Data Science

CHAPMAN & HALL/CRC DATA SCIENCE SERIES

Reflecting the interdisciplinary nature of the field, this book series brings together researchers, practitioners, and instructors from statistics, computer science, machine learning, and analytics. The series will publish cutting-edge research, industry applications, and textbooks in data science.

The inclusion of concrete examples, applications, and methods is highly encouraged. The scope of the series includes titles in the areas of machine learning, pattern recognition, predictive analytics, business analytics, Big Data, visualization, programming, software, learning analytics, data wrangling, interactive graphics, and reproducible research.

Published Titles

Statistical Foundations of Data Science
Jianqing Fan, Runze Li, Cun-Hui Zhang, and Hui Zou

A Tour of Data Science: Learn R and Python in Parallel
Nailong Zhang

Explanatory Model Analysis
Explore, Explain, and Examine Predictive Models
Przemyslaw Biecek and *Tomasz Burzykowski*

An Introduction to IoT Analytics
Harry G. Perros

Data Analytics
A Small Data Approach
Shuai Huang and Houtao Deng

Public Policy Analytics
Code and Context for Data Science in Government
Ken Steif

Supervised Machine Learning for Text Analysis in R
Emil Hvitfeldt and Julia Silge

Massive Graph Analytics
Edited by David Bader

Data Science
A First Introduction
Tiffany Timbers, Trevor Campbell, and Melissa Lee

Tree-Based Methods
A Practical Introduction with Applications in R
Brandon M. Greenwell

Urban Informatics
Using Big Data to Understand and Serve Communities
Daniel T. O'Brien

For more information about this series, please visit: https://www.routledge.com/Chapman--HallCRC-Data-Science-Series/book-series/CHDSS

Data Science

A First Introduction

Tiffany Timbers
Trevor Campbell
Melissa Lee

CRC Press
Taylor & Francis Group
Boca Raton London New York

CRC Press is an imprint of the
Taylor & Francis Group, an **informa** business
A CHAPMAN & HALL BOOK

First edition published 2022
by CRC Press
6000 Broken Sound Parkway NW, Suite 300, Boca Raton, FL 33487-2742

and by CRC Press
4 Park Square, Milton Park, Abingdon, Oxon, OX14 4RN

CRC Press is an imprint of Taylor & Francis Group, LLC

Library of Congress Cataloging-in-Publication Data

Names: Timbers, Tiffany, author. | Campbell, Trevor, author. | Lee, Melissa, author.
Title: Data science : a first introduction / Tiffany Timbers, Trevor Campbell and Melissa Lee.
Description: First edition. | Boca Raton : CRC Press, 2022. | Series: Statistics | Includes bibliographical references and index.
Identifiers: LCCN 2021054754 (print) | LCCN 2021054755 (ebook) | ISBN 9780367532178 (hardback) | ISBN 9780367524685 (paperback) | ISBN 9781003080978 (ebook)
Subjects: LCSH: Mathematical statistics--Data processing--Textbooks. | R (Computer program language)--Textbooks. | Quantitative research--Data processing--Textbooks.
Classification: LCC QA276.45.R3 T56 2022 (print) | LCC QA276.45.R3 (ebook) | DDC 519.50285/5133--dc23/eng20220301
LC record available at https://lccn.loc.gov/2021054754
LC ebook record available at https://lccn.loc.gov/2021054755

ISBN: 978-0-367-53217-8 (hbk)
ISBN: 978-0-367-52468-5 (pbk)
ISBN: 978-1-003-08097-8 (ebk)

DOI: 10.1201/9781003080978

Typeset in Latin Modern font
by KnowledgeWorks Global Ltd.

Publisher's note: This book has been prepared from camera-ready copy provided by the authors.

For my husband Curtis and daughter Rowan. Thank-you for your love (and patience with my late night writing).
– Tiffany

To mom and dad: here's a book. Pretty neat, eh? Love you guys.
– Trevor

To mom and dad, thank you for all your love and support.
– Melissa

Contents

Foreword

Roger D. Peng

Johns Hopkins Bloomberg School of Public Health

2022-01-04

The field of data science has expanded and grown significantly in recent years, attracting excitement and interest from many different directions. The demand for introductory educational materials has grown concurrently with the growth of the field itself, leading to a proliferation of textbooks, courses, blog posts, and tutorials. This book is an important contribution to this fast-growing literature, but given the wide availability of materials, a reader should be inclined to ask, "What is the unique contribution of *this* book?" In order to answer that question it is useful to step back for a moment and consider the development of the field of data science over the past few years.

When thinking about data science, it is important to consider two questions: "What is data science?" and "How should one do data science?" The former question is under active discussion amongst a broad community of researchers and practitioners and there does not appear to be much consensus to date. However, there seems a general understanding that data science focuses on the more "active" elements—data wrangling, cleaning, and analysis—of answering questions with data. These elements are often highly problem-specific and may seem difficult to generalize across applications. Nevertheless, over time we have seen some core elements emerge that appear to repeat themselves as useful concepts across different problems. Given the lack of clear agreement over the definition of data science, there is a strong need for a book like this one to propose a vision for what the field is and what the implications are for the activities in which members of the field engage.

The first important concept addressed by this book is tidy data, which is a format for tabular data formally introduced to the statistical community in a 2014 paper by Hadley Wickham. The tidy data organization strategy has proven a powerful abstract concept for conducting data analysis, in large part because of the vast toolchain implemented in the Tidyverse collection of R packages. The second key concept is the development of workflows for reproducible and auditable data analyses. Modern data analyses have only grown in complexity due to the availability of data and the ease with which we can implement complex data analysis procedures. Furthermore, these data analyses are often part of decision-making processes that may have significant impacts on people and communities. Therefore, there is a critical need to build reproducible analyses that can be studied and repeated by others in a reliable manner. Statistical methods clearly represent an important element of data science for building prediction and classification models and for making inferences about unobserved populations. Finally, because a field can succeed only if it fosters an active and collaborative community, it has become clear that being fluent in the tools of collaboration is a core element of data science.

This book takes these core concepts and focuses on how one can apply them to *do* data science in a rigorous manner. Students who learn from this book will be well-versed in the techniques and principles behind producing reliable evidence from data. This book is centered around the use of the R programming language within the tidy data framework, and as such employs the most recent advances in data analysis coding. The use of Jupyter notebooks for exercises immediately places the student in an environment that encourages auditability and reproducibility of analyses. The integration of git and GitHub into the course is a key tool for teaching about collaboration and community, key concepts that are critical to data science.

The demand for training in data science continues to increase. The availability of large quantities of data to answer a variety of questions, the computational power available to many more people than ever before, and the public awareness of the importance of data for decision-making have all contributed to the need for high-quality data science work. This book provides a sophisticated first introduction to the field of data science and provides a balanced mix of practical skills along with generalizable principles. As we continue to introduce students to data science and train them to confront an expanding array of data science problems, they will be well-served by the ideas presented here.

Preface

This textbook aims to be an approachable introduction to the world of data science. In this book, we define **data science** as the process of generating insight from data through **reproducible** and **auditable** processes. If you analyze some data and give your analysis to a friend or colleague, they should be able to re-run the analysis from start to finish and get the same result you did (*reproducibility*). They should also be able to see and understand all the steps in the analysis, as well as the history of how the analysis developed (*auditability*). Creating reproducible and auditable analyses allows both you and others to easily double-check and validate your work.

At a high level, in this book, you will learn how to

(1) identify common problems in data science, and
(2) solve those problems with reproducible and auditable workflows.

Figure 1 summarizes what you will learn in each chapter of this book. Throughout, you will learn how to use the R programming language [R Core Team, 2021] to perform all the tasks associated with data analysis. You will spend the first four chapters learning how to use R to load, clean, wrangle (i.e., restructure the data into a usable format) and visualize data while answering descriptive and exploratory data analysis questions. In the next six chapters, you will learn how to answer predictive, exploratory, and inferential data analysis questions with common methods in data science, including classification, regression, clustering, and estimation. In the final chapters (11–13), you will learn how to combine R code, formatted text, and images in a single coherent document with Jupyter, use version control for collaboration, and install and configure the software needed for data science on your own computer. If you are reading this book as part of a course that you are taking, the instructor may have set up all of these tools already for you; in this case, you can continue on through the book reading the chapters in order. But if you are reading this independently, you may want to jump to these last three chapters early before going on to make sure your computer is set up in such a way that you can try out the example code that we include throughout the book.

Each chapter in the book has an accompanying worksheet that provides exercises to help you practice the concepts you will learn. We strongly recommend

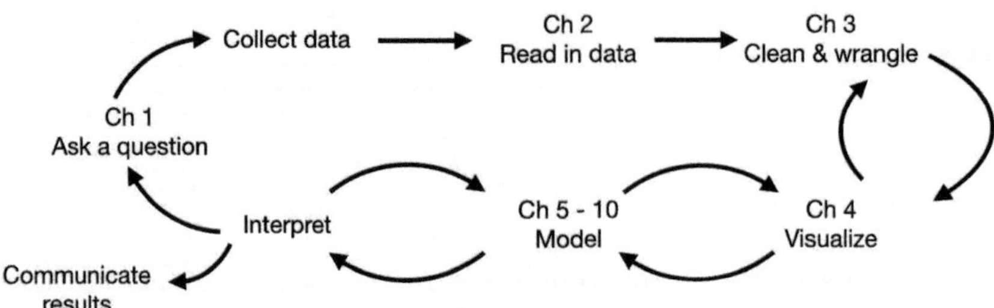

FIGURE 1: Where are we going?

that you work through the worksheet when you finish reading each chapter before moving on to the next chapter. All of the worksheets are available at `https://github.com/UBC-DSCI/data-science-a-first-intro-worksheets#readme`; the "Exercises" section at the end of each chapter points you to the right worksheet for that chapter. For each worksheet, you can either launch an interactive version of the worksheet in your browser by clicking the "launch binder" button, or preview a non-interactive version of the worksheet by clicking "view worksheet." If you instead decide to download the worksheet and run it on your own machine, make sure to follow the instructions for computer setup found in Chapter 13. This will ensure that the automated feedback and guidance that the worksheets provide will function as intended.

Acknowledgments

We'd like to thank everyone that has contributed to the development of *Data Science: A First Introduction*[1]. This is an open source textbook that began as a collection of course readings for DSCI 100, a new introductory data science course at the University of British Columbia (UBC). Several faculty members in the UBC Department of Statistics were pivotal in shaping the direction of that course, and as such, contributed greatly to the broad structure and list of topics in this book. We would especially like to thank Matías Salibían-Barrera for his mentorship during the initial development and roll-out of both DSCI 100 and this book. His door was always open when we needed to chat about how to best introduce and teach data science to our first-year students.

We would also like to thank all those who contributed to the process of publishing this book. In particular, we would like to thank all of our reviewers for their feedback and suggestions: Rohan Alexander, Isabella Ghement, Virgilio Gómez Rubio, Albert Kim, Adam Loy, Maria Prokofieva, Emily Riederer, and Greg Wilson. The book was improved substantially by their insights. We would like to give special thanks to Jim Zidek for his support and encouragement throughout the process, and to Roger Peng for graciously offering to write the Foreword.

Finally, we owe a debt of gratitude to all of the students of DSCI 100 over the past few years. They provided invaluable feedback on the book and worksheets; they found bugs for us (and stood by very patiently in class while we frantically fixed those bugs); and they brought a level of enthusiasm to the class that sustained us during the hard work of creating a new course and writing a textbook. Our interactions with them taught us how to teach data science, and that learning is reflected in the content of this book.

[1]`https://datasciencebook.ca`

About the authors

Tiffany Timbers is an Assistant Professor of Teaching in the Department of Statistics and Co-Director for the Master of Data Science program (Vancouver Option) at the University of British Columbia. In these roles she teaches and develops curriculum around the responsible application of Data Science to solve real-world problems. One of her favorite courses she teaches is a graduate course on collaborative software development, which focuses on teaching how to create R and Python packages using modern tools and workflows.

Trevor Campbell is an Assistant Professor in the Department of Statistics at the University of British Columbia. His research focuses on automated, scalable Bayesian inference algorithms, Bayesian nonparametrics, streaming data, and Bayesian theory. He was previously a postdoctoral associate advised by Tamara Broderick in the Computer Science and Artificial Intelligence Laboratory (CSAIL) and Institute for Data, Systems, and Society (IDSS) at MIT, a Ph.D. candidate under Jonathan How in the Laboratory for Information and Decision Systems (LIDS) at MIT, and before that he was in the Engineering Science program at the University of Toronto.

Melissa Lee is an Assistant Professor of Teaching in the Department of Statistics at the University of British Columbia. She teaches and develops curriculum for undergraduate statistics and data science courses. Her work focuses on student-centered approaches to teaching, developing and assessing open educational resources, and promoting equity, diversity, and inclusion initiatives.

1

R and the Tidyverse

1.1 Overview

This chapter provides an introduction to data science and the R programming language. The goal here is to get your hands dirty right from the start! We will walk through an entire data analysis, and along the way introduce different types of data analysis question, some fundamental programming concepts in R, and the basics of loading, cleaning, and visualizing data. In the following chapters, we will dig into each of these steps in much more detail; but for now, let's jump in to see how much we can do with data science!

1.2 Chapter learning objectives

By the end of the chapter, readers will be able to do the following:

- Identify the different types of data analysis question and categorize a question into the correct type.
- Load the `tidyverse` package into R.
- Read tabular data with `read_csv`.
- Create new variables and objects in R using the assignment symbol.
- Create and organize subsets of tabular data using `filter`, `select`, `arrange`, and `slice`.
- Visualize data with a `ggplot` bar plot.
- Use `?` to access help and documentation tools in R.

1.3 Canadian languages data set

In this chapter, we will walk through a full analysis of a data set relating to languages spoken at home by Canadian residents. Many Indigenous peoples exist in Canada with their own cultures and languages; these languages are often unique to Canada and not spoken anywhere else in the world [Statistics Canada, 2018]. Sadly, colonization has led to the loss of many of these languages. For instance, generations of children were not allowed to speak their mother tongue (the first language an individual learns in childhood) in Canadian residential schools. Colonizers also renamed places they had "discovered" [Wilson, 2018]. Acts such as these have significantly harmed the continuity of Indigenous languages in Canada, and some languages are considered "endangered" as few people report speaking them. To learn more, please see *Canadian Geographic*'s article, "Mapping Indigenous Languages in Canada" [Walker, 2017], *They Came for the Children: Canada, Aboriginal peoples, and Residential Schools* [Truth and Reconciliation Commission of Canada, 2012] and the *Truth and Reconciliation Commission of Canada's Calls to Action* [Truth and Reconciliation Commission of Canada, 2015].

The data set we will study in this chapter is taken from the `canlang` R data package[1] [Timbers, 2020], which has population language data collected during the 2016 Canadian census [Statistics Canada, 2016a]. In this data, there are 214 languages recorded, each having six different properties:

1. `category`: Higher-level language category, describing whether the language is an Official Canadian language, an Aboriginal (i.e., Indigenous) language, or a Non-Official and Non-Aboriginal language.
2. `language`: The name of the language.
3. `mother_tongue`: Number of Canadian residents who reported the language as their mother tongue. Mother tongue is generally defined as the language someone was exposed to since birth.
4. `most_at_home`: Number of Canadian residents who reported the language as being spoken most often at home.
5. `most_at_work`: Number of Canadian residents who reported the language as being used most often at work.
6. `lang_known`: Number of Canadian residents who reported knowledge of the language.

According to the census, more than 60 Aboriginal languages were reported as being spoken in Canada. Suppose we want to know which are the most common; then we might ask the following question, which we wish to answer using our data:

Which ten Aboriginal languages were most often reported in 2016 as mother tongues in Canada, and how many people speak each of them?

[1] `https://ttimbers.github.io/canlang/`

Note: Data science cannot be done without a deep understanding of the data and problem domain. In this book, we have simplified the data sets used in our examples to concentrate on methods and fundamental concepts. But in real life, you cannot and should not do data science without a domain expert. Alternatively, it is common to practice data science in your own domain of expertise! Remember that when you work with data, it is essential to think about *how* the data were collected, which affects the conclusions you can draw. If your data are biased, then your results will be biased!

1.4 Asking a question

Every good data analysis begins with a *question*—like the above—that you aim to answer using data. As it turns out, there are actually a number of different *types* of question regarding data: descriptive, exploratory, inferential, predictive, causal, and mechanistic, all of which are defined in Table 1.1. Carefully formulating a question as early as possible in your analysis—and correctly identifying which type of question it is—will guide your overall approach to the analysis as well as the selection of appropriate tools.

TABLE 1.1: Types of data analysis question [Leek and Peng, 2015, Peng and Matsui, 2015].

Question type	Description	Example
Descriptive	A question that asks about summarized characteristics of a data set without interpretation (i.e., report a fact).	How many people live in each province and territory in Canada?
Exploratory	A question that asks if there are patterns, trends, or relationships within a single data set. Often used to propose hypotheses for future study.	Does political party voting change with indicators of wealth in a set of data collected on 2,000 people living in Canada?

Question type	Description	Example
Predictive	A question that asks about predicting measurements or labels for individuals (people or things). The focus is on what things predict some outcome, but not what causes the outcome.	What political party will someone vote for in the next Canadian election?
Inferential	A question that looks for patterns, trends, or relationships in a single data set **and** also asks for quantification of how applicable these findings are to the wider population.	Does political party voting change with indicators of wealth for all people living in Canada?
Causal	A question that asks about whether changing one factor will lead to a change in another factor, on average, in the wider population.	Does wealth lead to voting for a certain political party in Canadian elections?
Mechanistic	A question that asks about the underlying mechanism of the observed patterns, trends, or relationships (i.e., how does it happen?)	How does wealth lead to voting for a certain political party in Canadian elections?

In this book, you will learn techniques to answer the first four types of question: descriptive, exploratory, predictive, and inferential; causal and mechanistic questions are beyond the scope of this book. In particular, you will learn how to apply the following analysis tools:

1. **Summarization:** computing and reporting aggregated values pertaining to a data set. Summarization is most often used to answer descriptive questions, and can occasionally help with answering exploratory questions. For example, you might use summarization to answer the following question: *What is the average race time for runners in this data set?* Tools for summarization are covered in detail in Chapters 2 and 3, but appear regularly throughout the text.
2. **Visualization:** plotting data graphically. Visualization is typically used to answer descriptive and exploratory questions, but plays a critical supporting role in answering all of the types of question in

Table 1.1. For example, you might use visualization to answer the following question: *Is there any relationship between race time and age for runners in this data set?* This is covered in detail in Chapter 4, but again appears regularly throughout the book.

3. **Classification:** predicting a class or category for a new observation. Classification is used to answer predictive questions. For example, you might use classification to answer the following question: *Given measurements of a tumor's average cell area and perimeter, is the tumor benign or malignant?* Classification is covered in Chapters 5 and 6.
4. **Regression:** predicting a quantitative value for a new observation. Regression is also used to answer predictive questions. For example, you might use regression to answer the following question: *What will be the race time for a 20-year-old runner who weighs 50kg?* Regression is covered in Chapters 7 and 8.
5. **Clustering:** finding previously unknown/unlabeled subgroups in a data set. Clustering is often used to answer exploratory questions. For example, you might use clustering to answer the following question: *What products are commonly bought together on Amazon?* Clustering is covered in Chapter 9.
6. **Estimation:** taking measurements for a small number of items from a large group and making a good guess for the average or proportion for the large group. Estimation is used to answer inferential questions. For example, you might use estimation to answer the following question: *Given a survey of cellphone ownership of 100 Canadians, what proportion of the entire Canadian population own Android phones?* Estimation is covered in Chapter 10.

Referring to Table 1.1, our question about Aboriginal languages is an example of a *descriptive question*: we are summarizing the characteristics of a data set without further interpretation. And referring to the list above, it looks like we should use visualization and perhaps some summarization to answer the question. So in the remainder of this chapter, we will work towards making a visualization that shows us the ten most common Aboriginal languages in Canada and their associated counts, according to the 2016 census.

1.5 Loading a tabular data set

A data set is, at its core essence, a structured collection of numbers and characters. Aside from that, there are really no strict rules; data sets can come in many different forms! Perhaps the most common form of data set that you will find in the wild, however, is *tabular data*. Think spreadsheets in Microsoft Excel: tabular data are rectangular-shaped and spreadsheet-like, as shown in Figure 1.1. In this book, we will focus primarily on tabular data.

Since we are using R for data analysis in this book, the first step for us is to load the data into R. When we load tabular data into R, it is represented as a *data frame* object. Figure 1.1 shows that an R data frame is very similar to a spreadsheet. We refer to the rows as **observations**; these are the things that we collect the data on, e.g., voters, cities, etc. We refer to the columns as **variables**; these are the characteristics of those observations, e.g., voters' political affiliations, cities' populations, etc.

Spreadsheet

	A	B	C	D	E	F
1	category	language	mother_tongue	most_at_home	most_at_work	lang_known
2	Aboriginal languages	Aboriginal languages, n.o.s.	590	235	30	665
3	Non-Official & Non-Aboriginal languages	Afrikaans	10260	4785	85	23415
4	Non-Official & Non-Aboriginal languages	Afro-Asiatic languages, n.i.e.	1150	445	10	2775
5	Non-Official & Non-Aboriginal languages	Akan (Twi)	13460	5985	25	22150
6	Non-Official & Non-Aboriginal languages	Albanian	26895	13135	345	31930
7	Aboriginal languages	Algonquian languages, n.i.e.	45	10	0	120
8	Aboriginal languages	Algonquin	1260	370	40	2480
9	Non-Official & Non-Aboriginal languages	American Sign Language	2685	3020	1145	21930
10	Non-Official & Non-Aboriginal languages	Amharic	22465	12785	200	33670
11	Non-Official & Non-Aboriginal languages	Arabic	419890	223535	5585	629055

Data frame in R

```
# A tibble: 214 x 6
   category                                language                        mother_tongue most_at_home most_at_work lang_known
   <chr>                                   <chr>                                   <dbl>        <dbl>        <dbl>      <dbl>
 1 Aboriginal languages                    Aboriginal languages, n.o.s.              590          235           30        665
 2 Non-Official & Non-Aboriginal languages Afrikaans                               10260         4785           85      23415
 3 Non-Official & Non-Aboriginal languages Afro-Asiatic languages, n.i.e.           1150          445           10       2775
 4 Non-Official & Non-Aboriginal languages Akan (Twi)                              13460         5985           25      22150
 5 Non-Official & Non-Aboriginal languages Albanian                                26895        13135          345      31930
 6 Aboriginal languages                    Algonquian languages, n.i.e.               45           10            0        120
 7 Aboriginal languages                    Algonquin                                1260          370           40       2480
 8 Non-Official & Non-Aboriginal languages American Sign Language                   2685         3020         1145      21930
 9 Non-Official & Non-Aboriginal languages Amharic                                 22465        12785          200      33670
10 Non-Official & Non-Aboriginal languages Arabic                                 419890       223535         5585     629055
# … with 204 more rows
```

FIGURE 1.1: A spreadsheet versus a data frame in R.

The first kind of data file that we will learn how to load into R as a data frame is the *comma-separated values* format (`.csv` for short). These files have names ending in `.csv`, and can be opened and saved using common spreadsheet programs like Microsoft Excel and Google Sheets. For example, the `.csv` file named `can_lang.csv` is included with the code for this book[2]. If we were to open this data in a plain text editor (a program like Notepad that just shows text with no formatting), we would see each row on its own line, and each entry in the table separated by a comma:

[2] https://github.com/UBC-DSCI/introduction-to-datascience/tree/master/data

```
category,language,mother_tongue,most_at_home,most_at_work,lang_known
Aboriginal languages,"Aboriginal languages, n.o.s.",590,235,30,665
Non-Official & Non-Aboriginal languages,Afrikaans,10260,4785,85,23415
Non-Official & Non-Aboriginal languages,"Afro-Asiatic languages, n.i.e.",1150,44
Non-Official & Non-Aboriginal languages,Akan (Twi),13460,5985,25,22150
Non-Official & Non-Aboriginal languages,Albanian,26895,13135,345,31930
Aboriginal languages,"Algonquian languages, n.i.e.",45,10,0,120
Aboriginal languages,Algonquin,1260,370,40,2480
Non-Official & Non-Aboriginal languages,American Sign Language,2685,3020,1145,21
Non-Official & Non-Aboriginal languages,Amharic,22465,12785,200,33670
```

To load this data into R so that we can do things with it (e.g., perform analyses or create data visualizations), we will need to use a *function.* A function is a special word in R that takes instructions (we call these *arguments*) and does something. The function we will use to load a `.csv` file into R is called `read_csv`. In its most basic use-case, `read_csv` expects that the data file:

- has column names (or *headers*),
- uses a comma (,) to separate the columns, and
- does not have row names.

Below you'll see the code used to load the data into R using the `read_csv` function. Note that the `read_csv` function is not included in the base installation of R, meaning that it is not one of the primary functions ready to use when you install R. Therefore, you need to load it from somewhere else before you can use it. The place from which we will load it is called an R *package.* An R package is a collection of functions that can be used in addition to the built-in R package functions once loaded. The `read_csv` function, in particular, can be made accessible by loading the `tidyverse` R package[3] [Wickham, 2021b, Wickham et al., 2019] using the `library` function. The `tidyverse` package contains many functions that we will use throughout this book to load, clean, wrangle, and visualize data.

```
library(tidyverse)
```

```
##     --     Attaching     packages     ---------------------------------------
tidyverse 1.3.1 --

## v ggplot2 3.3.5     v purrr   0.3.4
## v tibble  3.1.5     v dplyr   1.0.7
## v tidyr   1.1.4     v stringr 1.4.0
## v readr   2.0.2
```

[3]https://tidyverse.tidyverse.org/

```
##        --        Conflicts        ------------------------------------------
tidyverse_conflicts() --
## x dplyr::filter() masks stats::filter()
## x dplyr::lag()    masks stats::lag()
```

Note: You may have noticed that we got some extra output from R saying `Attaching packages` and `Conflicts` below our code line. These are examples of *messages* in R, which give the user more information that might be handy to know. The `Attaching packages` message is natural when loading `tidyverse`, since `tidyverse` actually automatically causes other packages to be imported too, such as `dplyr`. In the future, when we load `tidyverse` in this book, we will silence these messages to help with the readability of the book. The `Conflicts` message is also totally normal in this circumstance. This message tells you if functions from different packages share the same name, which is confusing to R. For example, in this case, the `dplyr` package and the `stats` package both provide a function called `filter`. The message above (`dplyr::filter() masks stats::filter()`) is R telling you that it is going to default to the `dplyr` package version of this function. So if you use the `filter` function, you will be using the `dplyr` version. In order to use the `stats` version, you need to use its full name `stats::filter`. Messages are not errors, so generally you don't need to take action when you see a message; but you should always read the message and critically think about what it means and whether you need to do anything about it.

After loading the `tidyverse` package, we can call the `read_csv` function and pass it a single argument: the name of the file, `"can_lang.csv"`. We have to put quotes around file names and other letters and words that we use in our code to distinguish it from the special words (like functions!) that make up the R programming language. The file's name is the only argument we need to provide because our file satisfies everything else that the `read_csv` function expects in the default use-case. Figure 1.2 describes how we use the `read_csv` to read data into R.

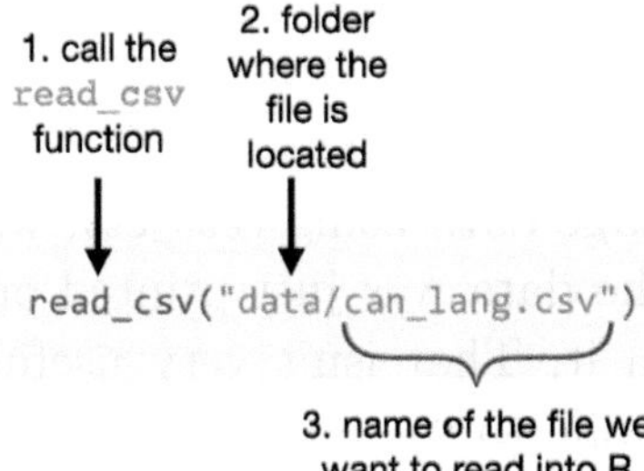

FIGURE 1.2: Syntax for the `read_csv` function.

```
read_csv("data/can_lang.csv")
```

```
## # A tibble: 214 x 6
##    category        language    mother_tongue most_at_home most_at_work lang_known
##     <chr>           <chr>              <dbl>        <dbl>        <dbl>      <dbl>
##  1 Aboriginal la~ Aboriginal~            590          235           30        665
##  2 Non-Official ~ Afrikaans            10260         4785           85      23415
##  3 Non-Official ~ Afro-Asiat~           1150          445           10       2775
##  4 Non-Official ~ Akan (Twi)           13460         5985           25      22150
##  5 Non-Official ~ Albanian             26895        13135          345      31930
##  6 Aboriginal la~ Algonquian~             45           10            0        120
##  7 Aboriginal la~ Algonquin             1260          370           40       2480
##  8 Non-Official ~ American S~           2685         3020         1145      21930
##  9 Non-Official ~ Amharic              22465        12785          200      33670
## 10 Non-Official ~ Arabic              419890       223535         5585     629055
## # ... with 204 more rows
```

Note: There is another function that also loads csv files named `read.csv`. We will *always* use `read_csv` in this book, as it is designed to play nicely with all of the other `tidyverse` functions, which we will use extensively. Be careful not to accidentally use `read.csv`, as it can cause some tricky errors to occur in your code that are hard to track down!

1.6 Naming things in R

When we loaded the 2016 Canadian census language data using `read_csv`, we did not give this data frame a name. Therefore the data was just printed on the screen, and we cannot do anything else with it. That isn't very useful. What would be more useful would be to give a name to the data frame that `read_csv` outputs, so that we can refer to it later for analysis and visualization.

The way to assign a name to a value in R is via the *assignment symbol* `<-`. On the left side of the assignment symbol you put the name that you want to use, and on the right side of the assignment symbol you put the value that you want the name to refer to. Names can be used to refer to almost anything in R, such as numbers, words (also known as *strings* of characters), and data frames! Below, we set `my_number` to `3` (the result of `1+2`) and we set `name` to the string `"Alice"`.

```
my_number <- 1 + 2
name <- "Alice"
```

Note that when we name something in R using the assignment symbol, `<-`, we do not need to surround the name we are creating with quotes. This is because we are formally telling R that this special word denotes the value of whatever is on the right-hand side. Only characters and words that act as *values* on the right-hand side of the assignment symbol—e.g., the file name `"data/can_lang.csv"` that we specified before, or `"Alice"` above—need to be surrounded by quotes.

After making the assignment, we can use the special name words we have created in place of their values. For example, if we want to do something with the value `3` later on, we can just use `my_number` instead. Let's try adding 2 to `my_number`; you will see that R just interprets this as adding 3 and 2:

```
my_number + 2
```

```
## [1] 5
```

Object names can consist of letters, numbers, periods `.` and underscores `_`. Other symbols won't work since they have their own meanings in R. For example, `+` is the addition symbol; if we try to assign a name with the `+` symbol, R will complain and we will get an error!

```
na + me <- 1
```

```
Error: unexpected assignment in "na+me <-"
```

There are certain conventions for naming objects in R. When naming an object we suggest using only lowercase letters, numbers and underscores `_` to separate the words in a name. R is case sensitive, which means that `Letter` and `letter` would be two different objects in R. You should also try to give your objects meaningful names. For instance, you *can* name a data frame `x`. However, using more meaningful terms, such as `language_data`, will help you remember what each name in your code represents. We recommend following the Tidyverse naming conventions outlined in the *Tidyverse Style Guide* [Wickham, 2020]. Let's now use the assignment symbol to give the name `can_lang` to the 2016 Canadian census language data frame that we get from `read_csv`.

```
can_lang <- read_csv("data/can_lang.csv")
```

Wait a minute, nothing happened this time! Where's our data? Actually, something did happen: the data was loaded in and now has the name `can_lang` associated with it. And we can use that name to access the data frame and do things with it. For example, we can type the name of the data frame to print the first few rows on the screen. You will also see at the top that the number of observations (i.e., rows) and variables (i.e., columns) are printed. Printing the first few rows of a data frame like this is a handy way to get a quick sense for what is contained in a data frame.

```
can_lang
```

```
## # A tibble: 214 x 6
##   category       language   mother_tongue most_at_home most_at_work lang_known
##    <chr>           <chr>              <dbl>        <dbl>        <dbl>      <dbl>
## 1 Aboriginal la~ Aboriginal~           590          235           30        665
## 2 Non-Official ~ Afrikaans           10260         4785           85      23415
## 3 Non-Official ~ Afro-Asiat~          1150          445           10       2775
## 4 Non-Official ~ Akan (Twi)          13460         5985           25      22150
## 5 Non-Official ~ Albanian            26895        13135          345      31930
## 6 Aboriginal la~ Algonquian~            45           10            0        120
## 7 Aboriginal la~ Algonquin            1260          370           40       2480
## 8 Non-Official ~ American S~          2685         3020         1145      21930
## 9 Non-Official ~ Amharic             22465        12785          200      33670
```

```
## 10 Non-Official ~ Arabic            419890       223535          5585     629055
## # ... with 204 more rows
```

1.7 Creating subsets of data frames with `filter` & `select`

Now that we've loaded our data into R, we can start wrangling the data to find the ten Aboriginal languages that were most often reported in 2016 as mother tongues in Canada. In particular, we will construct a table with the ten Aboriginal languages that have the largest counts in the `mother_tongue` column. The `filter` and `select` functions from the `tidyverse` package will help us here. The `filter` function allows you to obtain a subset of the rows with specific values, while the `select` function allows you to obtain a subset of the columns. Therefore, we can `filter` the rows to extract the Aboriginal languages in the data set, and then use `select` to obtain only the columns we want to include in our table.

1.7.1 Using `filter` to extract rows

Looking at the `can_lang` data above, we see the `category` column contains different high-level categories of languages, which include "Aboriginal languages", "Non-Official & Non-Aboriginal languages" and "Official languages". To answer our question we want to filter our data set so we restrict our attention to only those languages in the "Aboriginal languages" category.

We can use the `filter` function to obtain the subset of rows with desired values from a data frame. Figure 1.3 outlines what arguments we need to specify to use `filter`. The first argument to `filter` is the name of the data frame object, `can_lang`. The second argument is a *logical statement* to use when filtering the rows. A logical statement evaluates to either `TRUE` or `FALSE`; `filter` keeps only those rows for which the logical statement evaluates to `TRUE`. For example, in our analysis, we are interested in keeping only languages in the "Aboriginal languages" higher-level category. We can use the *equivalency operator* `==` to compare the values of the `category` column with the value `"Aboriginal languages"`; you will learn about many other kinds of logical statements in Chapter 3. Similar to when we loaded the data file and put quotes around the file name, here we need to put quotes around `"Aboriginal languages"`. Using quotes tells R that this is a string *value* and not one of the special words that make up R programming language, or one of the names we have given to data frames in the code we have already written.

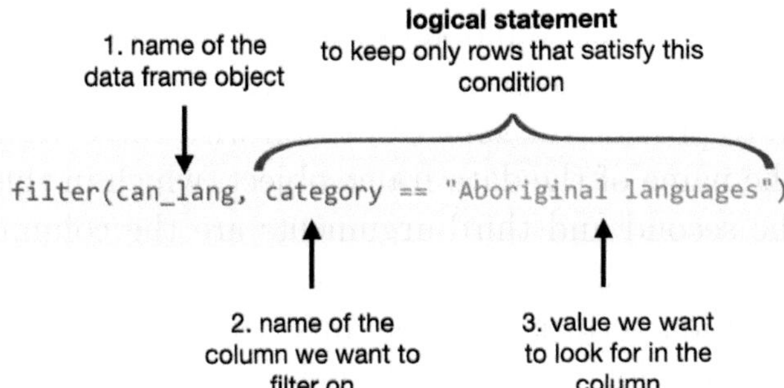

FIGURE 1.3: Syntax for the `filter` function.

With these arguments, `filter` returns a data frame that has all the columns of the input data frame, but only those rows we asked for in our logical filter statement.

```
aboriginal_lang <- filter(can_lang, category == "Aboriginal languages")
aboriginal_lang
```

```
## # A tibble: 67 x 6
##    category     language     mother_tongue most_at_home most_at_work lang_known
##    <chr>        <chr>                <dbl>        <dbl>        <dbl>      <dbl>
##  1 Aboriginal ~ Aboriginal l~          590          235           30        665
##  2 Aboriginal ~ Algonquian l~           45           10            0        120
##  3 Aboriginal ~ Algonquin             1260          370           40       2480
##  4 Aboriginal ~ Athabaskan l~           50           10            0         85
##  5 Aboriginal ~ Atikamekw             6150         5465         1100       6645
##  6 Aboriginal ~ Babine (Wets~          110           20           10        210
##  7 Aboriginal ~ Beaver                 190           50            0        340
##  8 Aboriginal ~ Blackfoot             2815         1110           85       5645
##  9 Aboriginal ~ Carrier               1025          250           15       2100
## 10 Aboriginal ~ Cayuga                  45           10           10        125
## # ... with 57 more rows
```

It's good practice to check the output after using a function in R. We can see the original `can_lang` data set contained 214 rows with multiple kinds of `category`. The data frame `aboriginal_lang` contains only 67 rows, and looks like it only contains languages in the "Aboriginal languages" in the `category` column. So it looks like the function gave us the result we wanted!

1.7.2 Using `select` to extract columns

Now let's use `select` to extract the `language` and `mother_tongue` columns from this data frame. Figure 1.4 shows us the syntax for the `select` function. To extract these columns, we need to provide the `select` function with three arguments. The first argument is the name of the data frame object, which in this example is `aboriginal_lang`. The second and third arguments are the column names that we want to select: `language` and `mother_tongue`. After passing these three arguments, the `select` function returns two columns (the `language` and `mother_tongue` columns that we asked for) as a data frame. This code is also a great example of why being able to name things in R is useful: you can see that we are using the result of our earlier `filter` step (which we named `aboriginal_lang`) here in the next step of the analysis!

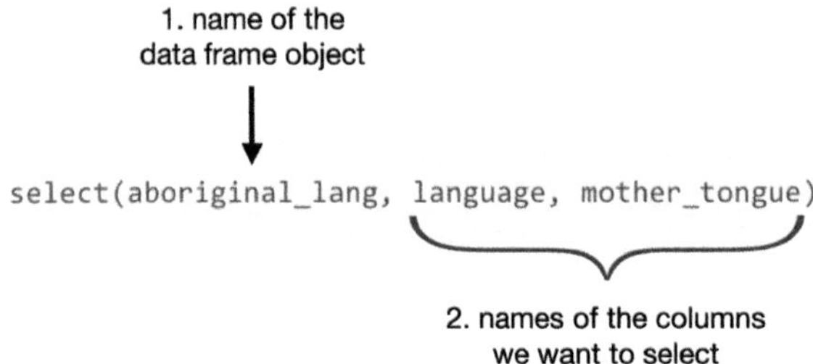

FIGURE 1.4: Syntax for the `select` function.

```
selected_lang <- select(aboriginal_lang, language, mother_tongue)
selected_lang
```

```
## # A tibble: 67 x 2
##    language                       mother_tongue
##    <chr>                                  <dbl>
##  1 Aboriginal languages, n.o.s.             590
##  2 Algonquian languages, n.i.e.              45
##  3 Algonquin                               1260
##  4 Athabaskan languages, n.i.e.              50
##  5 Atikamekw                               6150
##  6 Babine (Wetsuwet'en)                     110
##  7 Beaver                                   190
##  8 Blackfoot                               2815
##  9 Carrier                                 1025
```

```
## 10 Cayuga                              45
## # ... with 57 more rows
```

1.7.3 Using arrange to order and slice to select rows by index number

We have used `filter` and `select` to obtain a table with only the Aboriginal languages in the data set and their associated counts. However, we want to know the **ten** languages that are spoken most often. As a next step, we could order the `mother_tongue` column from greatest to least and then extract only the top ten rows. This is where the `arrange` and `slice` functions come to the rescue!

The `arrange` function allows us to order the rows of a data frame by the values of a particular column. Figure 1.5 details what arguments we need to specify to use the `arrange` function. We need to pass the data frame as the first argument to this function, and the variable to order by as the second argument. Since we want to choose the ten Aboriginal languages most often reported as a mother tongue language, we will use the `arrange` function to order the rows in our `selected_lang` data frame by the `mother_tongue` column. We want to arrange the rows in descending order (from largest to smallest), so we pass the column to the `desc` function before using it as an argument.

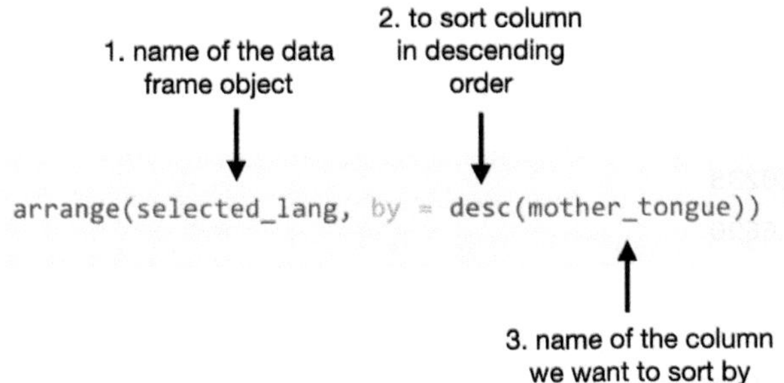

FIGURE 1.5: Syntax for the `arrange` function.

```
arranged_lang <- arrange(selected_lang, by = desc(mother_tongue))
arranged_lang
```

```
## # A tibble: 67 x 2
##    language             mother_tongue
##    <chr>                        <dbl>
##  1 Cree, n.o.s.                 64050
##  2 Inuktitut                    35210
```

```
##   3 Ojibway                 17885
##   4 Oji-Cree                12855
##   5 Dene                    10700
##   6 Montagnais (Innu)       10235
##   7 Mi'kmaq                  6690
##   8 Atikamekw                6150
##   9 Plains Cree              3065
##  10 Stoney                   3025
## # ... with 57 more rows
```

Next we will use the `slice` function, which selects rows according to their row number. Since we want to choose the most common ten languages, we will indicate we want the rows 1 to 10 using the argument `1:10`.

```
ten_lang <- slice(arranged_lang, 1:10)
ten_lang
```

```
## # A tibble: 10 x 2
##     language          mother_tongue
##     <chr>                     <dbl>
##   1 Cree, n.o.s.              64050
##   2 Inuktitut                 35210
##   3 Ojibway                   17885
##   4 Oji-Cree                  12855
##   5 Dene                      10700
##   6 Montagnais (Innu)         10235
##   7 Mi'kmaq                    6690
##   8 Atikamekw                  6150
##   9 Plains Cree                3065
##  10 Stoney                     3025
```

We have now answered our initial question by generating this table! Are we done? Well, not quite; tables are almost never the best way to present the result of your analysis to your audience. Even the simple table above with only two columns presents some difficulty: for example, you have to scrutinize the table quite closely to get a sense for the relative numbers of speakers of each language. When you move on to more complicated analyses, this issue only gets worse. In contrast, a *visualization* would convey this information in a much more easily understood format. Visualizations are a great tool for summarizing information to help you effectively communicate with your audience.

1.8 Exploring data with visualizations

Creating effective data visualizations is an essential component of any data analysis. In this section we will develop a visualization of the ten Aboriginal languages that were most often reported in 2016 as mother tongues in Canada, as well as the number of people that speak each of them.

1.8.1 Using `ggplot` to create a bar plot

In our data set, we can see that `language` and `mother_tongue` are in separate columns (or variables). In addition, there is a single row (or observation) for each language. The data are, therefore, in what we call a *tidy data* format. Tidy data is a fundamental concept and will be a significant focus in the remainder of this book: many of the functions from `tidyverse` require tidy data, including the `ggplot` function that we will use shortly for our visualization. We will formally introduce tidy data in Chapter 3.

We will make a bar plot to visualize our data. A bar plot is a chart where the heights of the bars represent certain values, like counts or proportions. We will make a bar plot using the `mother_tongue` and `language` columns from our `ten_lang` data frame. To create a bar plot of these two variables using the `ggplot` function, we must specify the data frame, which variables to put on the x and y axes, and what kind of plot to create. The `ggplot` function and its common usage is illustrated in Figure 1.6. Figure 1.7 shows the resulting bar plot generated by following the instructions in Figure 1.6.

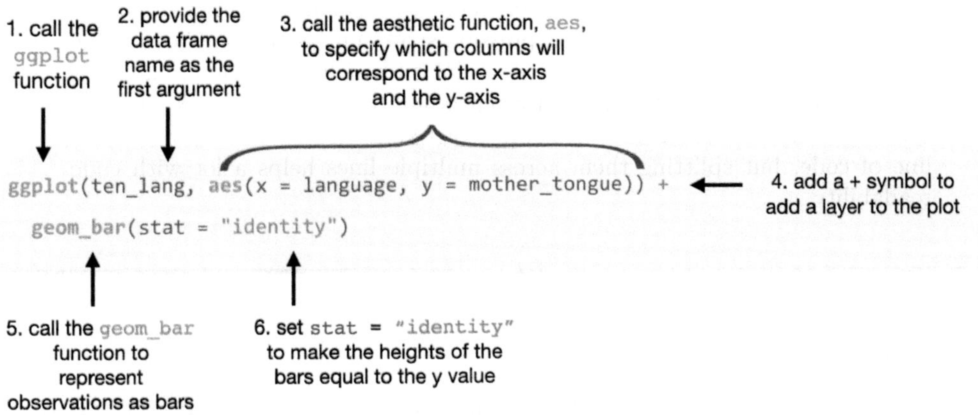

FIGURE 1.6: Creating a bar plot with the `ggplot` function.

```
ggplot(ten_lang, aes(x = language, y = mother_tongue)) +
  geom_bar(stat = "identity")
```

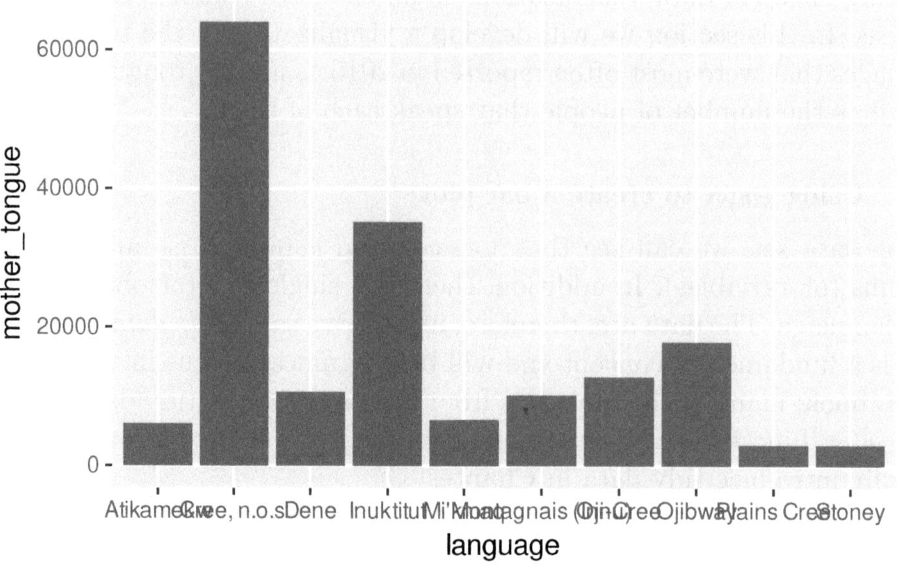

FIGURE 1.7: Bar plot of the ten Aboriginal languages most often reported by Canadian residents as their mother tongue. Note that this visualization is not done yet; there are still improvements to be made.

Note: The vast majority of the time, a single expression in R must be contained in a single line of code. However, there *are* a small number of situations in which you can have a single R expression span multiple lines. Above is one such case: here, R knows that a line cannot end with a `+` symbol, and so it keeps reading the next line to figure out what the right-hand side of the `+` symbol should be. We could, of course, put all of the added layers on one line of code, but splitting them across multiple lines helps a lot with code readability.

1.8.2 Formatting ggplot objects

It is exciting that we can already visualize our data to help answer our question, but we are not done yet! We can (and should) do more to improve the interpretability of the data visualization that we created. For example, by default, R uses the column names as the axis labels. Usually these column names do not have enough information about the variable in the column. We really should replace this default with a more informative label. For the example above, R uses the column name `mother_tongue` as the label for the y axis, but most people will not know what that is. And even if they did, they will not know how we measured this variable, or the group of people on which the measurements were taken. An axis label that reads "Mother Tongue (Number of Canadian Residents)" would be much more informative.

Adding additional layers to our visualizations that we create in `ggplot` is one common and easy way to improve and refine our data visualizations. New layers are added to `ggplot` objects using the `+` symbol. For example, we can use the `xlab` (short for x axis label) and `ylab` (short for y axis label) functions to add layers where we specify meaningful and informative labels for the x and y axes. Again, since we are specifying words (e.g. `"Mother Tongue (Number of Canadian Residents)"`) as arguments to `xlab` and `ylab`, we surround them with double quotation marks. We can add many more layers to format the plot further, and we will explore these in Chapter 4.

```
ggplot(ten_lang, aes(x = language, y = mother_tongue)) +
  geom_bar(stat = "identity") +
  xlab("Language") +
  ylab("Mother Tongue (Number of Canadian Residents)")
```

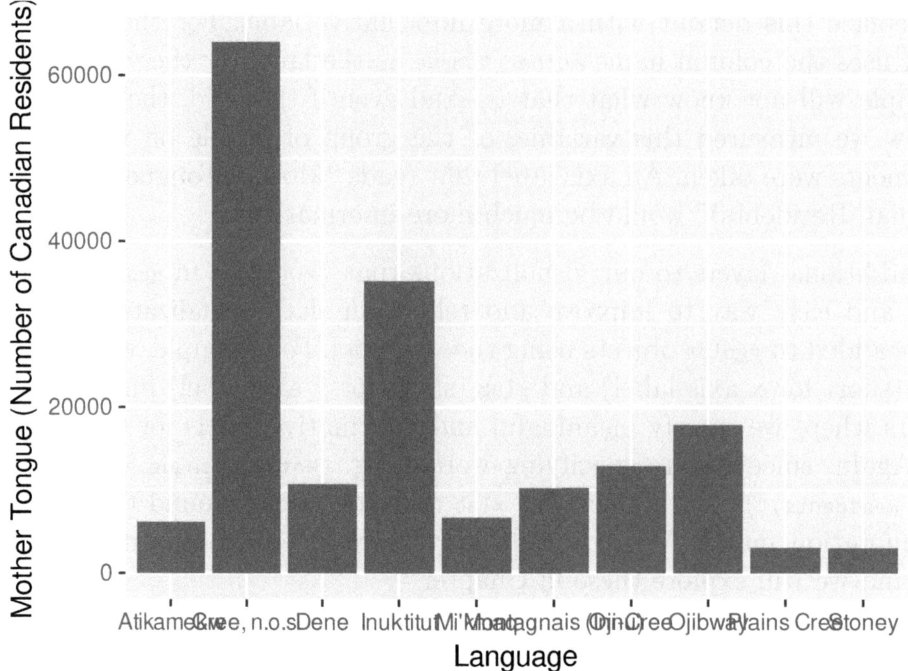

FIGURE 1.8: Bar plot of the ten Aboriginal languages most often reported by Canadian residents as their mother tongue with x and y labels. Note that this visualization is not done yet; there are still improvements to be made.

The result is shown in Figure 1.8. This is already quite an improvement! Let's tackle the next major issue with the visualization in Figure 1.8: the overlapping x axis labels, which are currently making it difficult to read the different language names. One solution is to rotate the plot such that the bars are horizontal rather than vertical. To accomplish this, we will swap the x and y coordinate axes:

```
ggplot(ten_lang, aes(x = mother_tongue, y = language)) +
  geom_bar(stat = "identity") +
  xlab("Mother Tongue (Number of Canadian Residents)") +
  ylab("Language")
```

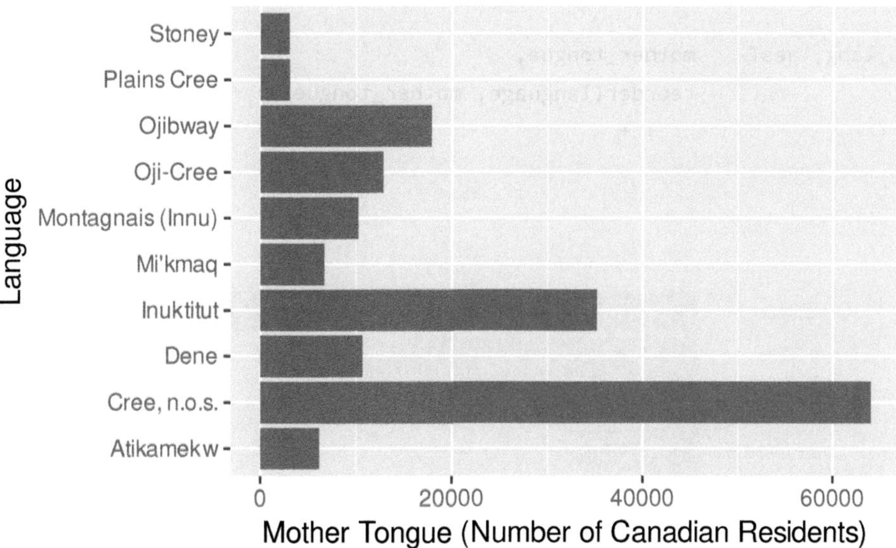

FIGURE 1.9: Horizontal bar plot of the ten Aboriginal languages most often reported by Canadian residents as their mother tongue. There are no more serious issues with this visualization, but it could be refined further.

Another big step forward, as shown in Figure 1.9! There are no more serious issues with the visualization. Now comes time to refine the visualization to make it even more well-suited to answering the question we asked earlier in this chapter. For example, the visualization could be made more transparent by organizing the bars according to the number of Canadian residents reporting each language, rather than in alphabetical order. We can reorder the bars using the `reorder` function, which orders a variable (here `language`) based on the values of the second variable (`mother_tongue`).

```
ggplot(ten_lang, aes(x = mother_tongue,
                     y = reorder(language, mother_tongue))) +
  geom_bar(stat = "identity") +
  xlab("Mother Tongue (Number of Canadian Residents)") +
  ylab("Language")
```

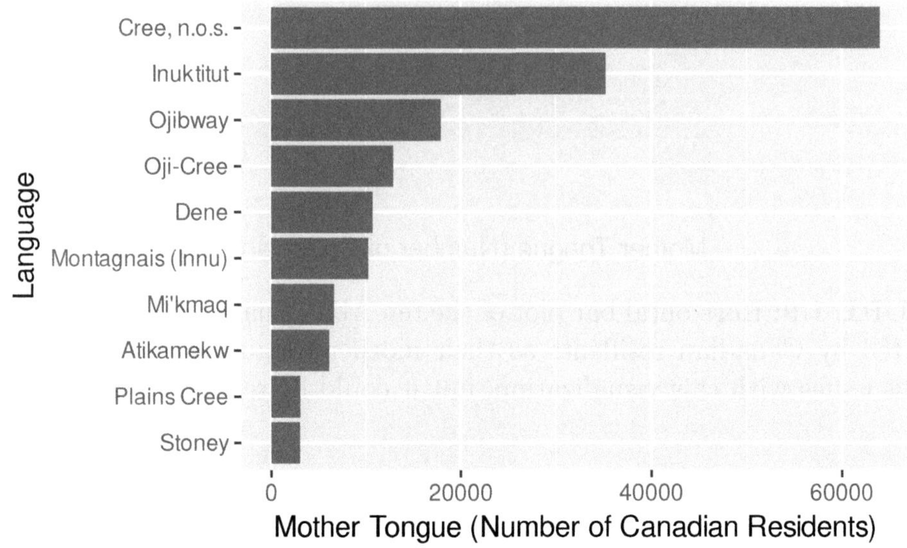

FIGURE 1.10: Bar plot of the ten Aboriginal languages most often reported by Canadian residents as their mother tongue with bars reordered.

Figure 1.10 provides a very clear and well-organized answer to our original question; we can see what the ten most often reported Aboriginal languages were, according to the 2016 Canadian census, and how many people speak each of them. For instance, we can see that the Aboriginal language most often reported was Cree n.o.s. with over 60,000 Canadian residents reporting it as their mother tongue.

Note: "n.o.s." means "not otherwise specified", so Cree n.o.s. refers to individuals who reported Cree as their mother tongue. In this data set, the Cree languages include the following categories: Cree n.o.s., Swampy Cree, Plains Cree, Woods Cree, and a 'Cree not included elsewhere' category (which includes Moose Cree, Northern East Cree and Southern East Cree) [Statistics Canada, 2016b].

1.8.3 Putting it all together

In the block of code below, we put everything from this chapter together, with a few modifications. In particular, we have actually skipped the `select` step that we did above; since you specify the variable names to plot in the `ggplot` function, you don't actually need to `select` the columns in advance when creating a visualization. We have also provided *comments* next to many of the lines of code below using the hash symbol `#`. When R sees a `#` sign, it will ignore all of the text that comes after the symbol on that line. So you can use comments to explain lines of code for others, and perhaps more importantly, your future self! It's good practice to get in the habit of commenting your code to improve its readability.

This exercise demonstrates the power of R. In relatively few lines of code, we performed an entire data science workflow with a highly effective data visualization! We asked a question, loaded the data into R, wrangled the data (using `filter`, `arrange` and `slice`) and created a data visualization to help answer our question. In this chapter, you got a quick taste of the data science workflow; continue on with the next few chapters to learn each of these steps in much more detail!

```
library(tidyverse)

# load the data set
can_lang <- read_csv("data/can_lang.csv")

# obtain the 10 most common Aboriginal languages
aboriginal_lang <- filter(can_lang, category == "Aboriginal languages")
arranged_lang <- arrange(aboriginal_lang, by = desc(mother_tongue))
ten_lang <- slice(arranged_lang, 1:10)

# create the visualization
ggplot(ten_lang, aes(x = mother_tongue,
                     y = reorder(language, mother_tongue))) +
  geom_bar(stat = "identity") +
  xlab("Mother Tongue (Number of Canadian Residents)") +
  ylab("Language")
```

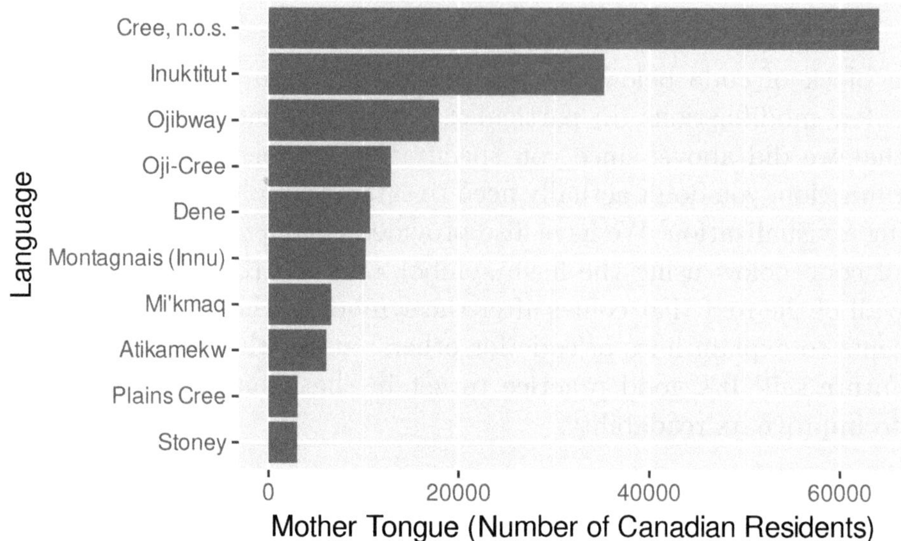

FIGURE 1.11: Putting it all together: bar plot of the ten Aboriginal languages most often reported by Canadian residents as their mother tongue.

1.9 Accessing documentation

There are many R functions in the `tidyverse` package (and beyond!), and nobody can be expected to remember what every one of them does or all of the arguments we have to give them. Fortunately, R provides the `?` symbol, which provides an easy way to pull up the documentation for most functions quickly. To use the `?` symbol to access documentation, you just put the name of the function you are curious about after the `?` symbol. For example, if you had forgotten what the `filter` function did or exactly what arguments to pass in, you could run the following code:

```
?filter
```

Figure 1.12 shows the documentation that will pop up, including a high-level description of the function, its arguments, a description of each, and more. Note that you may find some of the text in the documentation a bit too technical right now (for example, what is `dbplyr`, and what is grouped data?). Fear not: as you work through this book, many of these terms will be introduced to you, and slowly but surely you will become more adept at understanding and navigating documentation like that shown in Figure 1.12. But do keep in mind that the documentation is not written to *teach* you about a function; it is just there as a reference to *remind* you about the different arguments and usage of functions that you have already learned about elsewhere.

filter {dplyr} R Documentation

Subset rows using column values

Description

The `filter()` function is used to subset a data frame, retaining all rows that satisfy your conditions. To be retained, the row must produce a value of `TRUE` for all conditions. Note that when a condition evaluates to `NA` the row will be dropped, unlike base subsetting with `[`.

Usage

```
filter(.data, ..., .preserve = FALSE)
```

Arguments

`.data`	A data frame, data frame extension (e.g. a tibble), or a lazy data frame (e.g. from dbplyr or dtplyr). See *Methods*, below, for more details.
`...`	< `data-masking` > Expressions that return a logical value, and are defined in terms of the variables in `.data`. If multiple expressions are included, they are combined with the `&` operator. Only rows for which all conditions evaluate to `TRUE` are kept.

FIGURE 1.12: The documentation for the `filter` function, including a high-level description, a list of arguments and their meanings, and more.

1.10 Exercises

Practice exercises for the material covered in this chapter can be found in the accompanying worksheets repository[4] in the "R and the tidyverse" row. You can launch an interactive version of the worksheet in your browser by clicking the "launch binder" button. You can also preview a non-interactive version of the worksheet by clicking "view worksheet." If you instead decide to download the worksheet and run it on your own machine, make sure to follow the instructions for computer setup found in Chapter 13. This will ensure that the automated feedback and guidance that the worksheets provide will function as intended.

[4] `https://github.com/UBC-DSCI/data-science-a-first-intro-worksheets#readme`

2

Reading in data locally and from the web

2.1 Overview

In this chapter, you'll learn to read tabular data of various formats into R from your local device (e.g., your laptop) and the web. "Reading" (or "loading") is the process of converting data (stored as plain text, a database, HTML, etc.) into an object (e.g., a data frame) that R can easily access and manipulate. Thus reading data is the gateway to any data analysis; you won't be able to analyze data unless you've loaded it first. And because there are many ways to store data, there are similarly many ways to read data into R. The more time you spend upfront matching the data reading method to the type of data you have, the less time you will have to devote to re-formatting, cleaning and wrangling your data (the second step to all data analyses). It's like making sure your shoelaces are tied well before going for a run so that you don't trip later on!

2.2 Chapter learning objectives

By the end of the chapter, readers will be able to do the following:

- Define the following:
  - absolute file path
  - relative file path
  - **U**niform **R**esource **L**ocator (URL)
- Read data into R using a relative path and a URL.
- Compare and contrast the following functions:
  - `read_csv`
  - `read_tsv`
  - `read_csv2`
  - `read_delim`
  - `read_excel`

- Match the following `tidyverse` `read_*` function arguments to their descriptions:
  - `file`
  - `delim`
  - `col_names`
  - `skip`
- Choose the appropriate `tidyverse` `read_*` function and function arguments to load a given plain text tabular data set into R.
- Use `readxl` package's `read_excel` function and arguments to load a sheet from an excel file into R.
- Connect to a database using the `DBI` package's `dbConnect` function.
- List the tables in a database using the `DBI` package's `dbListTables` function.
- Create a reference to a database table that is queriable using the `tbl` from the `dbplyr` package.
- Retrieve data from a database query and bring it into R using the `collect` function from the `dbplyr` package.
- Use `write_csv` to save a data frame to a `.csv` file.
- (*Optional*) Obtain data using **a**pplication **p**rogramming **i**nterfaces (APIs) and web scraping.
  - Read HTML source code from a URL using the `rvest` package.
  - Read data from the Twitter API using the `rtweet` package.
  - Compare downloading tabular data from a plain text file (e.g., `.csv`), accessing data from an API, and scraping the HTML source code from a website.

2.3 Absolute and relative file paths

This chapter will discuss the different functions we can use to import data into R, but before we can talk about *how* we read the data into R with these functions, we first need to talk about *where* the data lives. When you load a data set into R, you first need to tell R where those files live. The file could live on your computer (*local*) or somewhere on the internet (*remote*).

The place where the file lives on your computer is called the "path". You can think of the path as directions to the file. There are two kinds of paths: *relative* paths and *absolute* paths. A relative path is where the file is with respect to where you currently are on the computer (e.g., where the file you're working in is). On the other hand, an absolute path is where the file is in respect to the computer's filesystem base (or root) folder.

Suppose our computer's filesystem looks like the picture in Figure 2.1, and we are working in a file titled `worksheet_02.ipynb`. If we want to read the `.csv` file named `happiness_report.csv` into R, we could do this using either a relative or an absolute path. We show both choices below.

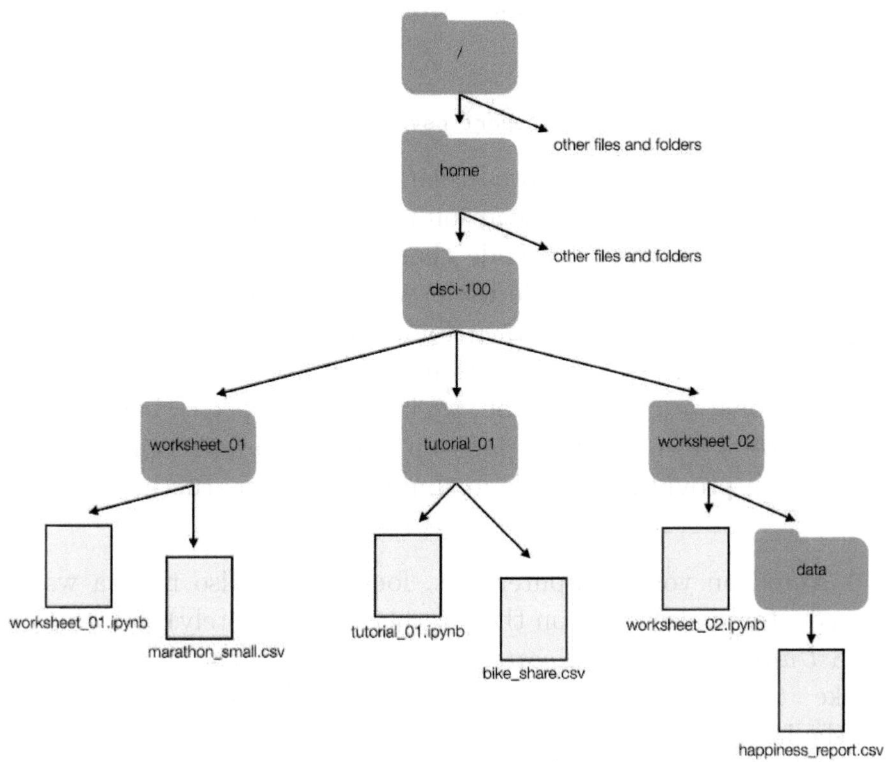

FIGURE 2.1: Example file system.

Reading `happiness_report.csv` using a relative path:

```
happy_data <- read_csv("data/happiness_report.csv")
```

Reading `happiness_report.csv` using an absolute path:

```
happy_data <- read_csv("/home/dsci-100/worksheet_02/data/happiness_report.csv")
```

So which one should you use? Generally speaking, you should use relative paths. Using a relative path helps ensure that your code can be run on a different computer (and as an added bonus, relative paths are often shorter—easier to type!). This is because a file's relative path is often the same across different computers, while a file's absolute path (the names of all of the folders between the computer's root, represented by `/`, and the file) isn't usually the same across different computers. For example, suppose Fatima and Jayden are working on a project together on the `happiness_report.csv` data. Fatima's file is stored at

`/home/Fatima/project/data/happiness_report.csv`,

while Jayden's is stored at

`/home/Jayden/project/data/happiness_report.csv`.

Even though Fatima and Jayden stored their files in the same place on their computers (in their home folders), the absolute paths are different due to their different usernames. If Jayden has code that loads the `happiness_report.csv` data using an absolute path, the code won't work on Fatima's computer. But the relative path from inside the `project` folder (`data/happiness_report.csv`) is the same on both computers; any code that uses relative paths will work on both! In the additional resources section, we include a link to a short video on the difference between absolute and relative paths. You can also check out the `here` package, which provides methods for finding and constructing file paths in R.

Beyond files stored on your computer (i.e., locally), we also need a way to locate resources stored elsewhere on the internet (i.e., remotely). For this purpose we use a *Uniform Resource Locator (URL)*, i.e., a web address that looks something like `https://datasciencebook.ca/`. URLs indicate the location of a resource on the internet and help us retrieve that resource.

2.4 Reading tabular data from a plain text file into R

2.4.1 `read_csv` to read in comma-separated files

Now that we have learned about *where* data could be, we will learn about *how* to import data into R using various functions. Specifically, we will learn how to *read* tabular data from a plain text file (a document containing only text) *into* R and *write* tabular data to a file *out of* R. The function we use to do this depends on the file's format. For example, in the last chapter, we learned about using the `tidyverse` `read_csv` function when reading `.csv` (**c**omma-**s**eparated **v**alues) files. In that case, the separator or *delimiter* that divided our columns was a comma (`,`). We only learned the case where the data matched the expected defaults of the `read_csv` function (column names are present, and commas are used as the delimiter between columns). In this section, we will learn how to read files that do not satisfy the default expectations of `read_csv`.

Before we jump into the cases where the data aren't in the expected default format for `tidyverse` and `read_csv`, let's revisit the more straightforward case where the defaults hold, and the only argument we need to give to the function is the path to the file, `data/can_lang.csv`. The `can_lang` data set contains language data from the 2016 Canadian census. We put `data/` before the file's name when we are loading the data set because this data set is located in a sub-folder, named `data`, relative to where we are running our R code. Here is what the text in the file `data/can_lang.csv` looks like.

```
category,language,mother_tongue,most_at_home,most_at_work,lang_known
Aboriginal languages,"Aboriginal languages, n.o.s.",590,235,30,665
Non-Official & Non-Aboriginal languages,Afrikaans,10260,4785,85,23415
Non-Official & Non-Aboriginal languages,"Afro-Asiatic languages, n.i.e.",1150,44
Non-Official & Non-Aboriginal languages,Akan (Twi),13460,5985,25,22150
Non-Official & Non-Aboriginal languages,Albanian,26895,13135,345,31930
Aboriginal languages,"Algonquian languages, n.i.e.",45,10,0,120
Aboriginal languages,Algonquin,1260,370,40,2480
Non-Official & Non-Aboriginal languages,American Sign Language,2685,3020,1145,21
Non-Official & Non-Aboriginal languages,Amharic,22465,12785,200,33670
```

And here is a review of how we can use `read_csv` to load it into R. First we load the `tidyverse` package to gain access to useful functions for reading the data.

```
library(tidyverse)
```

Next we use `read_csv` to load the data into R, and in that call we specify the relative path to the file.

```
canlang_data <- read_csv("data/can_lang.csv")
```

```
## Rows: 214 Columns: 6
## -- Column specification -----------------------------------------------------
-----
## Delimiter: ","
## chr (2): category, language
## dbl (4): mother_tongue, most_at_home, most_at_work, lang_known
##
## i Use `spec()` to retrieve the full column specification for this data.
## i Specify the column types or set `show_col_types = FALSE` to quiet this message.
```

Note: It is also normal and expected that a message is printed out after using the `read_csv` and related functions. This message lets you know the data types of each of the columns that R inferred while reading the data into R. In the future when we use this and related functions to load data in this book, we will silence these messages to help with the readability of the book.

Finally, to view the first 10 rows of the data frame, we must call it:

```
canlang_data
```

```
## # A tibble: 214 x 6
##    category          language          mother_tongue most_at_home most_at_work lang_known
##    <chr>             <chr>                     <dbl>        <dbl>        <dbl>      <dbl>
##  1 Aboriginal la~ Aboriginal~             590          235           30        665
##  2 Non-Official ~ Afrikaans             10260         4785           85      23415
##  3 Non-Official ~ Afro-Asiat~            1150          445           10       2775
##  4 Non-Official ~ Akan (Twi)            13460         5985           25      22150
##  5 Non-Official ~ Albanian              26895        13135          345      31930
##  6 Aboriginal la~ Algonquian~              45           10            0        120
##  7 Aboriginal la~ Algonquin              1260          370           40       2480
##  8 Non-Official ~ American S~            2685         3020         1145      21930
##  9 Non-Official ~ Amharic               22465        12785          200      33670
## 10 Non-Official ~ Arabic               419890       223535         5585     629055
## # ... with 204 more rows
```

2.4.2 Skipping rows when reading in data

Oftentimes, information about how data was collected, or other relevant information, is included at the top of the data file. This information is usually written in sentence and paragraph form, with no delimiter because it is not organized into columns. An example of this is shown below. This information gives the data scientist useful context and information about the data, however, it is not well formatted or intended to be read into a data frame cell along with the tabular data that follows later in the file.

```
Data source: https://ttimbers.github.io/canlang/
Data originally published in: Statistics Canada Census of Population 2016.
Reproduced and distributed on an as-is basis with their permission.
category,language,mother_tongue,most_at_home,most_at_work,lang_known
Aboriginal languages,"Aboriginal languages, n.o.s.",590,235,30,665
```

```
Non-Official & Non-Aboriginal languages,Afrikaans,10260,4785,85,23415
Non-Official & Non-Aboriginal languages,"Afro-Asiatic languages, n.i.e.",1150,44
Non-Official & Non-Aboriginal languages,Akan (Twi),13460,5985,25,22150
Non-Official & Non-Aboriginal languages,Albanian,26895,13135,345,31930
Aboriginal languages,"Algonquian languages, n.i.e.",45,10,0,120
Aboriginal languages,Algonquin,1260,370,40,2480
Non-Official & Non-Aboriginal languages,American Sign Language,2685,3020,1145,21
Non-Official & Non-Aboriginal languages,Amharic,22465,12785,200,33670
```

With this extra information being present at the top of the file, using `read_csv` as we did previously does not allow us to correctly load the data into R. In the case of this file we end up only reading in one column of the data set:

```
canlang_data <- read_csv("data/can_lang_meta-data.csv")
```

Note: In contrast to the normal and expected messages above, this time R printed out a warning for us indicating that there might be a problem with how our data is being read in.

```
canlang_data
```

```
## Warning: One or more parsing issues, see `problems()` for details

## # A tibble: 217 x 1
##    `Data source: https://ttimbers.github.io/canlang/`
##    <chr>
##  1 "Data originally published in: Statistics Canada Census of Population 2016."
##  2 "Reproduced and distributed on an as-is basis with their permission."
##  3 "category,language,mother_tongue,most_at_home,most_at_work,lang_known"
##  4 "Aboriginal languages,\"Aboriginal languages, n.o.s.\",590,235,30,665"
##  5 "Non-Official & Non-Aboriginal languages,Afrikaans,10260,4785,85,23415"
##          6      "Non-Official      &      Non-Aboriginal      languages,\"Afro-
Asiatic languages, n.i.e.\",~
##  7 "Non-Official & Non-Aboriginal languages,Akan (Twi),13460,5985,25,22150"
##  8 "Non-Official & Non-Aboriginal languages,Albanian,26895,13135,345,31930"
##  9 "Aboriginal languages,\"Algonquian languages, n.i.e.\",45,10,0,120"
## 10 "Aboriginal languages,Algonquin,1260,370,40,2480"
## # ... with 207 more rows
```

To successfully read data like this into R, the `skip` argument can be useful to tell R how many lines to skip before it should start reading in the data. In the example above, we would set this value to 3.

```
canlang_data <- read_csv("data/can_lang_meta-data.csv",
                         skip = 3)
canlang_data
```

```
## # A tibble: 214 x 6
##    category     language    mother_tongue most_at_home most_at_work lang_known
##    <chr>        <chr>               <dbl>        <dbl>        <dbl>      <dbl>
##  1 Aboriginal la~ Aboriginal~         590          235           30        665
##  2 Non-Official ~ Afrikaans         10260         4785           85      23415
##  3 Non-Official ~ Afro-Asiat~        1150          445           10       2775
##  4 Non-Official ~ Akan (Twi)        13460         5985           25      22150
##  5 Non-Official ~ Albanian          26895        13135          345      31930
##  6 Aboriginal la~ Algonquian~          45           10            0        120
##  7 Aboriginal la~ Algonquin          1260          370           40       2480
##  8 Non-Official ~ American S~        2685         3020         1145      21930
##  9 Non-Official ~ Amharic           22465        12785          200      33670
## 10 Non-Official ~ Arabic           419890       223535         5585     629055
## # ... with 204 more rows
```

How did we know to skip three lines? We looked at the data! The first three lines of the data had information we didn't need to import:

```
Data source: https://ttimbers.github.io/canlang/
Data originally published in: Statistics Canada Census of Population 2016.
Reproduced and distributed on an as-is basis with their permission.
```

The column names began at line 4, so we skipped the first three lines.

2.4.3 `read_tsv` to read in tab-separated files

Another common way data is stored is with tabs as the delimiter. Notice the data file, `can_lang_tab.tsv`, has tabs in between the columns instead of commas.

```
category	language	mother_tongue	most_at_home	most_at_work	lang_kno
Aboriginal languages	Aboriginal languages, n.o.s.	590	235	30	665
Non-Official & Non-Aboriginal languages	Afrikaans	10260	4785	85	23415
Non-Official & Non-Aboriginal languages	Afro-Asiatic languages, n.i.e.	1150
Non-Official & Non-Aboriginal languages	Akan (Twi)	13460	5985	25	22150
Non-Official & Non-Aboriginal languages	Albanian	26895	13135	345	31930
Aboriginal languages	Algonquian languages, n.i.e.	45	10	0	120
```

```
Aboriginal languages    Algonquin   1260    370 40  2480
Non-Official & Non-Aboriginal languages American Sign Language  2685    3020
Non-Official & Non-Aboriginal languages Amharic 22465   12785   200 33670
```

To read in this type of data, we can use the `read_tsv` to read in `.tsv` (tab separated values) files.

```
canlang_data <- read_tsv("data/can_lang_tab.tsv")
canlang_data
```

```
## # A tibble: 214 x 6
##    category     language    mother_tongue most_at_home most_at_work lang_known
##    <chr>        <chr>               <dbl>        <dbl>        <dbl>      <dbl>
##  1 Aboriginal la~ Aboriginal~         590          235           30        665
##  2 Non-Official ~ Afrikaans         10260         4785           85      23415
##  3 Non-Official ~ Afro-Asiat~        1150          445           10       2775
##  4 Non-Official ~ Akan (Twi)        13460         5985           25      22150
##  5 Non-Official ~ Albanian          26895        13135          345      31930
##  6 Aboriginal la~ Algonquian~          45           10            0        120
##  7 Aboriginal la~ Algonquin          1260          370           40       2480
##  8 Non-Official ~ American S~        2685         3020         1145      21930
##  9 Non-Official ~ Amharic           22465        12785          200      33670
## 10 Non-Official ~ Arabic           419890       223535         5585     629055
## # ... with 204 more rows
```

Let's compare the data frame here to the resulting data frame in Section 2.4.1 after using `read_csv`. Notice anything? They look the same! The same number of columns/rows and column names! So we needed to use different tools for the job depending on the file format and our resulting table (`canlang_data`) in both cases was the same!

2.4.4 `read_delim` as a more flexible method to get tabular data into R

`read_csv` and `read_tsv` are actually just special cases of the more general `read_delim` function. We can use `read_delim` to import both comma and tab-separated files (and more), we just have to specify the delimiter. The `can_lang.tsv` is a different version of this same data set with no column names and uses tabs as the delimiter instead of commas.

Here is how the file would look in a plain text editor:

```
Aboriginal languages    Aboriginal languages, n.o.s.    590 235 30  665
Non-Official & Non-Aboriginal languages Afrikaans   10260   4785    85  23415
Non-Official & Non-Aboriginal languages Afro-Asiatic languages, n.i.e.  1150
```

```
Non-Official & Non-Aboriginal languages Akan (Twi)  13460    5985     25  22150
Non-Official & Non-Aboriginal languages Albanian    26895    13135    345 31930
Aboriginal languages    Algonquian languages, n.i.e.    45  10  0   120
Aboriginal languages    Algonquin   1260    370 40  2480
Non-Official & Non-Aboriginal languages American Sign Language  2685    3020
Non-Official & Non-Aboriginal languages Amharic 22465   12785   200 33670
Non-Official & Non-Aboriginal languages Arabic  419890  223535  5585    629055
```

To get this into R using the `read_delim` function, we specify the first argument as the path to the file (as done with `read_csv`), and then provide values to the `delim` argument (here a tab, which we represent by `"\t"`) and the `col_names` argument (here we specify that there are no column names to assign, and give it the value of `FALSE`). `read_csv`, `read_tsv` and `read_delim` have a `col_names` argument and the default is `TRUE`.

Note: `\t` is an example of an *escaped character*, which always starts with a backslash (\). Escaped characters are used to represent non-printing characters (like the tab) or those with special meanings (such as quotation marks).

```
canlang_data <- read_delim("data/can_lang.tsv",
                           delim = "\t",
                           col_names = FALSE)
canlang_data
```

```
## # A tibble: 214 x 6
##    X1                                      X2             X3     X4    X5     X6
##    <chr>                                   <chr>       <dbl>  <dbl> <dbl>  <dbl>
##  1 Aboriginal languages                    Aborigina~    590    235    30    665
##  2 Non-Official & Non-Aboriginal languages Afrikaans   10260   4785    85  23415
##  3 Non-Official & Non-Aboriginal languages Afro-Asia~   1150    445    10   2775
##  4 Non-Official & Non-Aboriginal languages Akan (Twi)  13460   5985    25  22150
##  5 Non-Official & Non-Aboriginal languages Albanian    26895  13135   345  31930
##  6 Aboriginal languages                    Algonquia~     45     10     0    120
##  7 Aboriginal languages                    Algonquin    1260    370    40   2480
##  8 Non-Official & Non-Aboriginal languages American ~   2685   3020  1145  21930
##  9 Non-Official & Non-Aboriginal languages Amharic     22465  12785   200  33670
## 10 Non-Official & Non-Aboriginal languages Arabic     419890 223535  5585 629055
```

```
## # ... with 204 more rows
```

Data frames in R need to have column names. Thus if you read in data that don't have column names, R will assign names automatically. In the example above, R assigns each column a name of `X1, X2, X3, X4, X5, X6`.

It is best to rename your columns to help differentiate between them (e.g., `X1`, `X2`, etc., are not very descriptive names and will make it more confusing as you code). To rename your columns, you can use the `rename` function from the `dplyr` R package[1] [Wickham et al., 2021b] (one of the packages loaded with `tidyverse`, so we don't need to load it separately). The first argument is the data set, and in the subsequent arguments you write `new_name = old_name` for the selected variables to rename. We rename the `X1, X2, ..., X6` columns in the `canlang_data` data frame to more descriptive names below.

```
canlang_data <- rename(canlang_data,
       category = X1,
       language = X2,
       mother_tongue = X3,
       most_at_home = X4,
       most_at_work = X5,
       lang_known = X6)
canlang_data
```

```
## # A tibble: 214 x 6
##    category       language   mother_tongue most_at_home most_at_work lang_known
##    <chr>          <chr>              <dbl>        <dbl>        <dbl>      <dbl>
##  1 Aboriginal la~ Aboriginal~          590          235           30        665
##  2 Non-Official ~ Afrikaans          10260         4785           85      23415
##  3 Non-Official ~ Afro-Asiat~         1150          445           10       2775
##  4 Non-Official ~ Akan (Twi)         13460         5985           25      22150
##  5 Non-Official ~ Albanian           26895        13135          345      31930
##  6 Aboriginal la~ Algonquian~           45           10            0        120
##  7 Aboriginal la~ Algonquin           1260          370           40       2480
##  8 Non-Official ~ American S~         2685         3020         1145      21930
##  9 Non-Official ~ Amharic            22465        12785          200      33670
## 10 Non-Official ~ Arabic            419890       223535         5585     629055
## # ... with 204 more rows
```

[1] https://dplyr.tidyverse.org/

2.4.5 Reading tabular data directly from a URL

We can also use `read_csv`, `read_tsv` or `read_delim` (and related functions) to read in data directly from a **U**niform **R**esource **L**ocator (URL) that contains tabular data. Here, we provide the URL to `read_*` as the path to the file instead of a path to a local file on our computer. We need to surround the URL with quotes similar to when we specify a path on our local computer. All other arguments that we use are the same as when using these functions with a local file on our computer.

```
url <- "https://raw.githubusercontent.com/UBC-DSCI/data/main/can_lang.csv"
canlang_data <- read_csv(url)

canlang_data
```

```
## # A tibble: 214 x 6
##    category       language   mother_tongue most_at_home most_at_work lang_known
##    <chr>          <chr>              <dbl>        <dbl>        <dbl>      <dbl>
##  1 Aboriginal la~ Aboriginal~          590          235           30        665
##  2 Non-Official ~ Afrikaans          10260         4785           85      23415
##  3 Non-Official ~ Afro-Asiat~         1150          445           10       2775
##  4 Non-Official ~ Akan (Twi)         13460         5985           25      22150
##  5 Non-Official ~ Albanian           26895        13135          345      31930
##  6 Aboriginal la~ Algonquian~           45           10            0        120
##  7 Aboriginal la~ Algonquin           1260          370           40       2480
##  8 Non-Official ~ American S~         2685         3020         1145      21930
##  9 Non-Official ~ Amharic            22465        12785          200      33670
## 10 Non-Official ~ Arabic            419890       223535         5585     629055
## # ... with 204 more rows
```

2.4.6 Previewing a data file before reading it into R

In all the examples above, we gave you previews of the data file before we read it into R. Previewing data is essential to see whether or not there are column names, what the delimiters are, and if there are lines you need to skip. You should do this yourself when trying to read in data files. You can preview files in a plain text editor by right-clicking on the file, selecting "Open With," and choosing a plain text editor (e.g., Notepad).

2.5 Reading tabular data from a Microsoft Excel file

There are many other ways to store tabular data sets beyond plain text files, and similarly, many ways to load those data sets into R. For example, it is very common to encounter, and need to load into R, data stored as a Microsoft Excel spreadsheet (with the file name extension `.xlsx`). To be able to do this, a key thing to know is that even though `.csv` and `.xlsx` files look almost identical when loaded into Excel, the data themselves are stored completely differently. While `.csv` files are plain text files, where the characters you see when you open the file in a text editor are exactly the data they represent, this is not the case for `.xlsx` files. Take a look at a snippet of what a `.xlsx` file would look like in a text editor:

```
,?'O
    _rels/.rels???J1??>E?{7?
<?V????w8?'J???'QrJ???Tf?d??d?o?wZ'???@>?4'?|??hlIo??F
t                                                            8f??3wn
????t??u"/
          %~Ed2??<?w??
                         ?Pd(??J-?E???7?'t(?-GZ?????y???c~N?g[^_r?4
                                                                  yG?O
                                                                      ?K??G?



     ]TUEe??O??c[???????6q??s??d?m???\???H?^????3} ?rZY? ?:L60?^?????XTP+?|?
X?a??4VT?,D?Jq
```

This type of file representation allows Excel files to store additional things that you cannot store in a `.csv` file, such as fonts, text formatting, graphics, multiple sheets and more. And despite looking odd in a plain text editor, we can read Excel spreadsheets into R using the `readxl` package developed specifically for this purpose.

```
library(readxl)

canlang_data <- read_excel("data/can_lang.xlsx")
canlang_data
```

```
## # A tibble: 214 x 6
##   category        language   mother_tongue most_at_home most_at_work lang_known
##   <chr>            <chr>             <dbl>        <dbl>        <dbl>      <dbl>
```

```
## 1 Aboriginal la~ Aboriginal~       590       235       30      665
## 2 Non-Official ~ Afrikaans       10260      4785       85    23415
## 3 Non-Official ~ Afro-Asiat~      1150       445       10     2775
## 4 Non-Official ~ Akan (Twi)      13460      5985       25    22150
## 5 Non-Official ~ Albanian        26895     13135      345    31930
## 6 Aboriginal la~ Algonquian~        45        10        0      120
## 7 Aboriginal la~ Algonquin        1260       370       40     2480
## 8 Non-Official ~ American S~      2685      3020     1145    21930
## 9 Non-Official ~ Amharic         22465     12785      200    33670
## 10 Non-Official ~ Arabic        419890    223535     5585   629055
## # ... with 204 more rows
```

If the `.xlsx` file has multiple sheets, you have to use the `sheet` argument to specify the sheet number or name. You can also specify cell ranges using the `range` argument. This functionality is useful when a single sheet contains multiple tables (a sad thing that happens to many Excel spreadsheets since this makes reading in data more difficult).

As with plain text files, you should always explore the data file before importing it into R. Exploring the data beforehand helps you decide which arguments you need to load the data into R successfully. If you do not have the Excel program on your computer, you can use other programs to preview the file. Examples include Google Sheets and Libre Office.

In Table 2.1 we summarize the `read_*` functions we covered in this chapter. We also include the `read_csv2` function for data separated by semicolons ;, which you may run into with data sets where the decimal is represented by a comma instead of a period (as with some data sets from European countries).

TABLE 2.1: Summary of `read_*` functions

Data File Type	R Function	R Package
Comma (,) separated files	`read_csv`	`readr`
Tab (`\t`) separated files	`read_tsv`	`readr`
Semicolon (;) separated files	`read_csv2`	`readr`
Various formats (`.csv`, `.tsv`)	`read_delim`	`readr`
Excel files (`.xlsx`)	`read_excel`	`readxl`

Note: `readr` is a part of the `tidyverse` package so we did not need to load this package separately since we loaded `tidyverse`.

2.6 Reading data from a database

Another very common form of data storage is the relational database. Databases are great when you have large data sets or multiple users working on a project. There are many relational database management systems, such as SQLite, MySQL, PostgreSQL, Oracle, and many more. These different relational database management systems each have their own advantages and limitations. Almost all employ SQL (*structured query language*) to obtain data from the database. But you don't need to know SQL to analyze data from a database; several packages have been written that allow you to connect to relational databases and use the R programming language to obtain data. In this book, we will give examples of how to do this using R with SQLite and PostgreSQL databases.

2.6.1 Reading data from a SQLite database

SQLite is probably the simplest relational database system that one can use in combination with R. SQLite databases are self-contained and usually stored and accessed locally on one computer. Data is usually stored in a file with a `.db` extension. Similar to Excel files, these are not plain text files and cannot be read in a plain text editor.

The first thing you need to do to read data into R from a database is to connect to the database. We do that using the `dbConnect` function from the `DBI` (database interface) package. This does not read in the data, but simply tells R where the database is and opens up a communication channel that R can use to send SQL commands to the database.

```
library(DBI)

conn_lang_data <- dbConnect(RSQLite::SQLite(), "data/can_lang.db")
```

Often relational databases have many tables; thus, in order to retrieve data from a database, you need to know the name of the table in which the data is stored. You can get the names of all the tables in the database using the `dbListTables` function:

```
tables <- dbListTables(conn_lang_data)
tables
```

```
## [1] "lang"
```

The `dbListTables` function returned only one name, which tells us that there is only one table in this database. To reference a table in the database (so that we can perform operations like selecting columns and filtering rows), we use the `tbl` function from the `dbplyr` package. The object returned by the `tbl` function allows us to work with data stored in databases as if they were just regular data frames; but secretly, behind the scenes, `dbplyr` is turning your function calls (e.g., `select` and `filter`) into SQL queries!

```
library(dbplyr)

lang_db <- tbl(conn_lang_data, "lang")
lang_db
```

```
## # Source:   table<lang> [?? x 6]
## # Database: sqlite 3.36.0
## #   [/home/rstudio/introduction-to-datascience/pdf/data/can_lang.db]
##    category     language   mother_tongue most_at_home most_at_work lang_known
##    <chr>        <chr>              <dbl>        <dbl>        <dbl>      <dbl>
##  1 Aboriginal la~ Aboriginal~        590          235           30        665
##  2 Non-Official ~ Afrikaans        10260         4785           85      23415
##  3 Non-Official ~ Afro-Asiat~       1150          445           10       2775
##  4 Non-Official ~ Akan (Twi)       13460         5985           25      22150
##  5 Non-Official ~ Albanian         26895        13135          345      31930
##  6 Aboriginal la~ Algonquian~         45           10            0        120
##  7 Aboriginal la~ Algonquin         1260          370           40       2480
##  8 Non-Official ~ American S~       2685         3020         1145      21930
##  9 Non-Official ~ Amharic          22465        12785          200      33670
## 10 Non-Official ~ Arabic          419890       223535         5585     629055
## # ... with more rows
```

Although it looks like we just got a data frame from the database, we didn't! It's a *reference*; the data is still stored only in the SQLite database. The `dbplyr` package works this way because databases are often more efficient at selecting, filtering and joining large data sets than R. And typically the database will not even be stored on your computer, but rather a more powerful machine somewhere on the web. So R is lazy and waits to bring this data into memory until you explicitly tell it to using the `collect` function. Figure 2.2 highlights the difference between a `tibble` object in R and the output we just created. Notice in the table on the right, the first two lines of the output indicate the source is SQL. The last line doesn't show how many rows there are (R is trying to avoid performing expensive query operations), whereas the output for the `tibble` object does.

tibble object

```
## # A tibble: 214 x 6
##    category          language        mother_tongue most_at_home most_at_work lang_known
##    <chr>             <chr>                   <dbl>        <dbl>        <dbl>      <dbl>
##  1 Aboriginal la… Aboriginal…           590          235           30        665
##  2 Non-Official … Afrikaans            10260         4785           85      23415
##  3 Non-Official … Afro-Asiat…           1150          445           10       2775
##  4 Non-Official … Akan (Twi)           13460         5985           25      22150
##  5 Non-Official … Albanian             26895        13135          345      31930
##  6 Aboriginal la… Algonquian…             45           10            0        120
##  7 Aboriginal la… Algonquin             1260          370           40       2480
##  8 Non-Official … American S…           2685         3020         1145      21930
##  9 Non-Official … Amharic              22465        12785          200      33670
## 10 Non-Official … Arabic              419890       223535         5585     629055
## # … with 204 more rows
```

reference to data in database

```
## # Source:   table<lang> [?? x 6]
## # Database: sqlite 3.35.5
## #   [/home/rstudio/introduction-to-datascience/data/can_lang.db]
##    category          language        mother_tongue most_at_home most_at_work lang_known
##    <chr>             <chr>                   <dbl>        <dbl>        <dbl>      <dbl>
##  1 Aboriginal la… Aboriginal…           590          235           30        665
##  2 Non-Official … Afrikaans            10260         4785           85      23415
##  3 Non-Official … Afro-Asiat…           1150          445           10       2775
##  4 Non-Official … Akan (Twi)           13460         5985           25      22150
##  5 Non-Official … Albanian             26895        13135          345      31930
##  6 Aboriginal la… Algonquian…             45           10            0        120
##  7 Aboriginal la… Algonquin             1260          370           40       2480
##  8 Non-Official … American S…           2685         3020         1145      21930
##  9 Non-Official … Amharic              22465        12785          200      33670
## 10 Non-Official … Arabic              419890       223535         5585     629055
## # … with more rows
```

FIGURE 2.2: Comparison of a reference to data in a database and a tibble in R.

We can look at the SQL commands that are sent to the database when we write `tbl(conn_lang_data, "lang")` in R with the `show_query` function from the `dbplyr` package.

```
show_query(tbl(conn_lang_data, "lang"))
```

```
## <SQL>
## SELECT *
## FROM `lang`
```

The output above shows the SQL code that is sent to the database. When we write `tbl(conn_lang_data, "lang")` in R, in the background, the function is translating the R code into SQL, sending that SQL to the database, and then translating the response for us. So `dbplyr` does all the hard work of translating from R to SQL and back for us; we can just stick with R!

With our `lang_db` table reference for the 2016 Canadian Census data in hand, we can mostly continue onward as if it were a regular data frame. For example, we can use the `filter` function to obtain only certain rows. Below we filter the data to include only Aboriginal languages.

```
aboriginal_lang_db <- filter(lang_db, category == "Aboriginal languages")
aboriginal_lang_db
```

```
## # Source:   lazy query [?? x 6]
## # Database: sqlite 3.36.0
## #   [/home/rstudio/introduction-to-datascience/pdf/data/can_lang.db]
##    category     language     mother_tongue most_at_home most_at_work lang_known
##    <chr>        <chr>                <dbl>        <dbl>        <dbl>      <dbl>
##  1 Aboriginal ~ Aboriginal l~          590          235           30        665
##  2 Aboriginal ~ Algonquian l~           45           10            0        120
##  3 Aboriginal ~ Algonquin             1260          370           40       2480
##  4 Aboriginal ~ Athabaskan l~           50           10            0         85
##  5 Aboriginal ~ Atikamekw             6150         5465         1100       6645
##  6 Aboriginal ~ Babine (Wets~          110           20           10        210
##  7 Aboriginal ~ Beaver                 190           50            0        340
##  8 Aboriginal ~ Blackfoot             2815         1110           85       5645
##  9 Aboriginal ~ Carrier               1025          250           15       2100
## 10 Aboriginal ~ Cayuga                  45           10           10        125
## # ... with more rows
```

Above you can again see the hints that this data is not actually stored in R yet: the source is a `lazy query [?? x 6]` and the output says `... with more rows` at the end (both indicating that R does not know how many rows there are in total!), and a database type `sqlite 3.36.0` is listed. In order to actually retrieve this data in R as a data frame, we use the `collect` function. Below you will see that after running `collect`, R knows that the retrieved data has 67 rows, and there is no database listed any more.

```
aboriginal_lang_data <- collect(aboriginal_lang_db)
aboriginal_lang_data
```

```
## # A tibble: 67 x 6
##    category     language      mother_tongue most_at_home most_at_work lang_known
##    <chr>        <chr>                 <dbl>        <dbl>        <dbl>      <dbl>
##  1 Aboriginal ~ Aboriginal l~           590          235           30        665
##  2 Aboriginal ~ Algonquian l~            45           10            0        120
##  3 Aboriginal ~ Algonquin              1260          370           40       2480
##  4 Aboriginal ~ Athabaskan l~            50           10            0         85
##  5 Aboriginal ~ Atikamekw              6150         5465         1100       6645
##  6 Aboriginal ~ Babine (Wets~           110           20           10        210
##  7 Aboriginal ~ Beaver                  190           50            0        340
##  8 Aboriginal ~ Blackfoot              2815         1110           85       5645
##  9 Aboriginal ~ Carrier                1025          250           15       2100
## 10 Aboriginal ~ Cayuga                   45           10           10        125
## # ... with 57 more rows
```

Aside from knowing the number of rows, the data looks pretty similar in both outputs shown above. And `dbplyr` provides many more functions (not just `filter`) that you can use to directly feed the database reference (`lang_db`) into downstream analysis functions (e.g., `ggplot2` for data visualization). But `dbplyr` does not provide *every* function that we need for analysis; we do eventually need to call `collect`. For example, look what happens when we try to use `nrow` to count rows in a data frame:

```
nrow(aboriginal_lang_db)
```

```
## [1] NA
```

or `tail` to preview the last six rows of a data frame:

```
tail(aboriginal_lang_db)
```

```
## Error: tail() is not supported by sql sources
```

Additionally, some operations will not work to extract columns or single values from the reference given by the `tbl` function. Thus, once you have finished your data wrangling of the `tbl` database reference object, it is advisable to bring it into R as a data frame using `collect`. But be very careful using `collect`: databases are often *very* big, and reading an entire table into R might take a long time to run or even possibly crash your machine. So make sure you use `filter` and `select` on the database table to reduce the data to a reasonable size before using `collect` to read it into R!

2.6.2 Reading data from a PostgreSQL database

PostgreSQL (also called Postgres) is a very popular and open-source option for relational database software. Unlike SQLite, PostgreSQL uses a client–server database engine, as it was designed to be used and accessed on a network. This means that you have to provide more information to R when connecting to Postgres databases. The additional information that you need to include when you call the `dbConnect` function is listed below:

- `dbname`: the name of the database (a single PostgreSQL instance can host more than one database)
- `host`: the URL pointing to where the database is located
- `port`: the communication endpoint between R and the PostgreSQL database (usually `5432`)
- `user`: the username for accessing the database
- `password`: the password for accessing the database

Additionally, we must use the `RPostgres` package instead of `RSQLite` in the `dbConnect` function call. Below we demonstrate how to connect to a version of the `can_mov_db` database, which contains information about Canadian movies. Note that the `host` (`fakeserver.stat.ubc.ca`), `user` (`user0001`), and `password` (`abc123`) below are *not real*; you will not actually be able to connect to a database using this information.

```
library(RPostgres)
conn_mov_data <- dbConnect(RPostgres::Postgres(), dbname = "can_mov_db",
                           host = "fakeserver.stat.ubc.ca", port = 5432,
                           user = "user0001", password = "abc123")
```

After opening the connection, everything looks and behaves almost identically to when we were using an SQLite database in R. For example, we can again use `dbListTables` to find out what tables are in the `can_mov_db` database:

```
dbListTables(conn_mov_data)
```

```
 [1] "themes"            "medium"           "titles"     "title_aliases"       "forms"
 [6] "episodes"         "names"      "names_occupations" "occupation"       "ratings"
```

We see that there are 10 tables in this database. Let's first look at the `"ratings"` table to find the lowest rating that exists in the `can_mov_db` database:

```
ratings_db <- tbl(conn_mov_data, "ratings")
ratings_db
```

```
# Source:   table<ratings> [?? x 3]
# Database: postgres [user0001@fakeserver.stat.ubc.ca:5432/can_mov_db]
   title              average_rating num_votes
   <chr>                       <dbl>     <int>
 1 The Grand Seduction           6.6       150
 2 Rhymes for Young Ghouls       6.3      1685
 3 Mommy                         7.5      1060
 4 Incendies                     6.1      1101
 5 Bon Cop, Bad Cop              7.0       894
 6 Goon                          5.5      1111
 7 Monsieur Lazhar               5.6       610
 8 What if                       5.3      1401
 9 The Barbarian Invations       5.8        99
10 Away from Her                 6.9      2311
# … with more rows
```

To find the lowest rating that exists in the data base, we first need to extract the `average_rating` column using `select`:

```
avg_rating_db <- select(ratings_db, average_rating)
avg_rating_db
```

```
# Source:   lazy query [?? x 1]
# Database: postgres [user0001@fakeserver.stat.ubc.ca:5432/can_mov_db]
   average_rating
            <dbl>
 1            6.6
 2            6.3
 3            7.5
 4            6.1
```

```
 5            7.0
 6            5.5
 7            5.6
 8            5.3
 9            5.8
10            6.9
# … with more rows
```

Next we use `min` to find the minimum rating in that column:

```
min(avg_rating_db)
```

```
Error in min(avg_rating_db) : invalid 'type' (list) of argument
```

Instead of the minimum, we get an error! This is another example of when we need to use the `collect` function to bring the data into R for further computation:

```
avg_rating_data <- collect(avg_rating_db)
min(avg_rating_data)
```

```
[1] 1
```

We see the lowest rating given to a movie is 1, indicating that it must have been a really bad movie…

2.6.3 Why should we bother with databases at all?

Opening a database stored in a `.db` file involved a lot more effort than just opening a `.csv`, `.tsv`, or any of the other plain text or Excel formats. It was a bit of a pain to use a database in that setting since we had to use `dbplyr` to translate `tidyverse`-like commands (`filter`, `select` etc.) into SQL commands that the database understands. Not all `tidyverse` commands can currently be translated with SQLite databases. For example, we can compute a mean with an SQLite database but can't easily compute a median. So you might be wondering: why should we use databases at all?

Databases are beneficial in a large-scale setting:

- They enable storing large data sets across multiple computers with backups.
- They provide mechanisms for ensuring data integrity and validating input.
- They provide security and data access control.
- They allow multiple users to access data simultaneously and remotely without conflicts and errors. For example, there are billions of Google searches

conducted daily in 2021 [Real Time Statistics Project, 2021]. Can you imagine if Google stored all of the data from those searches in a single `.csv` file!? Chaos would ensue!

2.7 Writing data from R to a `.csv` file

At the middle and end of a data analysis, we often want to write a data frame that has changed (either through filtering, selecting, mutating or summarizing) to a file to share it with others or use it for another step in the analysis. The most straightforward way to do this is to use the `write_csv` function from the `tidyverse` package. The default arguments for this file are to use a comma (`,`) as the delimiter and include column names. Below we demonstrate creating a new version of the Canadian languages data set without the official languages category according to the Canadian 2016 Census, and then writing this to a `.csv` file:

```
no_official_lang_data <- filter(can_lang, category != "Official languages")
write_csv(no_official_lang_data, "data/no_official_languages.csv")
```

2.8 Obtaining data from the web

> **Note:** This section is not required reading for the remainder of the textbook. It is included for those readers interested in learning a little bit more about how to obtain different types of data from the web.

Data doesn't just magically appear on your computer; you need to get it from somewhere. Earlier in the chapter we showed you how to access data stored in a plain text, spreadsheet-like format (e.g., comma- or tab-separated) from a web URL using one of the `read_*` functions from the `tidyverse`. But as time goes on, it is increasingly uncommon to find data (especially large amounts of data) in this format available for download from a URL. Instead, websites now often offer something known as an **a**pplication **p**rogramming **i**nterface (API), which provides a programmatic way to ask for subsets of a data set. This allows the website owner to control *who* has access to the data, *what portion* of the data they have access to, and *how much* data they can access. Typically, the website owner will give you a *token* (a secret string of characters somewhat like a password) that you have to provide when accessing the API.

Another interesting thought: websites themselves *are* data! When you type a URL into your browser window, your browser asks the *web server* (another computer on the internet whose job it is to respond to requests for the website) to give it the website's data, and then your browser translates that data into something you can see. If the website shows you some information that you're interested in, you could *create* a data set for yourself by copying and pasting that information into a file. This process of taking information directly from what a website displays is called *web scraping* (or sometimes *screen scraping*). Now, of course, copying and pasting information manually is a painstaking and error-prone process, especially when there is a lot of information to gather. So instead of asking your browser to translate the information that the web server provides into something you can see, you can collect that data programmatically—in the form of **h**yper**t**ext **m**arkup **l**anguage (HTML) and **c**ascading **s**tyle **s**heet (CSS) code—and process it to extract useful information. HTML provides the basic structure of a site and tells the webpage how to display the content (e.g., titles, paragraphs, bullet lists etc.), whereas CSS helps style the content and tells the webpage how the HTML elements should be presented (e.g., colors, layouts, fonts etc.).

This subsection will show you the basics of both web scraping with the `rvest` R package[2] [Wickham, 2021a] and accessing the Twitter API using the `rtweet` R package[3] [Kearney, 2019].

2.8.1 Web scraping

HTML and CSS selectors

When you enter a URL into your browser, your browser connects to the web server at that URL and asks for the *source code* for the website. This is the data that the browser translates into something you can see; so if we are going to create our own data by scraping a website, we have to first understand what that data looks like! For example, let's say we are interested in knowing the average rental price (per square foot) of the most recently available one-bedroom apartments in Vancouver on Craiglist[4]. When we visit the Vancouver Craigslist website and search for one-bedroom apartments, we should see something similar to Figure 2.3.

[2]https://rvest.tidyverse.org/

[3]https://github.com/ropensci/rtweet

[4]https://vancouver.craigslist.org

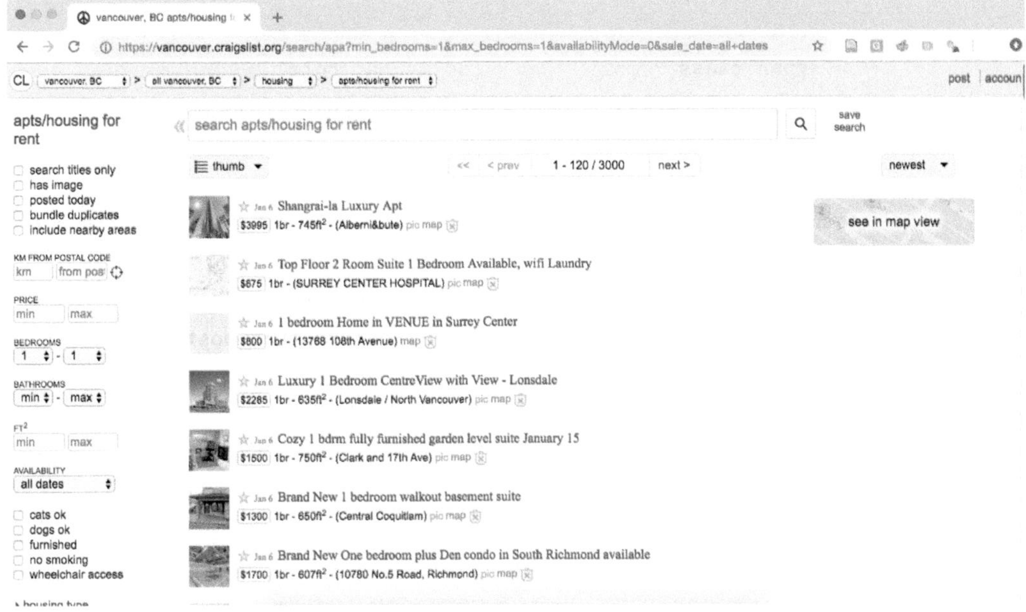

FIGURE 2.3: Craigslist webpage of advertisements for one-bedroom apartments.

Based on what our browser shows us, it's pretty easy to find the size and price for each apartment listed. But we would like to be able to obtain that information using R, without any manual human effort or copying and pasting. We do this by examining the *source code* that the web server actually sent our browser to display for us. We show a snippet of it below; the entire source is included with the code for this book[5]:

```
<span class="result-meta">
        <span class="result-price">$800</span>

        <span class="housing">
            1br -
        </span>

        <span class="result-hood"> (13768 108th Avenue)</span>

        <span class="result-tags">
            <span class="maptag" data-pid="6786042973">map</span>
        </span>
```

[5]https://github.com/UBC-DSCI/introduction-to-datascience/blob/master/img/website_source.txt

```
            <span class="banish icon icon-trash" role="button">
                <span class="screen-reader-text">hide this posting</span>
            </span>

        <span class="unbanish icon icon-trash red" role="button" aria-hidden
        <a href="#" class="restore-link">
            <span class="restore-narrow-text">restore</span>
            <span class="restore-wide-text">restore this posting</span>
        </a>

    </span>
  </p>
</li>
     <li class="result-row" data-pid="6788463837">

    <a href="https://vancouver.craigslist.org/nvn/apa/d/north-vancouver-luxu
            <span class="result-price">$2285</span>
    </a>
```

Oof...you can tell that the source code for a web page is not really designed for humans to understand easily. However, if you look through it closely, you will find that the information we're interested in is hidden among the muck. For example, near the top of the snippet above you can see a line that looks like

```
<span class="result-price">$800</span>
```

That is definitely storing the price of a particular apartment. With some more investigation, you should be able to find things like the date and time of the listing, the address of the listing, and more. So this source code most likely contains all the information we are interested in!

Let's dig into that line above a bit more. You can see that that bit of code has an *opening tag* (words between `<` and `>`, like `<span>`) and a *closing tag* (the same with a slash, like `</span>`). HTML source code generally stores its data between opening and closing tags like these. Tags are keywords that tell the web browser how to display or format the content. Above you can see that the information we want (`$800`) is stored between an opening and closing tag (`<span>` and `</span>`). In the opening tag, you can also see a very useful "class" (a special word that is sometimes included with opening tags): `class="result-price"`. Since we want R to programmatically sort through all of the source code for the website to find apartment prices, maybe we can look for all the tags with the `"result-price"` class, and grab the information between the opening and closing tag. Indeed, take a look at another line of the source snippet above:

```
<span class="result-price">$2285</span>
```

It's yet another price for an apartment listing, and the tags surrounding it have the `"result-price"` class. Wonderful! Now that we know what pattern we are looking for—a dollar amount between opening and closing tags that have the `"result-price"` class—we should be able to use code to pull out all of the matching patterns from the source code to obtain our data. This sort of "pattern" is known as a *CSS selector* (where CSS stands for **c**ascading **s**tyle **s**heet).

The above was a simple example of "finding the pattern to look for"; many websites are quite a bit larger and more complex, and so is their website source code. Fortunately, there are tools available to make this process easier. For example, SelectorGadget[6] is an open-source tool that simplifies identifying the generating and finding of CSS selectors. At the end of the chapter in the additional resources section, we include a link to a short video on how to install and use the SelectorGadget tool to obtain CSS selectors for use in web scraping. After installing and enabling the tool, you can click the website element for which you want an appropriate selector. For example, if we click the price of an apartment listing, we find that SelectorGadget shows us the selector `.result-price` in its toolbar, and highlights all the other apartment prices that would be obtained using that selector (Figure 2.4).

If we then click the size of an apartment listing, SelectorGadget shows us the `span` selector, and highlights many of the lines on the page; this indicates that the `span` selector is not specific enough to capture only apartment sizes (Figure 2.5).

To narrow the selector, we can click one of the highlighted elements that we *do not* want. For example, we can deselect the "pic/map" links, resulting in only the data we want highlighted using the `.housing` selector (Figure 2.6).

So to scrape information about the square footage and rental price of apartment listings, we need to use the two CSS selectors `.housing` and `.result-price`, respectively. The selector gadget returns them to us as a comma-separated list (here `.housing , .result-price`), which is exactly the format we need to provide to R if we are using more than one CSS selector.

[6] `https://selectorgadget.com/`

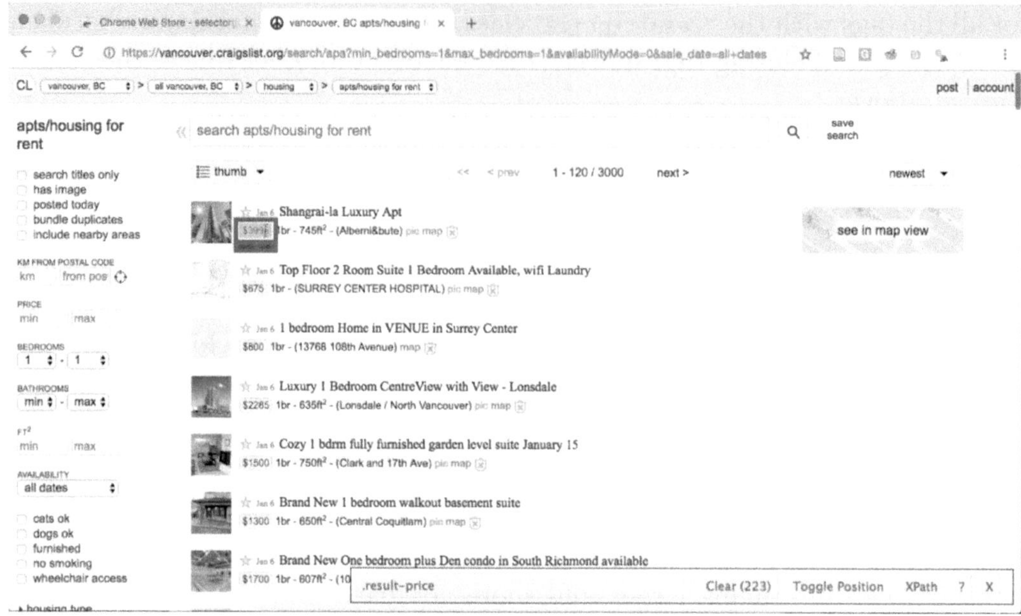

FIGURE 2.4: Using the SelectorGadget on a Craigslist webpage to obtain the CCS selector useful for obtaining apartment prices.

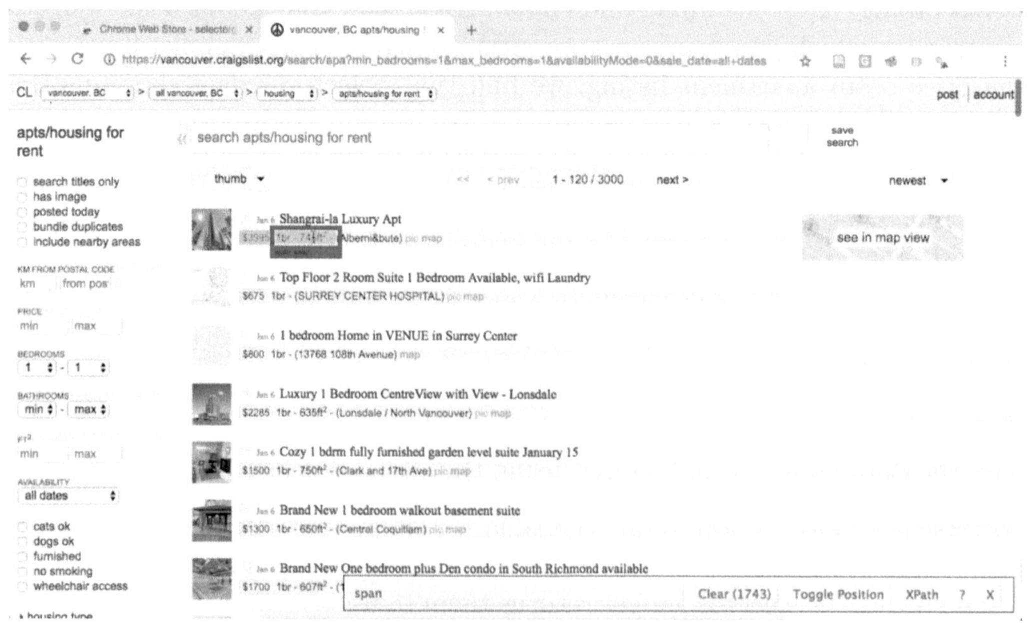

FIGURE 2.5: Using the SelectorGadget on a Craigslist webpage to obtain a CCS selector useful for obtaining apartment sizes.

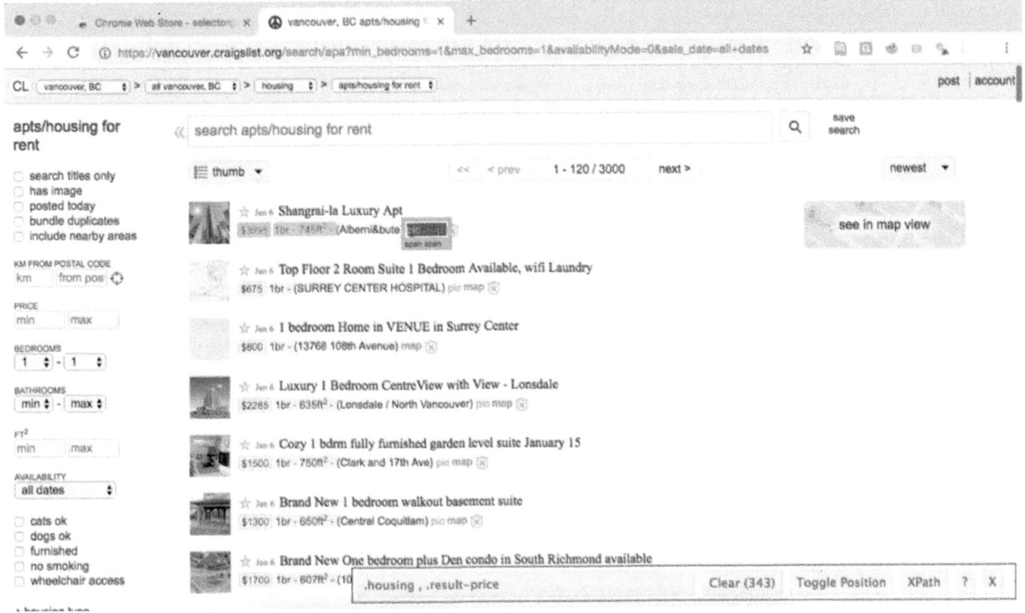

FIGURE 2.6: Using the SelectorGadget on a Craigslist webpage to refine the CCS selector to one that is most useful for obtaining apartment sizes.

Stop! Are you allowed to scrape that website? *Before* scraping data from the web, you should always check whether or not you are *allowed* to scrape it! There are two documents that are important for this: the `robots.txt` file and the Terms of Service document. If we take a look at Craigslist's Terms of Service document[7], we find the following text: *"You agree not to copy/collect CL content via robots, spiders, scripts, scrapers, crawlers, or any automated or manual equivalent (e.g., by hand)."* So unfortunately, without explicit permission, we are not allowed to scrape the website.

What to do now? Well, we *could* ask the owner of Craigslist for permission to scrape. However, we are not likely to get a response, and even if we did they would not likely give us permission. The more realistic answer is that we simply cannot scrape Craigslist. If we still want to find data about rental prices in Vancouver, we must go elsewhere. To continue learning how to scrape data from the web, let's instead scrape data on the population of Canadian cities from Wikipedia. We have checked the Terms of Service document[8], and it does not mention that web scraping is disallowed. We will use the SelectorGadget tool to pick elements that we are interested in (city names and population counts) and deselect others to indicate that we are not interested in them (province names), as shown in Figure 2.7.

[7] `https://www.craigslist.org/about/terms.of.use`

[8] `https://foundation.wikimedia.org/wiki/Terms_of_Use/en`

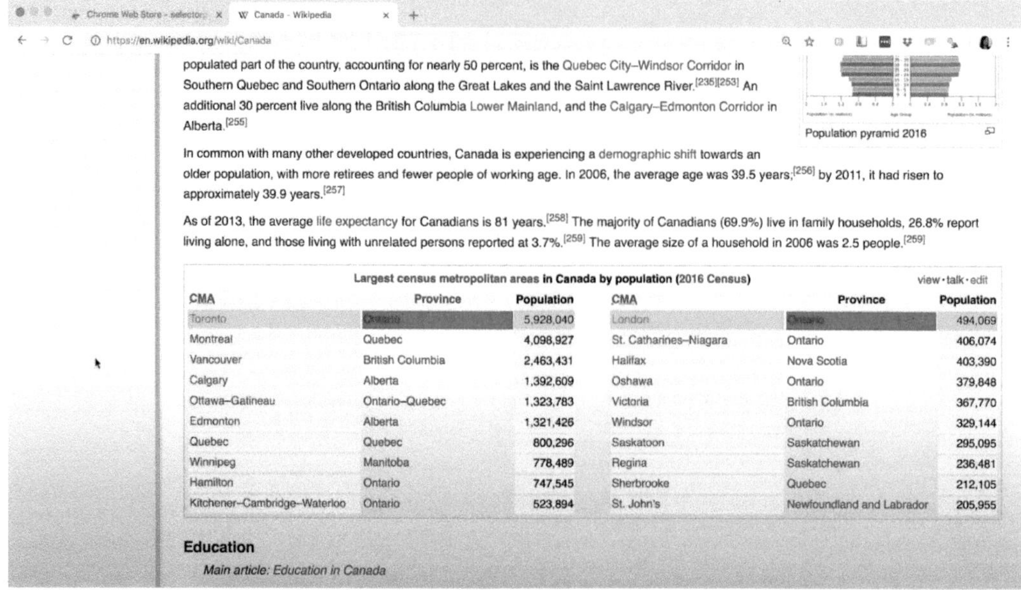

populated part of the country, accounting for nearly 50 percent, is the Quebec City–Windsor Corridor in Southern Quebec and Southern Ontario along the Great Lakes and the Saint Lawrence River.[235][253] An additional 30 percent live along the British Columbia Lower Mainland, and the Calgary–Edmonton Corridor in Alberta.[255]

In common with many other developed countries, Canada is experiencing a demographic shift towards an older population, with more retirees and fewer people of working age. In 2006, the average age was 39.5 years;[256] by 2011, it had risen to approximately 39.9 years.[257]

As of 2013, the average life expectancy for Canadians is 81 years.[258] The majority of Canadians (69.9%) live in family households, 26.8% report living alone, and those living with unrelated persons reported at 3.7%.[259] The average size of a household in 2006 was 2.5 people.[259]

Largest census metropolitan areas in Canada by population (2016 Census) view · talk · edit

CMA	Province	Population	CMA	Province	Population
Toronto	Ontario	5,928,040	London	Ontario	494,069
Montreal	Quebec	4,098,927	St. Catharines–Niagara	Ontario	406,074
Vancouver	British Columbia	2,463,431	Halifax	Nova Scotia	403,390
Calgary	Alberta	1,392,609	Oshawa	Ontario	379,848
Ottawa–Gatineau	Ontario–Quebec	1,323,783	Victoria	British Columbia	367,770
Edmonton	Alberta	1,321,426	Windsor	Ontario	329,144
Quebec	Quebec	800,296	Saskatoon	Saskatchewan	295,095
Winnipeg	Manitoba	778,489	Regina	Saskatchewan	236,481
Hamilton	Ontario	747,545	Sherbrooke	Quebec	212,105
Kitchener–Cambridge–Waterloo	Ontario	523,894	St. John's	Newfoundland and Labrador	205,955

Education

Main article: Education in Canada

FIGURE 2.7: Using the SelectorGadget on a Wikipedia webpage.

We include a link to a short video tutorial on this process at the end of the chapter in the additional resources section. SelectorGadget provides in its toolbar the following list of CSS selectors to use:

```
td:nth-child(5),
td:nth-child(7),
.infobox:nth-child(122) td:nth-child(1),
.infobox td:nth-child(3)
```

Now that we have the CSS selectors that describe the properties of the elements that we want to target, we can use them to find certain elements in web pages and extract data.

Using `rvest`

Now that we have our CSS selectors we can use the `rvest` R package to scrape our desired data from the website. We start by loading the `rvest` package:

```
library(rvest)
```

Next, we tell R what page we want to scrape by providing the webpage's URL in quotations to the function `read_html`:

```
page <- read_html("https://en.wikipedia.org/wiki/Canada")
```

The `read_html` function directly downloads the source code for the page at the URL you specify, just like your browser would if you navigated to that site. But instead of displaying the website to you, the `read_html` function just returns the HTML source code itself, which we have stored in the `page` variable. Next, we send the page object to the `html_nodes` function, along with the CSS selectors we obtained from the SelectorGadget tool. Make sure to surround the selectors with quotation marks; the function, `html_nodes`, expects that argument is a string. The `html_nodes` function then selects *nodes* from the HTML document that match the CSS selectors you specified. A *node* is an HTML tag pair (e.g., `<td>` and `</td>` which defines the cell of a table) combined with the content stored between the tags. For our CSS selector `td:nth-child(5)`, an example node that would be selected would be:

```
<td style="text-align:left;background:#f0f0f0;">
<a href="/wiki/London,_Ontario" title="London, Ontario">London</a>
</td>
```

We store the result of the `html_nodes` function in the `population_nodes` variable. Note that below we use the `paste` function with a comma separator (`sep=","`) to build the list of selectors. The `paste` function converts elements to characters and combines the values into a list. We use this function to build the list of selectors to maintain code readability; this avoids having one very long line of code with the string `"td:nth-child(5),td:nth-child(7),.infobox:nth-child(122) td:nth-child(1),.infobox td:nth-child(3)"` as the second argument of `html_nodes`:

```
selectors <- paste("td:nth-child(5)",
                   "td:nth-child(7)",
                   ".infobox:nth-child(122) td:nth-child(1)",
                   ".infobox td:nth-child(3)", sep = ",")

population_nodes <- html_nodes(page, selectors)
head(population_nodes)
```

```
## {xml_nodeset (6)}
## [1] <td style="text-align:left;background:#f0f0f0;"><a href="/wiki/London,_On .
## [2] <td style="text-align:right;">543,551\n</td>
## [3] <td style="text-align:left;background:#f0f0f0;"><a href="/wiki/Halifax,_N .
## [4] <td style="text-align:right;">465,703\n</td>
```

```
## [5] <td style="text-align:left;background:#f0f0f0;">\n<a href="/wiki/St._Cath .
## [6] <td style="text-align:right;">433,604\n</td>
```

Note: `head` is a function that is often useful for viewing only a short summary of an R object, rather than the whole thing (which may be quite a lot to look at). For example, here `head` shows us only the first 6 items in the `population_nodes` object. Note that some R objects by default print only a small summary. For example, `tibble` data frames only show you the first 10 rows. But not *all* R objects do this, and that's where the `head` function helps summarize things for you.

Next we extract the meaningful data—in other words, we get rid of the HTML code syntax and tags—from the nodes using the `html_text` function. In the case of the example node above, `html_text` function returns `"London"`.

```
population_text <- html_text(population_nodes)
head(population_text)
```

```
## [1] "London"                  "543,551\n"               "Halifax"
## [4] "465,703\n"               "St. Catharines-Niagara" "433,604\n"
```

Fantastic! We seem to have extracted the data of interest from the raw HTML source code. But we are not quite done; the data is not yet in an optimal format for data analysis. Both the city names and population are encoded as characters in a single vector, instead of being in a data frame with one character column for city and one numeric column for population (like a spreadsheet). Additionally, the populations contain commas (not useful for programmatically dealing with numbers), and some even contain a line break character at the end (`\n`). In Chapter 3, we will learn more about how to *wrangle* data such as this into a more useful format for data analysis using R.

2.8.2 Using an API

Rather than posting a data file at a URL for you to download, many websites these days provide an API that must be accessed through a programming language like R. The benefit of this is that data owners have much more control over the data they provide to users. However, unlike web scraping, there is no consistent way to access an API across websites. Every website typically has its own API designed especially for its own use case. Therefore we will just provide one example of accessing data through an API in this book, with the hope that it gives you enough of a basic idea that you can learn how to use another API if needed.

In particular, in this book we will show you the basics of how to use the `rtweet` package in R to access data from the Twitter API. One nice feature of this particular API is that you don't need a special *token* to access it; you simply need to make an account with them. Your access to the data will then be authenticated and controlled through your account username and password. If you have a Twitter account already (or are willing to make one), you can follow along with the examples that we show here. To get started, load the `rtweet` package:

```
library(rtweet)
```

This package provides an extensive set of functions to search Twitter for tweets, users, their followers, and more. Let's construct a small data set of the last 400 tweets and retweets from the @tidyverse[9] account. A few of the most recent tweets are shown in Figure 2.8.

Stop! Think about your API usage carefully!

When you access an API, you are initiating a transfer of data from a web server to your computer. Web servers are expensive to run and do not have infinite resources. If you try to ask for *too much data* at once, you can use up a huge amount of the server's bandwidth. If you try to ask for data *too frequently*—e.g., if you make many requests to the server in quick succession—you can also bog the server down and make it unable to talk to anyone else. Most servers have mechanisms to revoke your access if you are not careful, but you should try to prevent issues from happening in the first place by being extra careful with how you write and run your code. You should also keep in mind that when a website owner grants you API access, they also usually specify a limit (or *quota*) of how much data you can ask for. Be careful not to overrun your quota! In this example, we should take a look at the Twitter website[10] to see what limits we should abide by when using the API.

Using `rtweet`

After checking the Twitter website, it seems like asking for 400 tweets one time is acceptable. So we can use the `get_timelines` function to ask for the last 400 tweets from the @tidyverse[11] account.

---

[9] `https://twitter.com/tidyverse`

[10] `https://developer.twitter.com/en/docs/twitter-api/rate-limits`

[11] `https://twitter.com/tidyverse`

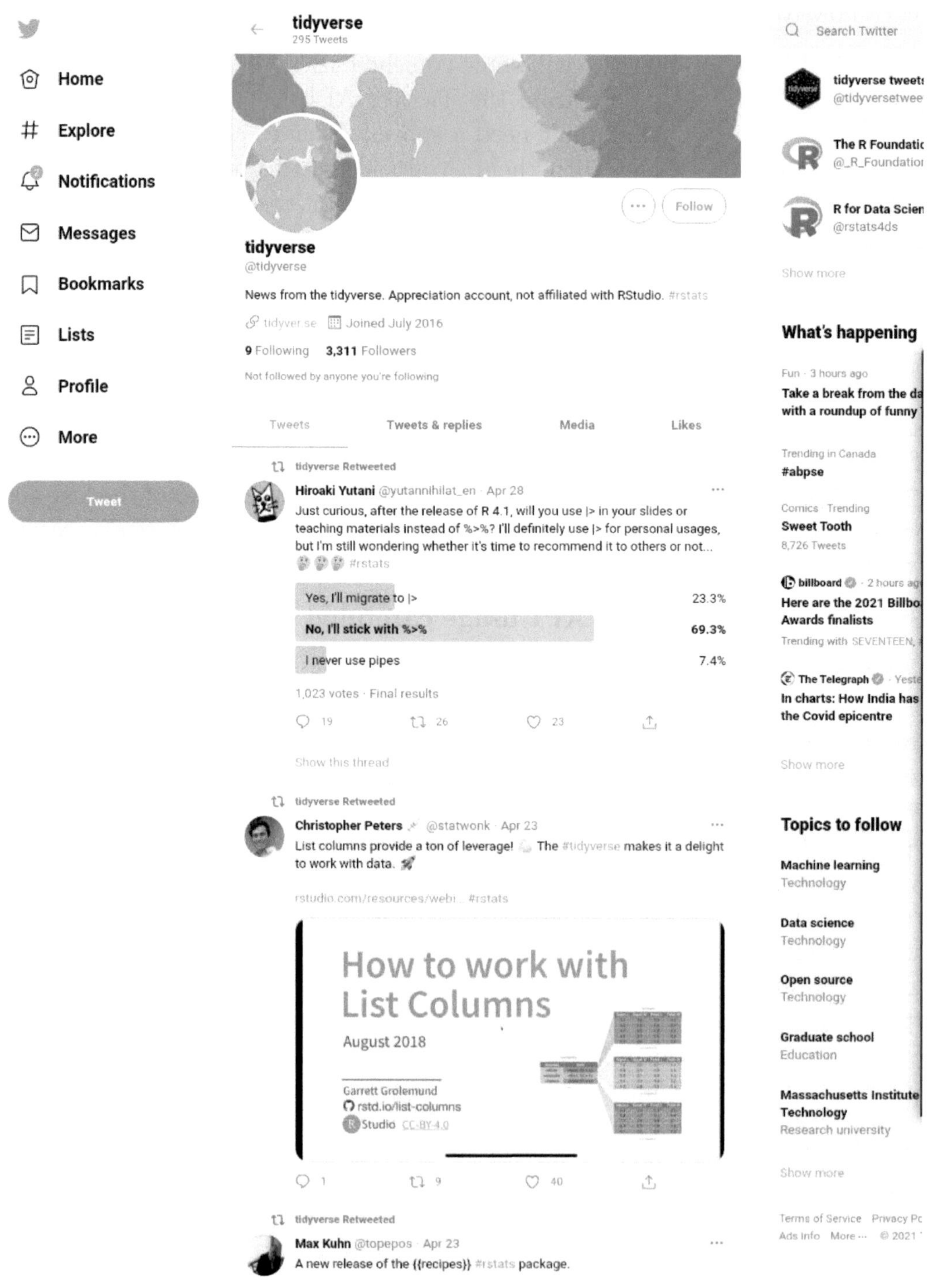

FIGURE 2.8: The tidyverse account Twitter feed.

```
tidyverse_tweets <- get_timelines('tidyverse', n=400)
```

When you call the `get_timelines` for the first time (or any other `rtweet` function that accesses the API), you will see a browser pop-up that looks something like Figure 2.9.

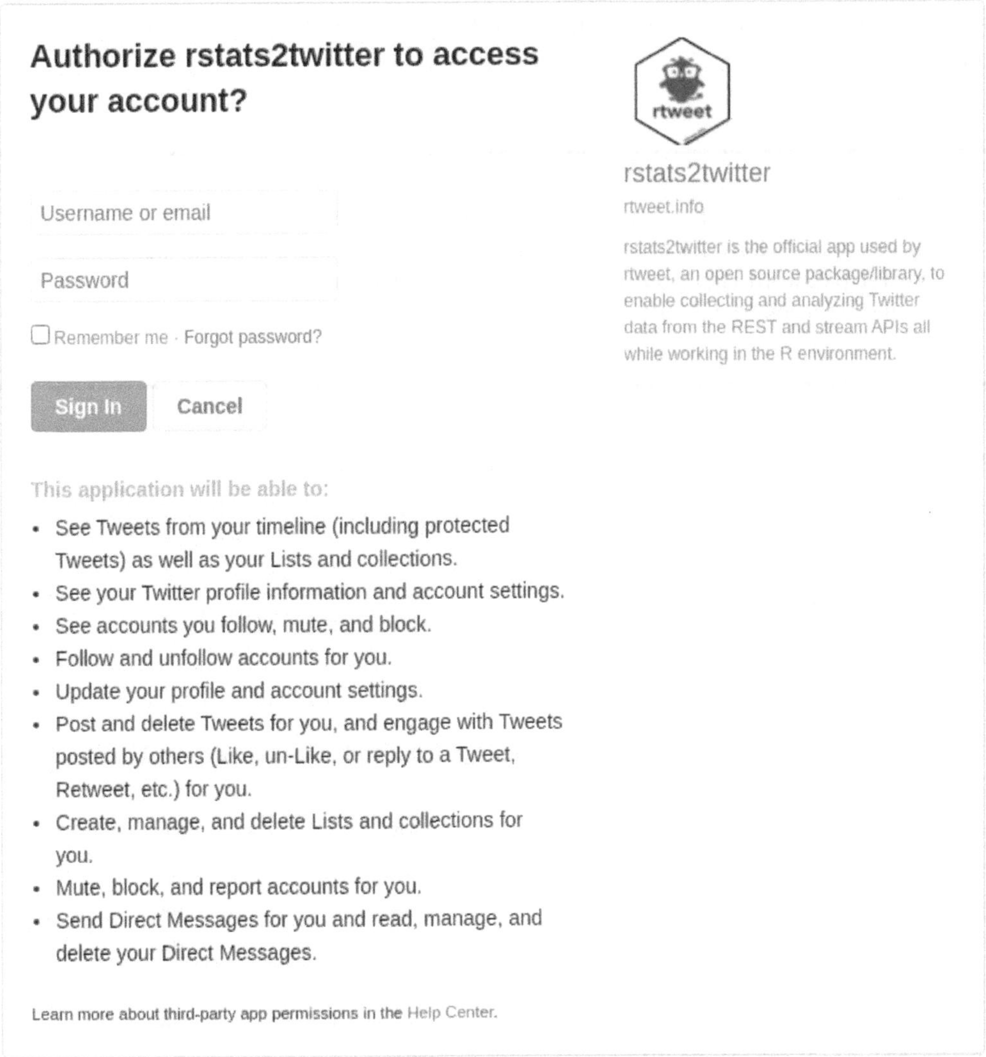

Authorize rstats2twitter to access your account?

Username or email

Password

Remember me · Forgot password?

Sign In Cancel

This application will be able to:

- See Tweets from your timeline (including protected Tweets) as well as your Lists and collections.
- See your Twitter profile information and account settings.
- See accounts you follow, mute, and block.
- Follow and unfollow accounts for you.
- Update your profile and account settings.
- Post and delete Tweets for you, and engage with Tweets posted by others (Like, un-Like, or reply to a Tweet, Retweet, etc.) for you.
- Create, manage, and delete Lists and collections for you.
- Mute, block, and report accounts for you.
- Send Direct Messages for you and read, manage, and delete your Direct Messages.

Learn more about third-party app permissions in the Help Center.

rtweet

rstats2twitter

rtweet.info

rstats2twitter is the official app used by rtweet, an open source package/library, to enable collecting and analyzing Twitter data from the REST and stream APIs all while working in the R environment.

FIGURE 2.9: The `rtweet` authorization prompt.

This is the `rtweet` package asking you to provide your own Twitter account's login information. When `rtweet` talks to the Twitter API, it uses your account information to authenticate requests; Twitter then can keep track of how much data you're asking for, and how frequently you're asking. If you want to follow along with this example using your own Twitter account, you should read over the list of permissions you are granting `rtweet` *very carefully* and make sure you are comfortable with it. Note that `rtweet` can be used to manage most aspects of your account (make posts, follow others, etc.), which is why `rtweet` asks for such extensive permissions. If you decide to allow `rtweet` to talk to the Twitter API using your account information, then input your username and password and hit "Sign In." Twitter will probably send you an email to say that there was an unusual login attempt on your account, and in that case you will have to take the one-time code they send you and provide that to the `rtweet` login page too.

Note: Every API has its own way to authenticate users when they try to access data. Many APIs require you to sign up to receive a *token*, which is a secret password that you input into the R package (like `rtweet`) that you are using to access the API.

With the authentication setup out of the way, let's run the `get_timelines` function again to actually access the API and take a look at what was returned:

```
tidyverse_tweets <- get_timelines('tidyverse', n = 400)
tidyverse_tweets
```

```
## # A tibble: 293 x 71
##    created_at          reply_to_status_id quoted_created_at reply_to_user_id
##    <dttm>              <lgl>              <dttm>            <lgl>
##  1 2021-04-29 11:59:04 NA                 NA                NA
##  2 2021-04-26 17:05:47 NA                 NA                NA
##  3 2021-04-24 09:13:12 NA                 NA                NA
##  4 2021-04-18 06:06:21 NA                 NA                NA
##  5 2021-04-12 05:48:33 NA                 NA                NA
##  6 2021-04-08 17:45:34 NA                 NA                NA
##  7 2021-04-01 05:01:38 NA                 NA                NA
##  8 2021-03-25 06:05:49 NA                 NA                NA
##  9 2021-03-18 17:16:21 NA                 NA                NA
## 10 2021-03-12 19:12:49 NA                 NA                NA
## # ... with 283 more rows, and 67 more variables: reply_to_screen_name <lgl>,
## #   is_quote <lgl>, is_retweet <lgl>, quoted_verified <lgl>,
## #   retweet_verified <lgl>, protected <lgl>, verified <lgl>,
## #   account_lang <lgl>, profile_background_url <lgl>, user_id <dbl>,
```

```
## #   status_id <dbl>, screen_name <chr>, text <chr>, source <chr>,
## #   ext_media_type <lgl>, lang <chr>, quoted_status_id <dbl>,
## #   quoted_text <chr>, quoted_source <chr>, quoted_user_id <dbl>, ...
```

The data has quite a few variables! (Notice that the output above shows that we have a data table with 293 rows and 71 columns). Let's reduce this down to a few variables of interest: `created_at`, `retweet_screen_name`, `is_retweet`, and `text`.

```
tidyverse_tweets <- select(tidyverse_tweets,
                           created_at,
                           retweet_screen_name,
                           is_retweet,
                           text)

tidyverse_tweets
```

```
## # A tibble: 293 x 4
##    created_at          retweet_screen_name is_retweet text
##    <dttm>              <chr>               <lgl>      <chr>
##  1 2021-04-29 11:59:04 yutannihilat_en     TRUE       "Just curious, after the ~
##  2 2021-04-26 17:05:47 statwonk            TRUE       "List columns provide a t~
##  3 2021-04-24 09:13:12 topepos             TRUE       "A new release of the {{r~
##  4 2021-04-18 06:06:21 ninarbrooks         TRUE       "Always typing `? pivot_l~
##  5 2021-04-12 05:48:33 rfunctionaday       TRUE       "If you are fluent in {dp~
##  6 2021-04-08 17:45:34 RhesusMaCassidy     TRUE       "R-Ladies of Göttingen! T~
##  7 2021-04-01 05:01:38 dvaughan32          TRUE       "I am ridiculously excite~
##  8 2021-03-25 06:05:49 rdpeng              TRUE       "New book out on using ti~
##  9 2021-03-18 17:16:21 SolomonKurz         TRUE       "The 0.2.0 version of my ~
## 10 2021-03-12 19:12:49 hadleywickham       TRUE       "rvest 1.0.0 out now! - h~
## # ... with 283 more rows
```

If you look back up at the image of the @tidyverse[12] Twitter page, you will recognize the text of the most recent few tweets in the above data frame. In other words, we have successfully created a small data set using the Twitter API—neat! This data is also quite different from what we obtained from web scraping; it is already well-organized into a `tidyverse` data frame (although not *every* API will provide data in such a nice format). From this point onward, the `tidyverse_tweets` data frame is stored on your machine, and you can play with it to your heart's content. For example, you can use `write_csv` to save it to a file and `read_csv` to read it into R again later; and after reading the next

[12] https://twitter.com/tidyverse

few chapters you will have the skills to compute the percentage of retweets versus tweets, find the most oft-retweeted account, make visualizations of the data, and much more! If you decide that you want to ask the Twitter API for more data (see the `rtweet` page[13] for more examples of what is possible), just be mindful as usual about how much data you are requesting and how frequently you are making requests.

2.9 Exercises

Practice exercises for the material covered in this chapter can be found in the accompanying worksheets repository[14] in the "Reading in data locally and from the web" row. You can launch an interactive version of the worksheet in your browser by clicking the "launch binder" button. You can also preview a non-interactive version of the worksheet by clicking "view worksheet." If you instead decide to download the worksheet and run it on your own machine, make sure to follow the instructions for computer setup found in Chapter 13. This will ensure that the automated feedback and guidance that the worksheets provide will function as intended.

2.10 Additional resources

- The `readr` documentation[15] provides the documentation for many of the reading functions we cover in this chapter. It is where you should look if you want to learn more about the functions in this chapter, the full set of arguments you can use, and other related functions. The site also provides a very nice cheat sheet that summarizes many of the data wrangling functions from this chapter.
- Sometimes you might run into data in such poor shape that none of the reading functions we cover in this chapter work. In that case, you can consult the data import chapter[16] from *R for Data Science* [Wickham and Grolemund, 2016], which goes into a lot more detail about how R parses text from files into data frames.

[13] `https://github.com/ropensci/rtweet`

[14] `https://github.com/UBC-DSCI/data-science-a-first-intro-worksheets#readme`

[15] `https://readr.tidyverse.org/`

[16] `https://r4ds.had.co.nz/data-import.html`

- The here R package[17] [Müller, 2020] provides a way for you to construct or find your files' paths.
- The readxl documentation[18] provides more details on reading data from Excel, such as reading in data with multiple sheets, or specifying the cells to read in.
- The rio R package[19] [Leeper, 2021] provides an alternative set of tools for reading and writing data in R. It aims to be a "Swiss army knife" for data reading/writing/converting, and supports a wide variety of data types (including data formats generated by other statistical software like SPSS and SAS).
- A video[20] from the Udacity course *Linux Command Line Basics* provides a good explanation of absolute versus relative paths.
- If you read the subsection on obtaining data from the web via scraping and APIs, we provide two companion tutorial video links for how to use the SelectorGadget tool to obtain desired CSS selectors for:
  - extracting the data for apartment listings on Craigslist[21], and
  - extracting Canadian city names and 2016 populations from Wikipedia[22].
- The polite R package[23] [Perepolkin, 2021] provides a set of tools for responsibly scraping data from websites.

---

[17] https://here.r-lib.org/
[18] https://readxl.tidyverse.org/
[19] https://github.com/leeper/rio
[20] https://www.youtube.com/embed/ephId3mYu9o
[21] https://www.youtube.com/embed/YdIWI6K64zo
[22] https://www.youtube.com/embed/O9HKbdhqYzk
[23] https://dmi3kno.github.io/polite/

3

Cleaning and wrangling data

3.1 Overview

This chapter is centered around defining tidy data—a data format that is suitable for analysis—and the tools needed to transform raw data into this format. This will be presented in the context of a real-world data science application, providing more practice working through a whole case study.

3.2 Chapter learning objectives

By the end of the chapter, readers will be able to do the following:

- Define the term "tidy data".
- Discuss the advantages of storing data in a tidy data format.
- Define what vectors, lists, and data frames are in R, and describe how they relate to each other.
- Describe the common types of data in R and their uses.
- Recall and use the following functions for their intended data wrangling tasks:
  - `across`
  - `c`
  - `filter`
  - `group_by`
  - `select`
  - `map`
  - `mutate`
  - `pull`
  - `pivot_longer`
  - `pivot_wider`
  - `rowwise`
  - `separate`
  - `summarize`

- Recall and use the following operators for their intended data wrangling tasks:
  - `==`
  - `%in%`
  - `!`
  - `&`
  - `|`
  - `|>` and `%>%`

3.3 Data frames, vectors, and lists

In Chapters 1 and 2, *data frames* were the focus: we learned how to import data into R as a data frame, and perform basic operations on data frames in R. In the remainder of this book, this pattern continues. The vast majority of tools we use will require that data are represented as a data frame in R. Therefore, in this section, we will dig more deeply into what data frames are and how they are represented in R. This knowledge will be helpful in effectively utilizing these objects in our data analyses.

3.3.1 What is a data frame?

A data frame is a table-like structure for storing data in R. Data frames are important to learn about because most data that you will encounter in practice can be naturally stored as a table. In order to define data frames precisely, we need to introduce a few technical terms:

- **variable:** a characteristic, number, or quantity that can be measured.
- **observation:** all of the measurements for a given entity.
- **value:** a single measurement of a single variable for a given entity.

Given these definitions, a **data frame** is a tabular data structure in R that is designed to store observations, variables, and their values. Most commonly, each column in a data frame corresponds to a variable, and each row corresponds to an observation. For example, Figure 3.1 displays a data set of city populations. Here, the variables are "region, year, population"; each of these are properties that can be collected or measured. The first observation is "Toronto, 2016, 2235145"; these are the values that the three variables take for the first entity in the data set. There are 13 entities in the data set in total, corresponding to the 13 rows in Figure 3.1.

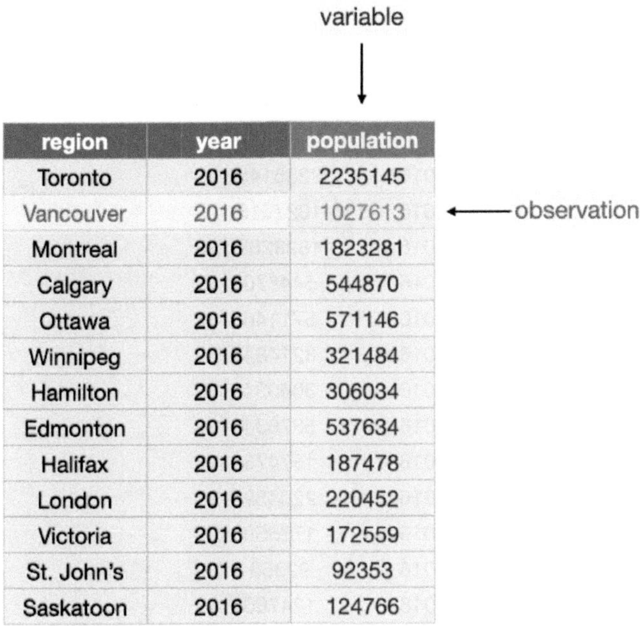

region	year	population
Toronto	2016	2235145
Vancouver	2016	1027613
Montreal	2016	1823281
Calgary	2016	544870
Ottawa	2016	571146
Winnipeg	2016	321484
Hamilton	2016	306034
Edmonton	2016	537634
Halifax	2016	187478
London	2016	220452
Victoria	2016	172559
St. John's	2016	92353
Saskatoon	2016	124766

FIGURE 3.1: A data frame storing data regarding the population of various regions in Canada. In this example data frame, the row that corresponds to the observation for the city of Vancouver is colored yellow, and the column that corresponds to the population variable is colored blue.

R stores the columns of a data frame as either *lists* or *vectors*. For example, the data frame in Figure 3.2 has three vectors whose names are `region`, `year` and `population`. The next two sections will explain what lists and vectors are.

3.3.2 What is a vector?

In R, **vectors** are objects that can contain one or more elements. The vector elements are ordered, and they must all be of the same **data type**; R has several different basic data types, as shown in Table 3.1. Figure 3.3 provides an example of a vector where all of the elements are of character type. You can create vectors in R using the `c` function (`c` stands for "concatenate"). For example, to create the vector `region` as shown in Figure 3.3, you would write:

```
year <- c("Toronto", "Montreal", "Vancouver", "Calgary", "Ottawa")
year
```

```
## [1] "Toronto"   "Montreal"  "Vancouver" "Calgary"   "Ottawa"
```

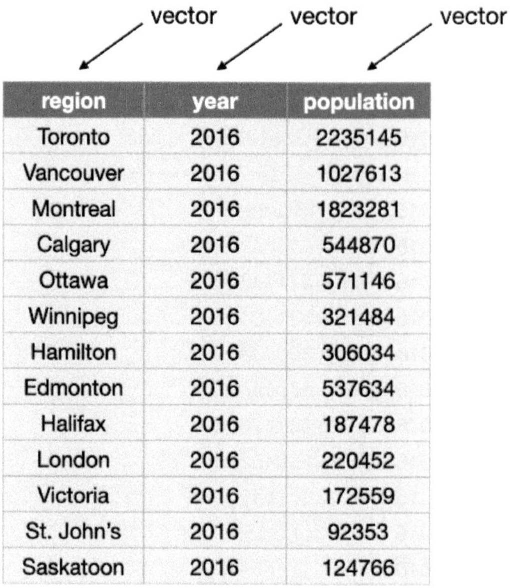

vector vector vector

region	year	population
Toronto	2016	2235145
Vancouver	2016	1027613
Montreal	2016	1823281
Calgary	2016	544870
Ottawa	2016	571146
Winnipeg	2016	321484
Hamilton	2016	306034
Edmonton	2016	537634
Halifax	2016	187478
London	2016	220452
Victoria	2016	172559
St. John's	2016	92353
Saskatoon	2016	124766

FIGURE 3.2: Data frame with three vectors.

Note: Technically, these objects are called "atomic vectors." In this book we have chosen to call them "vectors," which is how they are most commonly referred to in the R community. To be totally precise, "vector" is an umbrella term that encompasses both atomic vector and list objects in R. But this creates a confusing situation where the term "vector" could mean "atomic vector" *or* "the umbrella term for atomic vector and list," depending on context. Very confusing indeed! So to keep things simple, in this book we *always* use the term "vector" to refer to "atomic vector." We encourage readers who are enthusiastic to learn more to read the Vectors chapter of *Advanced R* [Wickham, 2019].

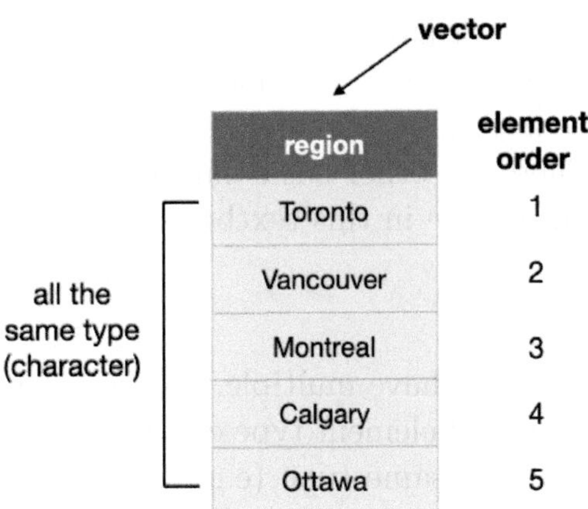

FIGURE 3.3: Example of a vector whose type is character.

TABLE 3.1: Basic data types in R

Data type	Abbreviation	Description	Example
character	chr	letters or numbers surrounded by quotes	"1" , "Hello world!"
double	dbl	numbers with decimals values	1.2333
integer	int	numbers that do not contain decimals	1L, 20L (where "L" tells R to store as an integer)
logical	lgl	either true or false	`TRUE`, `FALSE`
factor	fct	used to represent data with a limited number of values (usually categories)	a `color` variable with levels `red`, `green` and `orange`

It is important in R to make sure you represent your data with the correct type. Many of the `tidyverse` functions we use in this book treat the various data types differently. You should use integers and double types (which both fall under the "numeric" umbrella type) to represent numbers and perform arithmetic. Doubles are more common than integers in R, though; for instance, a double data type is the default when you create a vector of numbers using `c()`, and when you read in whole numbers via `read_csv`. Characters are used to represent data that should be thought of as "text", such as words, names, paths, URLs, and more. Factors help us encode variables that represent *categories*; a factor variable takes one of a discrete set of values known as *levels* (one for each category). The levels can be ordered or unordered. Even though factors can sometimes *look* like characters, they are not used to represent text, words, names, and paths in the way that characters are; in fact, R internally stores factors using integers! There are other basic data types in R, such as *raw* and *complex*, but we do not use these in this textbook.

3.3.3 What is a list?

Lists are also objects in R that have multiple, ordered elements. Vectors and lists differ by the requirement of element type consistency. All elements within a single vector must be of the same type (e.g., all elements are characters), whereas elements within a single list can be of different types (e.g., characters, integers, logicals, and even other lists). See Figure 3.4.

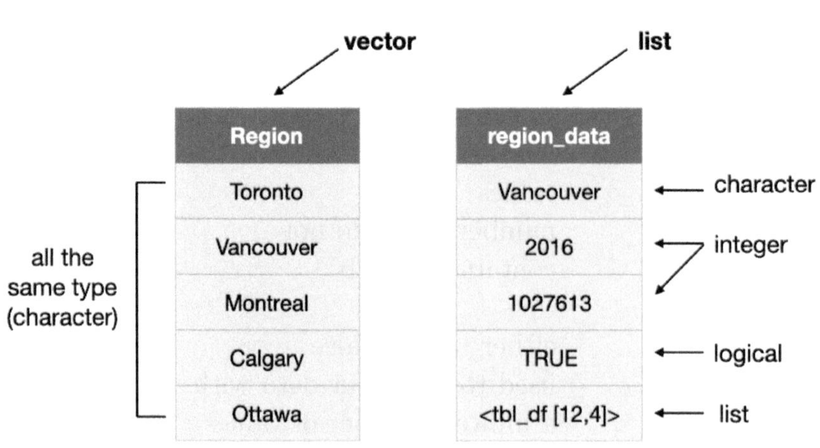

FIGURE 3.4: A vector versus a list.

3.3.4 What does this have to do with data frames?

A data frame is really a special kind of list that follows two rules:

1. Each element itself must either be a vector or a list.
2. Each element (vector or list) must have the same length.

Not all columns in a data frame need to be of the same type. Figure 3.5 shows a data frame where the columns are vectors of different types. But remember: because the columns in this example are *vectors*, the elements must be the same data type *within each column*. On the other hand, if our data frame had *list* columns, there would be no such requirement. It is generally much more common to use *vector* columns, though, as the values for a single variable are usually all of the same type.

vector of type character | vector of type integer | vector of type logical

region	year	voted
Toronto	2016	TRUE
Vancouver	2016	TRUE
Montreal	2016	TRUE
Calgary	2016	TRUE
Ottawa	2016	TRUE
Winnipeg	2016	TRUE
Hamilton	2016	TRUE
Edmonton	2016	TRUE
Halifax	2016	TRUE
London	2016	TRUE
Victoria	2016	TRUE
St. John's	2016	TRUE
Saskatoon	2016	TRUE

FIGURE 3.5: Data frame and vector types.

The functions from the `tidyverse` package that we use often give us a special class of data frame called a *tibble*. Tibbles have some additional features and benefits over the built-in data frame object. These include the ability to add useful attributes (such as grouping, which we will discuss later) and more predictable type preservation when subsetting. Because a tibble is just a data frame with some added features, we will collectively refer to both built-in R data frames and tibbles as data frames in this book.

Note: You can use the function `class` on a data object to assess whether a data frame is a built-in R data frame or a tibble. If the data object is a data frame, `class` will return `"data.frame"`. If the data object is a tibble it will return "tbl_df" "tbl" "data.frame". You can easily convert built-in R data frames to tibbles using the tidyverse as_tibble function. For example we can check the class of the Canadian languages data set, can_lang, we worked with in the previous chapters and we see it is a tibble.

```
class(can_lang)
```

```
## [1] "spec_tbl_df" "tbl_df"      "tbl"         "data.frame"
```

Vectors, data frames and lists are basic types of *data structure* in R, which are core to most data analyses. We summarize them in Table 3.2. There are several other data structures in the R programming language (*e.g.,* matrices), but these are beyond the scope of this book.

TABLE 3.2: Basic data structures in R

Data Structure	Description
vector	An ordered collection of one, or more, values of the *same data type.*
list	An ordered collection of one, or more, values of *possibly different data types.*
data frame	A list of either vectors or lists of the *same length*, with column names. We typically use a data frame to represent a data set.

3.4 Tidy data

There are many ways a tabular data set can be organized. This chapter will focus on introducing the **tidy data** format of organization and how to make your raw (and likely messy) data tidy. A tidy data frame satisfies the following three criteria [Wickham, 2014]:

- each row is a single observation,
- each column is a single variable, and
- each value is a single cell (i.e., its entry in the data frame is not shared with another value).

Figure 3.6 demonstrates a tidy data set that satisfies these three criteria.

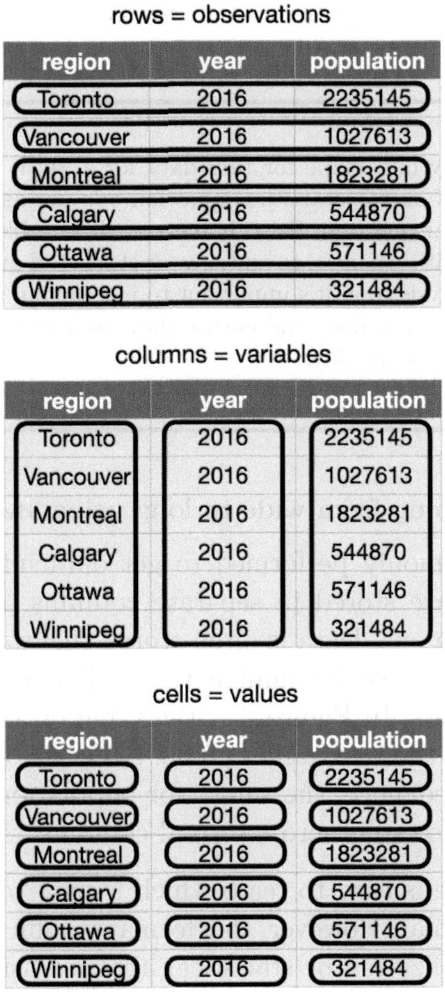

rows = observations

region	year	population
Toronto	2016	2235145
Vancouver	2016	1027613
Montreal	2016	1823281
Calgary	2016	544870
Ottawa	2016	571146
Winnipeg	2016	321484

columns = variables

region	year	population
Toronto	2016	2235145
Vancouver	2016	1027613
Montreal	2016	1823281
Calgary	2016	544870
Ottawa	2016	571146
Winnipeg	2016	321484

cells = values

region	year	population
Toronto	2016	2235145
Vancouver	2016	1027613
Montreal	2016	1823281
Calgary	2016	544870
Ottawa	2016	571146
Winnipeg	2016	321484

FIGURE 3.6: Tidy data satisfies three criteria.

There are many good reasons for making sure your data are tidy as a first step in your analysis. The most important is that it is a single, consistent format that nearly every function in the `tidyverse` recognizes. No matter what the variables and observations in your data represent, as long as the data frame is tidy, you can manipulate it, plot it, and analyze it using the same tools. If your data is *not* tidy, you will have to write special bespoke code in your analysis that will not only be error-prone, but hard for others to understand. Beyond making your analysis more accessible to others and less error-prone, tidy data is also typically easy for humans to interpret. Given these benefits, it is well worth spending the time to get your data into a tidy format upfront.

Fortunately, there are many well-designed `tidyverse` data cleaning/wrangling tools to help you easily tidy your data. Let's explore them below!

Note: Is there only one shape for tidy data for a given data set? Not necessarily! It depends on the statistical question you are asking and what the variables are for that question. For tidy data, each variable should be its own column. So, just as it's essential to match your statistical question with the appropriate data analysis tool, it's important to match your statistical question with the appropriate variables and ensure they are represented as individual columns to make the data tidy.

3.4.1 Tidying up: going from wide to long using `pivot_longer`

One task that is commonly performed to get data into a tidy format is to combine values that are stored in separate columns, but are really part of the same variable, into one. Data is often stored this way because this format is sometimes more intuitive for human readability and understanding, and humans create data sets. In Figure 3.7, the table on the left is in an untidy, "wide" format because the year values (2006, 2011, 2016) are stored as column names. And as a consequence, the values for population for the various cities over these years are also split across several columns.

For humans, this table is easy to read, which is why you will often find data stored in this wide format. However, this format is difficult to work with when performing data visualization or statistical analysis using R. For example, if we wanted to find the latest year it would be challenging because the year values are stored as column names instead of as values in a single column. So before we could apply a function to find the latest year (for example, by using `max`), we would have to first extract the column names to get them as a vector and then apply a function to extract the latest year. The problem only gets worse if you would like to find the value for the population for a given region for the latest year. Both of these tasks are greatly simplified once the data is tidied.

Another problem with data in this format is that we don't know what the numbers under each year actually represent. Do those numbers represent population size? Land area? It's not clear. To solve both of these problems, we can reshape this data set to a tidy data format by creating a column called "year" and a column called "population." This transformation—which makes the data "longer"—is shown as the right table in Figure 3.7.

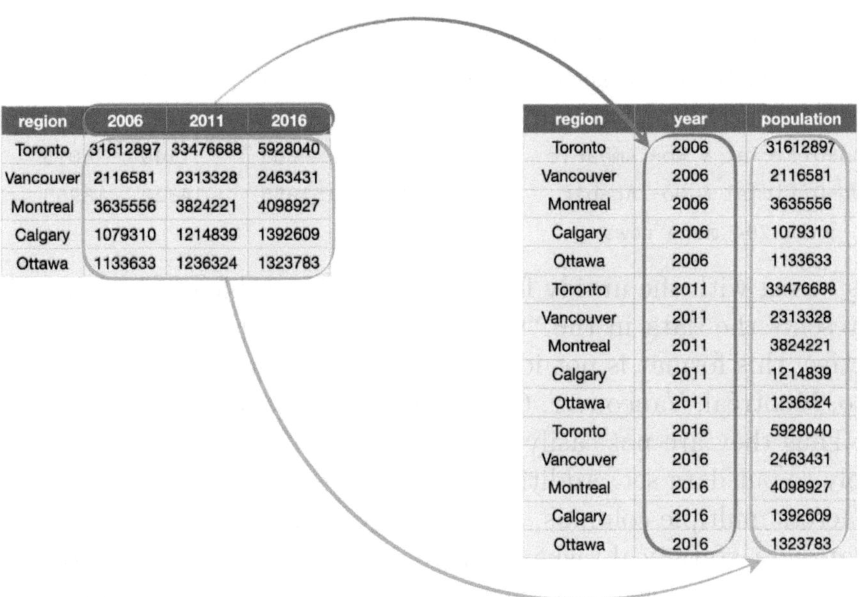

region	2006	2011	2016
Toronto	31612897	33476688	5928040
Vancouver	2116581	2313328	2463431
Montreal	3635556	3824221	4098927
Calgary	1079310	1214839	1392609
Ottawa	1133633	1236324	1323783

region	year	population
Toronto	2006	31612897
Vancouver	2006	2116581
Montreal	2006	3635556
Calgary	2006	1079310
Ottawa	2006	1133633
Toronto	2011	33476688
Vancouver	2011	2313328
Montreal	2011	3824221
Calgary	2011	1214839
Ottawa	2011	1236324
Toronto	2016	5928040
Vancouver	2016	2463431
Montreal	2016	4098927
Calgary	2016	1392609
Ottawa	2016	1323783

FIGURE 3.7: Pivoting data from a wide to long data format.

We can achieve this effect in R using the `pivot_longer` function from the `tidyverse` package. The `pivot_longer` function combines columns, and is usually used during tidying data when we need to make the data frame longer and narrower. To learn how to use `pivot_longer`, we will work through an example with the `region_lang_top5_cities_wide.csv` data set. This data set contains the counts of how many Canadians cited each language as their mother tongue for five major Canadian cities (Toronto, Montréal, Vancouver, Calgary and Edmonton) from the 2016 Canadian census. To get started, we will load the `tidyverse` package and use `read_csv` to load the (untidy) data.

```
library(tidyverse)

lang_wide <- read_csv("data/region_lang_top5_cities_wide.csv")
lang_wide
```

```
## # A tibble: 214 x 7
##   category             language          Toronto Montréal Vancouver Calgary Edmonton
##   <chr>                <chr>               <dbl>    <dbl>     <dbl>   <dbl>    <dbl>
## 1 Aboriginal langua~ Aboriginal la~       80       30        70      20       25
## 2 Non-Official & No~ Afrikaans           985       90      1435     960      575
## 3 Non-Official & No~ Afro-Asiatic ~      360      240        45      45       65
## 4 Non-Official & No~ Akan (Twi)         8485     1015       400     705      885
## 5 Non-Official & No~ Albanian          13260     2450      1090    1365      770
```

```
## 6 Aboriginal langua~ Algonquian la~      5      5     0     0     0
## 7 Aboriginal langua~ Algonquin           5     30     5     5     0
## 8 Non-Official & No~ American Sign~    470     50   265   100   180
## 9 Non-Official & No~ Amharic          7460    665  1140  4075  2515
## 10 Non-Official & No~ Arabic         85175 151955 14320 18965 17525
## # ... with 204 more rows
```

What is wrong with the untidy format above? The table on the left in Figure 3.8 represents the data in the "wide" (messy) format. From a data analysis perspective, this format is not ideal because the values of the variable *region* (Toronto, Montréal, Vancouver, Calgary and Edmonton) are stored as column names. Thus they are not easily accessible to the data analysis functions we will apply to our data set. Additionally, the *mother tongue* variable values are spread across multiple columns, which will prevent us from doing any desired visualization or statistical tasks until we combine them into one column. For instance, suppose we want to know the languages with the highest number of Canadians reporting it as their mother tongue among all five regions. This question would be tough to answer with the data in its current format. We *could* find the answer with the data in this format, though it would be much easier to answer if we tidy our data first. If mother tongue were instead stored as one column, as shown in the tidy data on the right in Figure 3.8, we could simply use the `max` function in one line of code to get the maximum value.

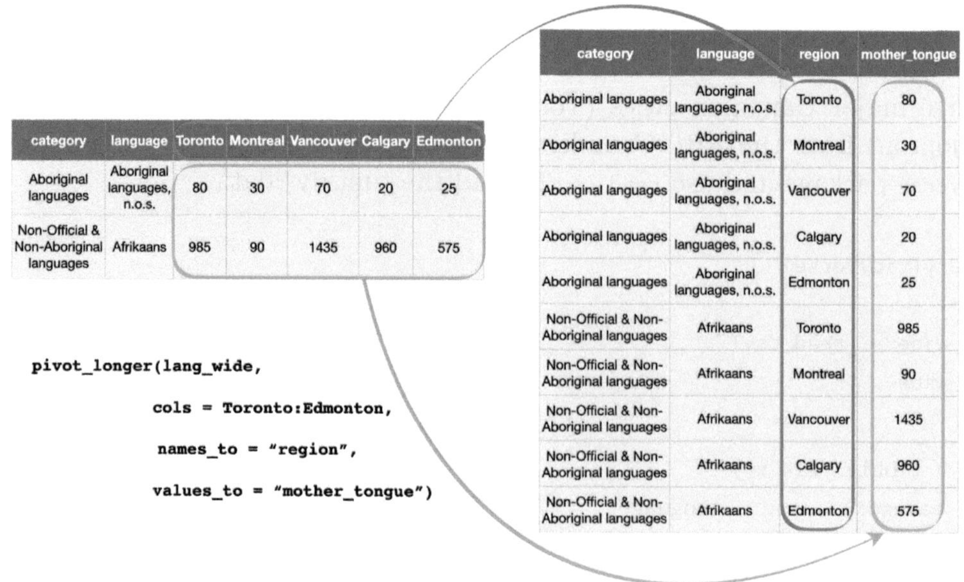

FIGURE 3.8: Going from wide to long with the `pivot_longer` function.

Figure 3.9 details the arguments that we need to specify in the `pivot_longer` function to accomplish this data transformation.

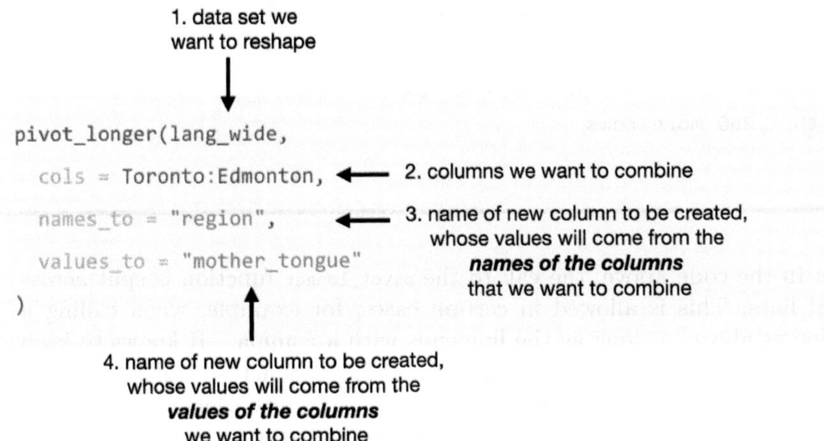

FIGURE 3.9: Syntax for the `pivot_longer` function.

We use `pivot_longer` to combine the Toronto, Montréal, Vancouver, Calgary, and Edmonton columns into a single column called `region`, and create a column called `mother_tongue` that contains the count of how many Canadians report each language as their mother tongue for each metropolitan area. We use a colon `:` between Toronto and Edmonton to tell R to select all the columns between Toronto and Edmonton:

```
lang_mother_tidy <- pivot_longer(lang_wide,
  cols = Toronto:Edmonton,
  names_to = "region",
  values_to = "mother_tongue"
)

lang_mother_tidy
```

```
## # A tibble: 1,070 x 4
##    category                                language          region  mother_tongue
##    <chr>                                    <chr>             <chr>           <dbl>
##  1 Aboriginal languages                     Aboriginal lan~ Toronto            80
##  2 Aboriginal languages                     Aboriginal lan~ Montré~            30
##  3 Aboriginal languages                     Aboriginal lan~ Vancou~            70
##  4 Aboriginal languages                     Aboriginal lan~ Calgary            20
##  5 Aboriginal languages                     Aboriginal lan~ Edmont~            25
##  6 Non-Official & Non-Aboriginal languages Afrikaans          Toronto           985
##  7 Non-Official & Non-Aboriginal languages Afrikaans          Montré~            90
```

```
##  8 Non-Official & Non-Aboriginal languages Afrikaans     Vancou~         1435
##  9 Non-Official & Non-Aboriginal languages Afrikaans     Calgary          960
## 10 Non-Official & Non-Aboriginal languages Afrikaans     Edmont~          575
## # ... with 1,060 more rows
```

Note: In the code above, the call to the `pivot_longer` function is split across several lines. This is allowed in certain cases; for example, when calling a function as above, as long as the line ends with a comma `,` R knows to keep reading on the next line. Splitting long lines like this across multiple lines is encouraged as it helps significantly with code readability. Generally speaking, you should limit each line of code to about 80 characters.

The data above is now tidy because all three criteria for tidy data have now been met:

1. All the variables (`category`, `language`, `region` and `mother_tongue`) are now their own columns in the data frame.
2. Each observation, (i.e., each language in a region) is in a single row.
3. Each value is a single cell, i.e., its row, column position in the data frame is not shared with another value.

3.4.2 Tidying up: going from long to wide using `pivot_wider`

Suppose we have observations spread across multiple rows rather than in a single row. For example, in Figure 3.10, the table on the left is in an untidy, long format because the `count` column contains three variables (population, commuter count, and year the city was incorporated) and information about each observation (here, population, commuter, and incorporated values for a region) is split across three rows. Remember: one of the criteria for tidy data is that each observation must be in a single row.

Using data in this format—where two or more variables are mixed together in a single column—makes it harder to apply many usual `tidyverse` functions. For example, finding the maximum number of commuters would require an additional step of filtering for the commuter values before the maximum can be computed. In comparison, if the data were tidy, all we would have to do is compute the maximum value for the commuter column. To reshape this untidy data set to a tidy (and in this case, wider) format, we need to create columns called "population", "commuters", and "incorporated." This is illustrated in the right table of Figure 3.10.

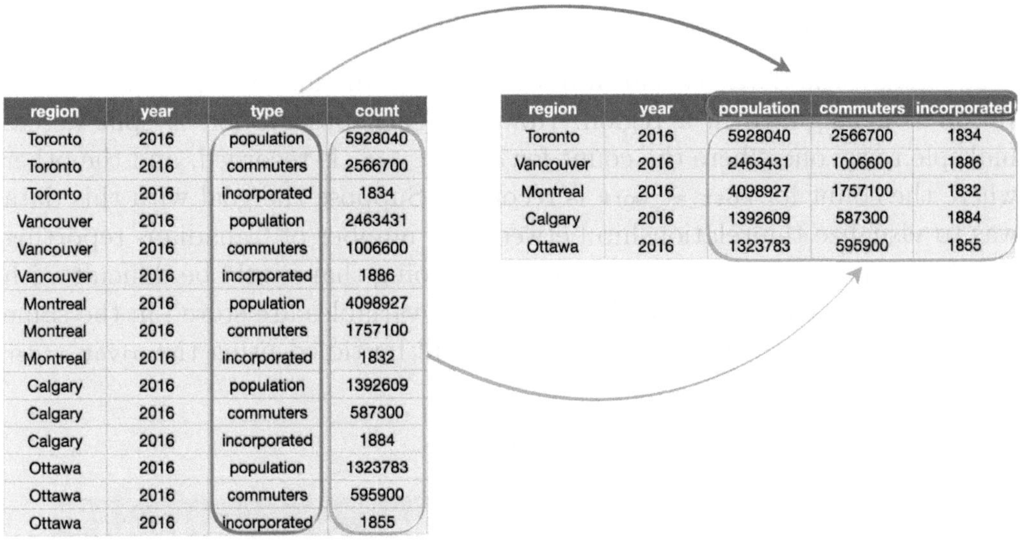

region	year	type	count
Toronto	2016	population	5928040
Toronto	2016	commuters	2566700
Toronto	2016	incorporated	1834
Vancouver	2016	population	2463431
Vancouver	2016	commuters	1006600
Vancouver	2016	incorporated	1886
Montreal	2016	population	4098927
Montreal	2016	commuters	1757100
Montreal	2016	incorporated	1832
Calgary	2016	population	1392609
Calgary	2016	commuters	587300
Calgary	2016	incorporated	1884
Ottawa	2016	population	1323783
Ottawa	2016	commuters	595900
Ottawa	2016	incorporated	1855

region	year	population	commuters	incorporated
Toronto	2016	5928040	2566700	1834
Vancouver	2016	2463431	1006600	1886
Montreal	2016	4098927	1757100	1832
Calgary	2016	1392609	587300	1884
Ottawa	2016	1323783	595900	1855

FIGURE 3.10: Going from long to wide data.

To tidy this type of data in R, we can use the `pivot_wider` function. The `pivot_wider` function generally increases the number of columns (widens) and decreases the number of rows in a data set. To learn how to use `pivot_wider`, we will work through an example with the `region_lang_top5_cities_long.csv` data set. This data set contains the number of Canadians reporting the primary language at home and work for five major cities (Toronto, Montréal, Vancouver, Calgary and Edmonton).

```
lang_long <- read_csv("data/region_lang_top5_cities_long.csv")
lang_long
```

```
## # A tibble: 2,140 x 5
##    region   category               language                       type         count
##    <chr>    <chr>                  <chr>                          <chr>        <dbl>
##  1 Montréal Aboriginal languages Aboriginal languages, n.o.s. most_at_home    15
##  2 Montréal Aboriginal languages Aboriginal languages, n.o.s. most_at_work     0
##  3 Toronto  Aboriginal languages Aboriginal languages, n.o.s. most_at_home    50
##  4 Toronto  Aboriginal languages Aboriginal languages, n.o.s. most_at_work     0
##  5 Calgary  Aboriginal languages Aboriginal languages, n.o.s. most_at_home     5
##  6 Calgary  Aboriginal languages Aboriginal languages, n.o.s. most_at_work     0
##  7 Edmonton Aboriginal languages Aboriginal languages, n.o.s. most_at_home    10
```

```
##  8 Edmonton  Aboriginal languages Aboriginal languages, n.o.s. most_at_work    0
##  9 Vancouver Aboriginal languages Aboriginal languages, n.o.s. most_at_home   15
## 10 Vancouver Aboriginal languages Aboriginal languages, n.o.s. most_at_work    0
## # ... with 2,130 more rows
```

What makes the data set shown above untidy? In this example, each observation is a language in a region. However, each observation is split across multiple rows: one where the count for `most_at_home` is recorded, and the other where the count for `most_at_work` is recorded. Suppose the goal with this data was to visualize the relationship between the number of Canadians reporting their primary language at home and work. Doing that would be difficult with this data in its current form, since these two variables are stored in the same column. Figure 3.11 shows how this data will be tidied using the `pivot_wider` function.

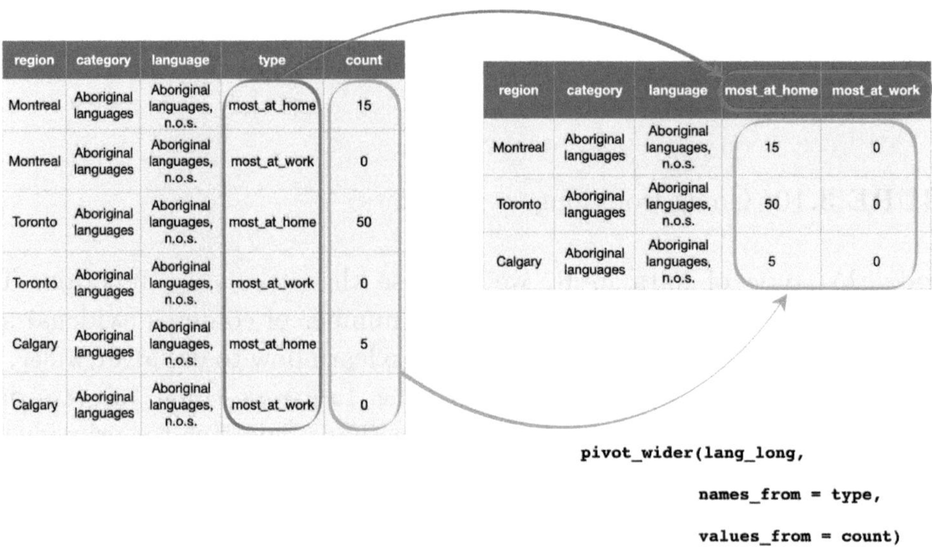

FIGURE 3.11: Going from long to wide with the `pivot_wider` function.

Figure 3.12 details the arguments that we need to specify in the `pivot_wider` function.

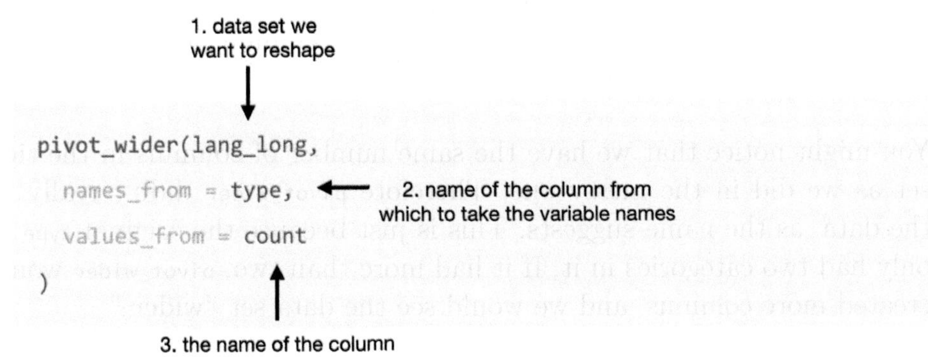

FIGURE 3.12: Syntax for the `pivot_wider` function.

We will apply the function as detailed in Figure 3.12.

```
lang_home_tidy <- pivot_wider(lang_long,
  names_from = type,
  values_from = count
)
lang_home_tidy
```

```
## # A tibble: 1,070 x 5
##    region    category                language          most_at_home most_at_work
##    <chr>     <chr>                   <chr>                    <dbl>        <dbl>
##  1 Montréal  Aboriginal languages    Aboriginal langu~           15            0
##  2 Toronto   Aboriginal languages    Aboriginal langu~           50            0
##  3 Calgary   Aboriginal languages    Aboriginal langu~            5            0
##  4 Edmonton  Aboriginal languages    Aboriginal langu~           10            0
##  5 Vancouver Aboriginal languages    Aboriginal langu~           15            0
##  6 Montréal  Non-Official & Non-Abo~ Afrikaans                   10            0
##  7 Toronto   Non-Official & Non-Abo~ Afrikaans                  265            0
##  8 Calgary   Non-Official & Non-Abo~ Afrikaans                  505           15
##  9 Edmonton  Non-Official & Non-Abo~ Afrikaans                  300            0
## 10 Vancouver Non-Official & Non-Abo~ Afrikaans                  520           10
## # ... with 1,060 more rows
```

The data above is now tidy! We can go through the three criteria again to check that this data is a tidy data set.

1. All the statistical variables are their own columns in the data frame (i.e., `most_at_home`, and `most_at_work` have been separated into their own columns in the data frame).
2. Each observation, (i.e., each language in a region) is in a single row.
3. Each value is a single cell (i.e., its row, column position in the data frame is not shared with another value).

You might notice that we have the same number of columns in the tidy data set as we did in the messy one. Therefore `pivot_wider` didn't really "widen" the data, as the name suggests. This is just because the original `type` column only had two categories in it. If it had more than two, `pivot_wider` would have created more columns, and we would see the data set "widen."

3.4.3 Tidying up: using `separate` to deal with multiple delimiters

Data are also not considered tidy when multiple values are stored in the same cell. The data set we show below is even messier than the ones we dealt with above: the `Toronto`, `Montréal`, `Vancouver`, `Calgary` and `Edmonton` columns contain the number of Canadians reporting their primary language at home and work in one column separated by the delimiter (`/`). The column names are the values of a variable, *and* each value does not have its own cell! To turn this messy data into tidy data, we'll have to fix these issues.

```
lang_messy <- read_csv("data/region_lang_top5_cities_messy.csv")
lang_messy
```

```
## # A tibble: 214 x 7
##    category           language          Toronto  Montréal Vancouver Calgary Edmonton
##    <chr>              <chr>             <chr>    <chr>    <chr>     <chr>   <chr>
##  1 Aboriginal langu~ Aboriginal la~ 50/0     15/0     15/0      5/0     10/0
##  2 Non-Official & N~ Afrikaans      265/0    10/0     520/10    505/15  300/0
##  3 Non-Official & N~ Afro-Asiatic ~ 185/10   65/0     10/0      15/0    20/0
##  4 Non-Official & N~ Akan (Twi)     4045/20  440/0    125/10    330/0   445/0
##  5 Non-Official & N~ Albanian       6380/215 1445/20  530/10    620/25  370/10
##  6 Aboriginal langu~ Algonquian la~ 5/0      0/0      0/0       0/0     0/0
##  7 Aboriginal langu~ Algonquin      0/0      10/0     0/0       0/0     0/0
##  8 Non-Official & N~ American Sign~ 720/245  70/0     300/140   85/25   190/85
##  9 Non-Official & N~ Amharic        3820/55  315/0    540/10    2730/50 1695/35
## 10 Non-Official & N~ Arabic         45025/1~ 72980/1~ 8680/275  11010/~ 10590/3~
## # ... with 204 more rows
```

First we'll use `pivot_longer` to create two columns, `region` and `value`, similar to what we did previously. The new `region` columns will contain the region names, and the new column `value` will be a temporary holding place for the data that we need to further separate, i.e., the number of Canadians reporting their primary language at home and work.

```
lang_messy_longer <- pivot_longer(lang_messy,
  cols = Toronto:Edmonton,
  names_to = "region",
  values_to = "value"
)

lang_messy_longer
```

```
## # A tibble: 1,070 x 4
##    category                                language                 region   value
##    <chr>                                   <chr>                    <chr>    <chr>
##  1 Aboriginal languages                    Aboriginal languages,~ Toronto  50/0
##  2 Aboriginal languages                    Aboriginal languages,~ Montréal 15/0
##  3 Aboriginal languages                    Aboriginal languages,~ Vancouv~ 15/0
##  4 Aboriginal languages                    Aboriginal languages,~ Calgary  5/0
##  5 Aboriginal languages                    Aboriginal languages,~ Edmonton 10/0
##  6 Non-Official & Non-Aboriginal languages Afrikaans              Toronto  265/0
##  7 Non-Official & Non-Aboriginal languages Afrikaans              Montréal 10/0
##  8 Non-Official & Non-Aboriginal languages Afrikaans              Vancouv~ 520/~
##  9 Non-Official & Non-Aboriginal languages Afrikaans              Calgary  505/~
## 10 Non-Official & Non-Aboriginal languages Afrikaans              Edmonton 300/0
## # ... with 1,060 more rows
```

Next we'll use `separate` to split the `value` column into two columns. One column will contain only the counts of Canadians that speak each language most at home, and the other will contain the counts of Canadians that speak each language most at work for each region. Figure 3.13 outlines what we need to specify to use `separate`.

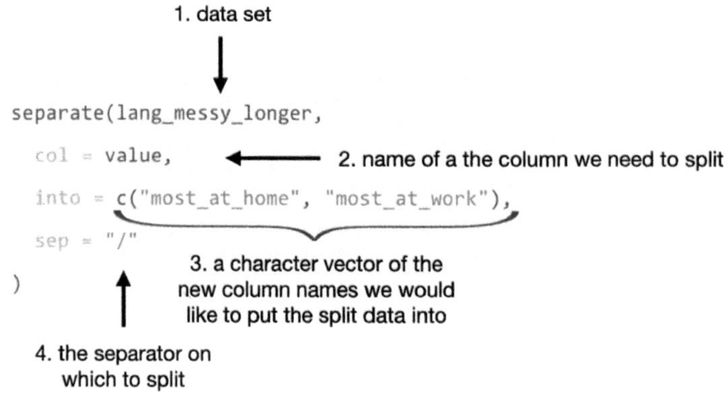

FIGURE 3.13: Syntax for the `separate` function.

```
tidy_lang <- separate(lang_messy_longer,
  col = value,
  into = c("most_at_home", "most_at_work"),
  sep = "/"
)

tidy_lang
```

```
## # A tibble: 1,070 x 5
##    category                 language          region  most_at_home most_at_work
##     <chr>                    <chr>             <chr>   <chr>        <chr>
##  1 Aboriginal languages     Aboriginal langua~ Toronto 50           0
##  2 Aboriginal languages     Aboriginal langua~ Montré~ 15           0
##  3 Aboriginal languages     Aboriginal langua~ Vancou~ 15           0
##  4 Aboriginal languages     Aboriginal langua~ Calgary 5            0
##  5 Aboriginal languages     Aboriginal langua~ Edmont~ 10           0
##  6 Non-Official & Non-Abor~ Afrikaans          Toronto 265          0
##  7 Non-Official & Non-Abor~ Afrikaans          Montré~ 10           0
##  8 Non-Official & Non-Abor~ Afrikaans          Vancou~ 520          10
##  9 Non-Official & Non-Abor~ Afrikaans          Calgary 505          15
## 10 Non-Official & Non-Abor~ Afrikaans          Edmont~ 300          0
## # ... with 1,060 more rows
```

Is this data set now tidy? If we recall the three criteria for tidy data:

- each row is a single observation,
- each column is a single variable, and
- each value is a single cell.

We can see that this data now satisfies all three criteria, making it easier to analyze. But we aren't done yet! Notice in the table above that the word `<chr>` appears beneath each of the column names. The word under the column name indicates the data type of each column. Here all of the variables are "character" data types. Recall, character data types are letter(s) or digits(s) surrounded by quotes. In the previous example in Section 3.4.2, the `most_at_home` and `most_at_work` variables were `<dbl>` (double)—you can verify this by looking at the tables in the previous sections—which is a type of numeric data. This change is due to the delimiter (`/`) when we read in this messy data set. R read these columns in as character types, and by default, `separate` will return columns as character data types.

It makes sense for `region`, `category`, and `language` to be stored as a character (or perhaps factor) type. However, suppose we want to apply any functions that treat the `most_at_home` and `most_at_work` columns as a number (e.g., finding rows above a numeric threshold of a column). In that case, it won't be possible to do if the variable is stored as a `character`. Fortunately, the `separate` function provides a natural way to fix problems like this: we can set `convert = TRUE` to convert the `most_at_home` and `most_at_work` columns to the correct data type.

```
tidy_lang <- separate(lang_messy_longer,
  col = value,
  into = c("most_at_home", "most_at_work"),
  sep = "/",
  convert = TRUE
)

tidy_lang
```

```
## # A tibble: 1,070 x 5
##    category                   language            region  most_at_home most_at_work
##    <chr>                      <chr>               <chr>          <int>        <int>
##  1 Aboriginal languages       Aboriginal langua~ Toronto            50            0
##  2 Aboriginal languages       Aboriginal langua~ Montré~            15            0
##  3 Aboriginal languages       Aboriginal langua~ Vancou~            15            0
##  4 Aboriginal languages       Aboriginal langua~ Calgary             5            0
##  5 Aboriginal languages       Aboriginal langua~ Edmont~            10            0
##  6 Non-Official & Non-Abor~ Afrikaans            Toronto           265            0
##  7 Non-Official & Non-Abor~ Afrikaans            Montré~            10            0
##  8 Non-Official & Non-Abor~ Afrikaans            Vancou~           520           10
##  9 Non-Official & Non-Abor~ Afrikaans            Calgary           505           15
## 10 Non-Official & Non-Abor~ Afrikaans            Edmont~           300            0
```

```
## # ... with 1,060 more rows
```

Now we see `<int>` appears under the `most_at_home` and `most_at_work` columns, indicating they are integer data types (i.e., numbers)!

3.5 Using `select` to extract a range of columns

Now that the `tidy_lang` data is indeed *tidy*, we can start manipulating it using the powerful suite of functions from the `tidyverse`. For the first example, recall the `select` function from Chapter 1, which lets us create a subset of columns from a data frame. Suppose we wanted to select only the columns `language`, `region`, `most_at_home` and `most_at_work` from the `tidy_lang` data set. Using what we learned in Chapter 1, we would pass the `tidy_lang` data frame as well as all of these column names into the `select` function:

```
selected_columns <- select(tidy_lang,
                           language,
                           region,
                           most_at_home,
                           most_at_work)
selected_columns
```

```
## # A tibble: 1,070 x 4
##    language                     region    most_at_home most_at_work
##    <chr>                        <chr>            <int>        <int>
##  1 Aboriginal languages, n.o.s. Toronto             50            0
##  2 Aboriginal languages, n.o.s. Montréal            15            0
##  3 Aboriginal languages, n.o.s. Vancouver           15            0
##  4 Aboriginal languages, n.o.s. Calgary              5            0
##  5 Aboriginal languages, n.o.s. Edmonton            10            0
##  6 Afrikaans                    Toronto            265            0
##  7 Afrikaans                    Montréal            10            0
##  8 Afrikaans                    Vancouver          520           10
##  9 Afrikaans                    Calgary            505           15
## 10 Afrikaans                    Edmonton           300            0
## # ... with 1,060 more rows
```

Here we wrote out the names of each of the columns. However, this method is time-consuming, especially if you have a lot of columns! Another approach is to use a "select helper". Select helpers are operators that make it easier for us to select columns. For instance, we can use a select helper to choose a range of columns rather than typing each column name out. To do this, we use the colon (:) operator to denote the range. For example, to get all the columns in the `tidy_lang` data frame from `language` to `most_at_work` we pass `language:most_at_work` as the second argument to the `select` function.

```
column_range <- select(tidy_lang, language:most_at_work)
column_range
```

```
## # A tibble: 1,070 x 4
##    language                       region    most_at_home most_at_work
##    <chr>                          <chr>            <int>        <int>
##  1 Aboriginal languages, n.o.s. Toronto             50            0
##  2 Aboriginal languages, n.o.s. Montréal            15            0
##  3 Aboriginal languages, n.o.s. Vancouver           15            0
##  4 Aboriginal languages, n.o.s. Calgary              5            0
##  5 Aboriginal languages, n.o.s. Edmonton            10            0
##  6 Afrikaans                      Toronto          265            0
##  7 Afrikaans                      Montréal          10            0
##  8 Afrikaans                      Vancouver        520           10
##  9 Afrikaans                      Calgary          505           15
## 10 Afrikaans                      Edmonton         300            0
## # ... with 1,060 more rows
```

Notice that we get the same output as we did above, but with less (and clearer!) code. This type of operator is especially handy for large data sets.

Suppose instead we wanted to extract columns that followed a particular pattern rather than just selecting a range. For example, let's say we wanted only to select the columns `most_at_home` and `most_at_work`. There are other helpers that allow us to select variables based on their names. In particular, we can use the `select` helper `starts_with` to choose only the columns that start with the word "most":

```
select(tidy_lang, starts_with("most"))
```

```
## # A tibble: 1,070 x 2
##    most_at_home most_at_work
##           <int>        <int>
##  1           50            0
##  2           15            0
##  3           15            0
##  4            5            0
```

```
## 5          10        0
## 6         265        0
## 7          10        0
## 8         520       10
## 9         505       15
## 10        300        0
## # ... with 1,060 more rows
```

We could also have chosen the columns containing an underscore _ by adding `contains("_")` as the second argument in the `select` function, since we notice the columns we want contain underscores and the others don't.

```
select(tidy_lang, contains("_"))
```

```
## # A tibble: 1,070 x 2
##    most_at_home most_at_work
##           <int>        <int>
##  1           50            0
##  2           15            0
##  3           15            0
##  4            5            0
##  5           10            0
##  6          265            0
##  7           10            0
##  8          520           10
##  9          505           15
## 10          300            0
## # ... with 1,060 more rows
```

There are many different `select` helpers that select variables based on certain criteria. The additional resources section at the end of this chapter provides a comprehensive resource on `select` helpers.

3.6 Using `filter` to extract rows

Next, we revisit the `filter` function from Chapter 1, which lets us create a subset of rows from a data frame. Recall the two main arguments to the `filter` function: the first is the name of the data frame object, and the second is a *logical statement* to use when filtering the rows. `filter` works by returning the rows where the logical statement evaluates to `TRUE`. This section will highlight more advanced usage of the `filter` function. In particular, this section provides an in-depth treatment of the variety of logical statements one can use in the filter function to select subsets of rows.

3.6.1 Extracting rows that have a certain value with ==

Suppose we are only interested in the subset of rows in tidy_lang corresponding to the official languages of Canada (English and French). We can filter for these rows by using the *equivalency operator* (==) to compare the values of the category column with the value "Official languages". With these arguments, filter returns a data frame with all the columns of the input data frame but only the rows we asked for in the logical statement, i.e., those where the category column holds the value "Official languages". We name this data frame official_langs.

```
official_langs <- filter(tidy_lang, category == "Official languages")
official_langs
```

```
## # A tibble: 10 x 5
##    category           language region    most_at_home most_at_work
##    <chr>              <chr>    <chr>            <int>        <int>
##  1 Official languages English  Toronto        3836770      3218725
##  2 Official languages English  Montréal        620510       412120
##  3 Official languages English  Vancouver      1622735      1330555
##  4 Official languages English  Calgary        1065070       844740
##  5 Official languages English  Edmonton       1050410       792700
##  6 Official languages French   Toronto          29800        11940
##  7 Official languages French   Montréal       2669195      1607550
##  8 Official languages French   Vancouver         8630         3245
##  9 Official languages French   Calgary           8630         2140
## 10 Official languages French   Edmonton         10950         2520
```

3.6.2 Extracting rows that do not have a certain value with !=

What if we want all the other language categories in the data set *except* for those in the "Official languages" category? We can accomplish this with the != operator, which means "not equal to". So if we want to find all the rows where the category does *not* equal "Official languages" we write the code below.

```
filter(tidy_lang, category != "Official languages")
```

```
## # A tibble: 1,060 x 5
##    category                    language             region  most_at_home most_at_work
```

```
##   <chr>                    <chr>              <chr>        <int>        <int>
##  1 Aboriginal languages     Aboriginal langua~ Toronto         50            0
##  2 Aboriginal languages     Aboriginal langua~ Montré~         15            0
##  3 Aboriginal languages     Aboriginal langua~ Vancou~         15            0
##  4 Aboriginal languages     Aboriginal langua~ Calgary          5            0
##  5 Aboriginal languages     Aboriginal langua~ Edmont~         10            0
##  6 Non-Official & Non-Abor~ Afrikaans          Toronto        265            0
##  7 Non-Official & Non-Abor~ Afrikaans          Montré~         10            0
##  8 Non-Official & Non-Abor~ Afrikaans          Vancou~        520           10
##  9 Non-Official & Non-Abor~ Afrikaans          Calgary        505           15
## 10 Non-Official & Non-Abor~ Afrikaans          Edmont~        300            0
## # ... with 1,050 more rows
```

3.6.3 Extracting rows satisfying multiple conditions using `,` or `&`

Suppose now we want to look at only the rows for the French language in Montréal. To do this, we need to filter the data set to find rows that satisfy multiple conditions simultaneously. We can do this with the comma symbol (`,`), which in the case of `filter` is interpreted by R as "and". We write the code as shown below to filter the `official_langs` data frame to subset the rows where `region == "Montréal"` *and* the `language == "French"`.

```
filter(official_langs, region == "Montréal", language == "French")
```

```
## # A tibble: 1 x 5
##   category             language region   most_at_home most_at_work
##   <chr>                <chr>    <chr>           <int>        <int>
## 1 Official languages French   Montréal      2669195      1607550
```

We can also use the ampersand (`&`) logical operator, which gives us cases where *both* one condition *and* another condition are satisfied. You can use either comma (`,`) or ampersand (`&`) in the `filter` function interchangeably.

```
filter(official_langs, region == "Montréal" & language == "French")
```

```
## # A tibble: 1 x 5
##   category             language region   most_at_home most_at_work
##   <chr>                <chr>    <chr>           <int>        <int>
## 1 Official languages French   Montréal      2669195      1607550
```

3.6.4 Extracting rows satisfying at least one condition using `|`

Suppose we were interested in only those rows corresponding to cities in Alberta in the `official_langs` data set (Edmonton and Calgary). We can't use `,` as we did above because `region` cannot be both Edmonton *and* Calgary simultaneously. Instead, we can use the vertical pipe (`|`) logical operator, which gives us the cases where one condition *or* another condition *or* both are satisfied. In the code below, we ask R to return the rows where the `region` columns are equal to "Calgary" *or* "Edmonton".

```
filter(official_langs, region == "Calgary" | region == "Edmonton")
```

```
## # A tibble: 4 x 5
##   category           language region   most_at_home most_at_work
##   <chr>              <chr>    <chr>           <int>        <int>
## 1 Official languages English  Calgary       1065070       844740
## 2 Official languages English  Edmonton      1050410       792700
## 3 Official languages French   Calgary          8630         2140
## 4 Official languages French   Edmonton        10950         2520
```

3.6.5 Extracting rows with values in a vector using `%in%`

Next, suppose we want to see the populations of our five cities. Let's read in the `region_data.csv` file that comes from the 2016 Canadian census, as it contains statistics for number of households, land area, population and number of dwellings for different regions.

```
region_data <- read_csv("data/region_data.csv")
region_data
```

```
## # A tibble: 35 x 5
##    region         households  area population dwellings
##    <chr>               <dbl> <dbl>      <dbl>     <dbl>
##  1 Belleville          43002 1355.     103472     45050
##  2 Lethbridge          45696 3047.     117394     48317
##  3 Thunder Bay         52545 2618.     121621     57146
##  4 Peterborough        50533 1637.     121721     55662
##  5 Saint John          52872 3793.     126202     58398
##  6 Brantford           52530 1086.     134203     54419
##  7 Moncton             61769 2625.     144810     66699
##  8 Guelph              59280  604.     151984     63324
##  9 Trois-Rivières      72502 1053.     156042     77734
```

```
## 10 Saguenay              72479 3079.     160980     77968
## # ... with 25 more rows
```

To get the population of the five cities we can filter the data set using the `%in%` operator. The `%in%` operator is used to see if an element belongs to a vector. Here we are filtering for rows where the value in the `region` column matches any of the five cities we are intersted in: Toronto, Montréal, Vancouver, Calgary, and Edmonton.

```
city_names <- c("Toronto", "Montréal", "Vancouver", "Calgary", "Edmonton")
five_cities <- filter(region_data,
                      region %in% city_names)
five_cities
```

```
## # A tibble: 5 x 5
##   region     households  area population dwellings
##   <chr>           <dbl> <dbl>      <dbl>     <dbl>
## 1 Edmonton       502143 9858.    1321426    537634
## 2 Calgary        519693 5242.    1392609    544870
## 3 Vancouver      960894 3040.    2463431   1027613
## 4 Montréal      1727310 4638.    4098927   1823281
## 5 Toronto       2135909 6270.    5928040   2235145
```

Note: What's the difference between `==` and `%in%`? Suppose we have two vectors, `vectorA` and `vectorB`. If you type `vectorA == vectorB` into R it will compare the vectors element by element. R checks if the first element of `vectorA` equals the first element of `vectorB`, the second element of `vectorA` equals the second element of `vectorB`, and so on. On the other hand, `vectorA %in% vectorB` compares the first element of `vectorA` to all the elements in `vectorB`. Then the second element of `vectorA` is compared to all the elements in `vectorB`, and so on. Notice the difference between `==` and `%in%` in the example below.

```
c("Vancouver", "Toronto") == c("Toronto", "Vancouver")
```

```
## [1] FALSE FALSE
```

```
c("Vancouver", "Toronto") %in% c("Toronto", "Vancouver")
```

```
## [1] TRUE TRUE
```

3.6.6 Extracting rows above or below a threshold using > and <

We saw in Section 3.6.3 that 2,669,195 people reported speaking French in Montréal as their primary language at home. If we are interested in finding the official languages in regions with higher numbers of people who speak it as their primary language at home compared to French in Montréal, then we can use `filter` to obtain rows where the value of `most_at_home` is greater than 2,669,195.

```
filter(official_langs, most_at_home > 2669195)
```

```
## # A tibble: 1 x 5
##   category             language region  most_at_home most_at_work
##   <chr>                <chr>    <chr>          <int>        <int>
## 1 Official languages English  Toronto      3836770      3218725
```

`filter` returns a data frame with only one row, indicating that when considering the official languages, only English in Toronto is reported by more people as their primary language at home than French in Montréal according to the 2016 Canadian census.

3.7 Using mutate to modify or add columns

3.7.1 Using mutate to modify columns

In Section 3.4.3, when we first read in the `"region_lang_top5_cities_messy.csv"` data, all of the variables were "character" data types. During the tidying process, we used the `convert` argument from the `separate` function to convert the `most_at_home` and `most_at_work` columns to the desired integer (i.e., numeric class) data types. But suppose we didn't use the `convert` argument, and needed to modify the column type some other way. Below we create such a situation so that we can demonstrate how to use `mutate` to change the column types of a data frame. `mutate` is a useful function to modify or create new data frame columns.

```
lang_messy <- read_csv("data/region_lang_top5_cities_messy.csv")
lang_messy_longer <- pivot_longer(lang_messy,
              cols = Toronto:Edmonton,
              names_to = "region",
              values_to = "value")
tidy_lang_chr <- separate(lang_messy_longer, col = value,
```

```
         into = c("most_at_home", "most_at_work"),
         sep = "/")
official_langs_chr <- filter(tidy_lang_chr, category == "Official languages")

official_langs_chr
```

```
## # A tibble: 10 x 5
##    category           language region    most_at_home most_at_work
##    <chr>              <chr>    <chr>     <chr>        <chr>
##  1 Official languages English  Toronto   3836770      3218725
##  2 Official languages English  Montréal  620510       412120
##  3 Official languages English  Vancouver 1622735      1330555
##  4 Official languages English  Calgary   1065070      844740
##  5 Official languages English  Edmonton  1050410      792700
##  6 Official languages French   Toronto   29800        11940
##  7 Official languages French   Montréal  2669195      1607550
##  8 Official languages French   Vancouver 8630         3245
##  9 Official languages French   Calgary   8630         2140
## 10 Official languages French   Edmonton  10950        2520
```

To use `mutate`, again we first specify the data set in the first argument, and in the following arguments, we specify the name of the column we want to modify or create (here `most_at_home` and `most_at_work`), an `=` sign, and then the function we want to apply (here `as.numeric`). In the function we want to apply, we refer directly to the column name upon which we want it to act (here `most_at_home` and `most_at_work`). In our example, we are naming the columns the same names as columns that already exist in the data frame ("most_at_home", "most_at_work") and this will cause `mutate` to *overwrite* those columns (also referred to as modifying those columns *in-place*). If we were to give the columns a new name, then `mutate` would create new columns with the names we specified. `mutate`'s general syntax is detailed in Figure 3.14.

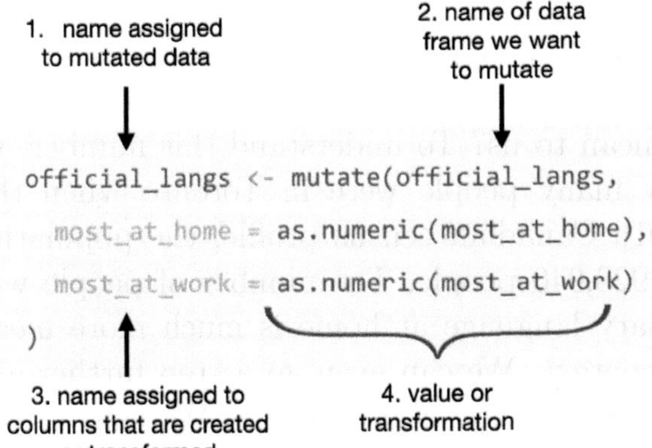

FIGURE 3.14: Syntax for the `mutate` function.

Below we use `mutate` to convert the columns `most_at_home` and `most_at_work` to numeric data types in the `official_langs` data set as described in Figure 3.14:

```
official_langs_numeric <- mutate(official_langs_chr,
  most_at_home = as.numeric(most_at_home),
  most_at_work = as.numeric(most_at_work)
)

official_langs_numeric
```

```
## # A tibble: 10 x 5
##    category           language region    most_at_home most_at_work
##    <chr>              <chr>    <chr>            <dbl>        <dbl>
##  1 Official languages English  Toronto        3836770      3218725
##  2 Official languages English  Montréal        620510       412120
##  3 Official languages English  Vancouver      1622735      1330555
##  4 Official languages English  Calgary        1065070       844740
##  5 Official languages English  Edmonton       1050410       792700
##  6 Official languages French   Toronto          29800        11940
##  7 Official languages French   Montréal       2669195      1607550
##  8 Official languages French   Vancouver         8630         3245
##  9 Official languages French   Calgary           8630         2140
## 10 Official languages French   Edmonton         10950         2520
```

Now we see `<dbl>` appears under the `most_at_home` and `most_at_work` columns, indicating they are double data types (which is a numeric data type)!

3.7.2 Using `mutate` to create new columns

We can see in the table that 3,836,770 people reported speaking English in Toronto as their primary language at home, according to the 2016 Canadian census. What does this number mean to us? To understand this number, we need context. In particular, how many people were in Toronto when this data was collected? From the 2016 Canadian census profile, the population of Toronto was reported to be 5,928,040 people. The number of people who report that English is their primary language at home is much more meaningful when we report it in this context. We can even go a step further and transform this count to a relative frequency or proportion. We can do this by dividing the number of people reporting a given language as their primary language at home by the number of people who live in Toronto. For example, the proportion of people who reported that their primary language at home was English in the 2016 Canadian census was 0.65 in Toronto.

Let's use `mutate` to create a new column in our data frame that holds the proportion of people who speak English for our five cities of focus in this chapter. To accomplish this, we will need to do two tasks beforehand:

1. Create a vector containing the population values for the cities.
2. Filter the `official_langs` data frame so that we only keep the rows where the language is English.

To create a vector containing the population values for the five cities (Toronto, Montréal, Vancouver, Calgary, Edmonton), we will use the `c` function (recall that `c` stands for "concatenate"):

```
city_pops <- c(5928040, 4098927, 2463431, 1392609, 1321426)
city_pops
```

```
## [1] 5928040 4098927 2463431 1392609 1321426
```

And next, we will filter the `official_langs` data frame so that we only keep the rows where the language is English. We will name the new data frame we get from this `english_langs`:

```
english_langs <- filter(official_langs, language == "English")
english_langs
```

```
## # A tibble: 5 x 5
##   category            language region    most_at_home most_at_work
##   <chr>               <chr>    <chr>            <int>        <int>
```

```
## 1 Official languages English  Toronto         3836770       3218725
## 2 Official languages English  Montréal         620510        412120
## 3 Official languages English  Vancouver       1622735       1330555
## 4 Official languages English  Calgary         1065070        844740
## 5 Official languages English  Edmonton        1050410        792700
```

Finally, we can use `mutate` to create a new column, named `most_at_home_proportion`, that will have value that corresponds to the proportion of people reporting English as their primary language at home. We will compute this by dividing the column by our vector of city populations.

```
english_langs <- mutate(english_langs,
                        most_at_home_proportion = most_at_home / city_pops)

english_langs
```

```
## # A tibble: 5 x 6
##   category           language region  most_at_home most_at_work most_at_home_pr~
##   <chr>              <chr>    <chr>          <int>        <int>            <dbl>
## 1 Official languages English  Toronto      3836770      3218725            0.647
## 2 Official languages English  Montré~       620510       412120            0.151
## 3 Official languages English  Vancou~      1622735      1330555            0.659
## 4 Official languages English  Calgary      1065070       844740            0.765
## 5 Official languages English  Edmont~      1050410       792700            0.795
```

In the computation above, we had to ensure that we ordered the `city_pops` vector in the same order as the cities were listed in the `english_langs` data frame. This is because R will perform the division computation we did by dividing each element of the `most_at_home` column by each element of the `city_pops` vector, matching them up by position. Failing to do this would have resulted in the incorrect math being performed.

Note: In more advanced data wrangling, one might solve this problem in a less error-prone way though using a technique called "joins." We link to resources that discuss this in the additional resources at the end of this chapter.

3.8 Combining functions using the pipe operator, |>

In R, we often have to call multiple functions in a sequence to process a data frame. The basic ways of doing this can become quickly unreadable if there are many steps. For example, suppose we need to perform three operations on a data frame called `data`:

1) add a new column `new_col` that is double another `old_col`,
2) filter for rows where another column, `other_col`, is more than 5, and
3) select only the new column `new_col` for those rows.

One way of performing these three steps is to just write multiple lines of code, storing temporary objects as you go:

```
output_1 <- mutate(data, new_col = old_col * 2)
output_2 <- filter(output_1, other_col > 5)
output <- select(output_2, new_col)
```

This is difficult to understand for multiple reasons. The reader may be tricked into thinking the named `output_1` and `output_2` objects are important for some reason, while they are just temporary intermediate computations. Further, the reader has to look through and find where `output_1` and `output_2` are used in each subsequent line.

Another option for doing this would be to *compose* the functions:

```
output <- select(filter(mutate(data, new_col = old_col * 2),
                        other_col > 5),
                 new_col)
```

Code like this can also be difficult to understand. Functions compose (reading from left to right) in the *opposite order* in which they are computed by R (above, `mutate` happens first, then `filter`, then `select`). It is also just a really long line of code to read in one go.

The *pipe operator* (`|>`) solves this problem, resulting in cleaner and easier-to-follow code. `|>` is built into R so you don't need to load any packages to use it. You can think of the pipe as a physical pipe. It takes the output from the function on the left-hand side of the pipe, and passes it as the first argument to the function on the right-hand side of the pipe. The code below accomplishes the same thing as the previous two code blocks:

```
output <- data |>
  mutate(new_col = old_col * 2) |>
  filter(other_col > 5) |>
  select(new_col)
```

Note: You might also have noticed that we split the function calls across lines after the pipe, similar to when we did this earlier in the chapter for long function calls. Again, this is allowed and recommended, especially when the piped function calls create a long line of code. Doing this makes your code more readable. When you do this, it is important to end each line with the pipe operator `|>` to tell R that your code is continuing onto the next line.

Note: In this textbook, we will be using the base R pipe operator syntax, `|>`. This base R `|>` pipe operator was inspired by a previous version of the pipe operator, `%>%`. The `%>%` pipe operator is not built into R and is from the `magrittr` R package. The `tidyverse` metapackage imports the `%>%` pipe operator via `dplyr` (which in turn imports the `magrittr` R package). There are some other differences between `%>%` and `|>` related to more advanced R uses, such as sharing and distributing code as R packages, however, these are beyond the scope of this textbook. We have this note in the book to make the reader aware that `%>%` exists as it is still commonly used in data analysis code and in many data science books and other resources. In most cases these two pipes are interchangeable and either can be used.

3.8.1 Using `|>` to combine `filter` and `select`

Let's work with the tidy `tidy_lang` data set from Section 3.4.3, which contains the number of Canadians reporting their primary language at home and work for five major cities (Toronto, Montréal, Vancouver, Calgary, and Edmonton):

```
tidy_lang
```

```
## # A tibble: 1,070 x 5
##    category                 language          region most_at_home most_at_work
##    <chr>                    <chr>             <chr>         <int>        <int>
```

```
##  1 Aboriginal languages    Aboriginal langua~ Toronto        50          0
##  2 Aboriginal languages    Aboriginal langua~ Montré~        15          0
##  3 Aboriginal languages    Aboriginal langua~ Vancou~        15          0
##  4 Aboriginal languages    Aboriginal langua~ Calgary         5          0
##  5 Aboriginal languages    Aboriginal langua~ Edmont~        10          0
##  6 Non-Official & Non-Abor~ Afrikaans         Toronto       265          0
##  7 Non-Official & Non-Abor~ Afrikaans         Montré~        10          0
##  8 Non-Official & Non-Abor~ Afrikaans         Vancou~       520         10
##  9 Non-Official & Non-Abor~ Afrikaans         Calgary       505         15
## 10 Non-Official & Non-Abor~ Afrikaans         Edmont~       300          0
## # ... with 1,060 more rows
```

Suppose we want to create a subset of the data with only the languages and counts of each language spoken most at home for the city of Vancouver. To do this, we can use the functions `filter` and `select`. First, we use `filter` to create a data frame called `van_data` that contains only values for Vancouver.

```
van_data <- filter(tidy_lang, region == "Vancouver")
van_data
```

```
## # A tibble: 214 x 5
##    category                language           region  most_at_home most_at_work
##    <chr>                   <chr>              <chr>          <int>        <int>
##  1 Aboriginal languages    Aboriginal languag~ Vancou~          15            0
##  2 Non-Official & Non-Abo~ Afrikaans          Vancou~          520           10
##  3 Non-Official & Non-Abo~ Afro-Asiatic langu~ Vancou~          10            0
##  4 Non-Official & Non-Abo~ Akan (Twi)         Vancou~          125           10
##  5 Non-Official & Non-Abo~ Albanian           Vancou~          530           10
##  6 Aboriginal languages    Algonquian languag~ Vancou~           0            0
##  7 Aboriginal languages    Algonquin          Vancou~            0            0
##  8 Non-Official & Non-Abo~ American Sign Lang~ Vancou~         300          140
##  9 Non-Official & Non-Abo~ Amharic            Vancou~          540           10
## 10 Non-Official & Non-Abo~ Arabic             Vancou~         8680          275
## # ... with 204 more rows
```

We then use `select` on this data frame to keep only the variables we want:

```
van_data_selected <- select(van_data, language, most_at_home)
van_data_selected
```

```
## # A tibble: 214 x 2
##    language                            most_at_home
```

```
##    <chr>                                   <int>
##  1 Aboriginal languages, n.o.s.               15
##  2 Afrikaans                                 520
##  3 Afro-Asiatic languages, n.i.e.             10
##  4 Akan (Twi)                                125
##  5 Albanian                                  530
##  6 Algonquian languages, n.i.e.                0
##  7 Algonquin                                   0
##  8 American Sign Language                    300
##  9 Amharic                                   540
## 10 Arabic                                   8680
## # ... with 204 more rows
```

Although this is valid code, there is a more readable approach we could take by using the pipe, `|>`. With the pipe, we do not need to create an intermediate object to store the output from `filter`. Instead, we can directly send the output of `filter` to the input of `select`:

```
van_data_selected <- filter(tidy_lang, region == "Vancouver") |>
    select(language, most_at_home)

van_data_selected
```

```
## # A tibble: 214 x 2
##    language                       most_at_home
##    <chr>                                 <int>
##  1 Aboriginal languages, n.o.s.             15
##  2 Afrikaans                               520
##  3 Afro-Asiatic languages, n.i.e.           10
##  4 Akan (Twi)                              125
##  5 Albanian                                530
##  6 Algonquian languages, n.i.e.              0
##  7 Algonquin                                 0
##  8 American Sign Language                  300
##  9 Amharic                                 540
## 10 Arabic                                 8680
## # ... with 204 more rows
```

But wait...Why do the `select` function calls look different in these two examples? Remember: when you use the pipe, the output of the first function is automatically provided as the first argument for the function that comes after it. Therefore you do not specify the first argument in that function call. In the code above, The pipe passes the left-hand side (the output of `filter`) to the first argument of the function on the right (`select`), so in the `select` function you only see the second argument (and beyond). As you can see, both of these approaches—with and without pipes—give us the same output, but the second approach is clearer and more readable.

3.8.2 Using `|>` with more than two functions

The pipe operator (|>) can be used with any function in R. Additionally, we can pipe together more than two functions. For example, we can pipe together three functions to:

- `filter` rows to include only those where the counts of the language most spoken at home are greater than 10,000,
- `select` only the columns corresponding to `region`, `language` and `most_at_home`, and
- `arrange` the data frame rows in order by counts of the language most spoken at home from smallest to largest.

As we saw in Chapter 1, we can use the `tidyverse` `arrange` function to order the rows in the data frame by the values of one or more columns. Here we pass the column name `most_at_home` to arrange the data frame rows by the values in that column, in ascending order.

```
large_region_lang <- filter(tidy_lang, most_at_home > 10000) |>
  select(region, language, most_at_home) |>
  arrange(most_at_home)

large_region_lang
```

```
## # A tibble: 67 x 3
##    region    language most_at_home
##    <chr>     <chr>           <int>
##  1 Edmonton  Arabic          10590
##  2 Montréal  Tamil           10670
##  3 Vancouver Russian         10795
##  4 Edmonton  Spanish         10880
##  5 Edmonton  French          10950
##  6 Calgary   Arabic          11010
##  7 Calgary   Urdu            11060
##  8 Vancouver Hindi           11235
##  9 Montréal  Armenian        11835
```

```
## 10 Toronto    Romanian          12200
## # ... with 57 more rows
```

You will notice above that we passed `tidy_lang` as the first argument of the `filter` function. We can also pipe the data frame into the same sequence of functions rather than using it as the first argument of the first function. These two choices are equivalent, and we get the same result.

```
large_region_lang <- tidy_lang |>
  filter(most_at_home > 10000) |>
  select(region, language, most_at_home) |>
  arrange(most_at_home)

large_region_lang
```

```
## # A tibble: 67 x 3
##    region    language most_at_home
##    <chr>     <chr>           <int>
##  1 Edmonton  Arabic          10590
##  2 Montréal  Tamil           10670
##  3 Vancouver Russian         10795
##  4 Edmonton  Spanish         10880
##  5 Edmonton  French          10950
##  6 Calgary   Arabic          11010
##  7 Calgary   Urdu            11060
##  8 Vancouver Hindi           11235
##  9 Montréal  Armenian        11835
## 10 Toronto   Romanian        12200
## # ... with 57 more rows
```

Now that we've shown you the pipe operator as an alternative to storing temporary objects and composing code, does this mean you should *never* store temporary objects or compose code? Not necessarily! There are times when you will still want to do these things. For example, you might store a temporary object before feeding it into a plot function so you can iteratively change the plot without having to redo all of your data transformations. Additionally, piping many functions can be overwhelming and difficult to debug; you may want to store a temporary object midway through to inspect your result before moving on with further steps.

3.9 Aggregating data with `summarize` and `map`

3.9.1 Calculating summary statistics on whole columns

As a part of many data analyses, we need to calculate a summary value for the data (a *summary statistic*). Examples of summary statistics we might want to calculate are the number of observations, the average/mean value for a column, the minimum value, etc. Oftentimes, this summary statistic is calculated from the values in a data frame column, or columns, as shown in Figure 3.15.

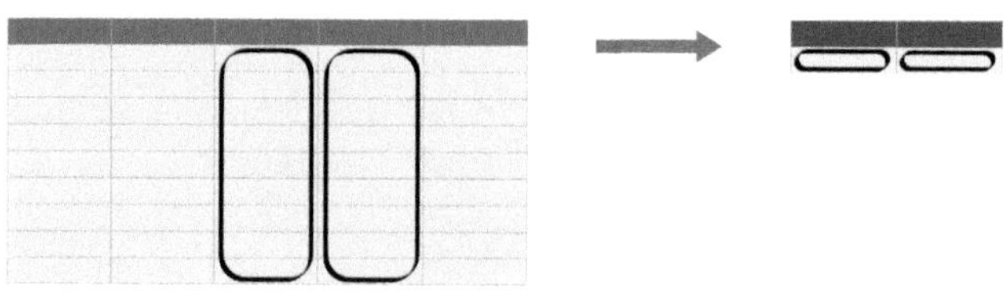

FIGURE 3.15: `summarize` is useful for calculating summary statistics on one or more column(s). In its simplest use case, it creates a new data frame with a single row containing the summary statistic(s) for each column being summarized. The darker, top row of each table represents the column headers.

A useful `dplyr` function for calculating summary statistics is `summarize`, where the first argument is the data frame and subsequent arguments are the summaries we want to perform. Here we show how to use the `summarize` function to calculate the minimum and maximum number of Canadians reporting a particular language as their primary language at home. First a reminder of what `region_lang` looks like:

```
region_lang
```

```
## # A tibble: 7,490 x 7
##    region  category language  mother_tongue most_at_home most_at_work lang_known
##    <chr>   <chr>    <chr>             <dbl>        <dbl>        <dbl>      <dbl>
##  1 St. Jo~ Aborigi~ Aborigin~             5            0            0          0
##  2 Halifax Aborigi~ Aborigin~             5            0            0          0
##  3 Moncton Aborigi~ Aborigin~             0            0            0          0
##  4 Saint ~ Aborigi~ Aborigin~             0            0            0          0
##  5 Saguen~ Aborigi~ Aborigin~             5            5            0          0
##  6 Québec  Aborigi~ Aborigin~             0            5            0         20
##  7 Sherbr~ Aborigi~ Aborigin~             0            0            0          0
##  8 Trois-~ Aborigi~ Aborigin~             0            0            0          0
```

```
## 9 Montré~ Aborigi~ Aborigin~              30          15           0         10
## 10 Kingst~ Aborigi~ Aborigin~              0           0           0          0
## # ... with 7,480 more rows
```

We apply `summarize` to calculate the minimum and maximum number of Canadians reporting a particular language as their primary language at home, for any region:

```
summarize(region_lang,
          min_most_at_home = min(most_at_home),
          max_most_at_home = max(most_at_home))
```

```
## # A tibble: 1 x 2
##   min_most_at_home max_most_at_home
##              <dbl>            <dbl>
## 1                0          3836770
```

From this we see that there are some languages in the data set that no one speaks as their primary language at home. We also see that the most commonly spoken primary language at home is spoken by 3,836,770 people.

3.9.2 Calculating summary statistics when there are NAs

In data frames in R, the value `NA` is often used to denote missing data. Many of the base R statistical summary functions (e.g., `max`, `min`, `mean`, `sum`, etc) will return `NA` when applied to columns containing `NA` values. Usually that is not what we want to happen; instead, we would usually like R to ignore the missing entries and calculate the summary statistic using all of the other non-`NA` values in the column. Fortunately many of these functions provide an argument `na.rm` that lets us tell the function what to do when it encounters `NA` values. In particular, if we specify `na.rm = TRUE`, the function will ignore missing values and return a summary of all the non-missing entries. We show an example of this combined with `summarize` below.

First we create a new version of the `region_lang` data frame, named `region_lang_na`, that has a seemingly innocuous `NA` in the first row of the `most_at_home` column:

```
region_lang_na
```

```
## # A tibble: 7,490 x 7
##   region  category language  mother_tongue most_at_home most_at_work lang_known
##    <chr>   <chr>    <chr>            <dbl>        <dbl>        <dbl>      <dbl>
```

```
## 1 St. Jo~ Aborigi~ Aborigin~          5          NA          0          0
## 2 Halifax Aborigi~ Aborigin~          5           0          0          0
## 3 Moncton Aborigi~ Aborigin~          0           0          0          0
## 4 Saint ~ Aborigi~ Aborigin~          0           0          0          0
## 5 Saguen~ Aborigi~ Aborigin~          5           5          0          0
## 6 Québec  Aborigi~ Aborigin~          0           5          0         20
## 7 Sherbr~ Aborigi~ Aborigin~          0           0          0          0
## 8 Trois-~ Aborigi~ Aborigin~          0           0          0          0
## 9 Montré~ Aborigi~ Aborigin~         30          15          0         10
## 10 Kingst~ Aborigi~ Aborigin~         0           0          0          0
## # ... with 7,480 more rows
```

Now if we apply the `summarize` function as above, we see that we no longer get the minimum and maximum returned, but just an `NA` instead!

```
summarize(region_lang_na,
          min_most_at_home = min(most_at_home),
          max_most_at_home = max(most_at_home))
```

```
## # A tibble: 1 x 2
##   min_most_at_home max_most_at_home
##              <dbl>            <dbl>
## 1               NA               NA
```

We can fix this by adding the `na.rm = TRUE` as explained above:

```
summarize(region_lang_na,
          min_most_at_home = min(most_at_home, na.rm = TRUE),
          max_most_at_home = max(most_at_home, na.rm = TRUE))
```

```
## # A tibble: 1 x 2
##   min_most_at_home max_most_at_home
##              <dbl>            <dbl>
## 1                0          3836770
```

3.9.3 Calculating summary statistics for groups of rows

A common pairing with `summarize` is `group_by`. Pairing these functions together can let you summarize values for subgroups within a data set, as illustrated in Figure 3.16. For example, we can use `group_by` to group the regions of the `region_lang` data frame and then calculate the minimum and maximum number of Canadians reporting the language as the primary language at home for each of the regions in the data set.

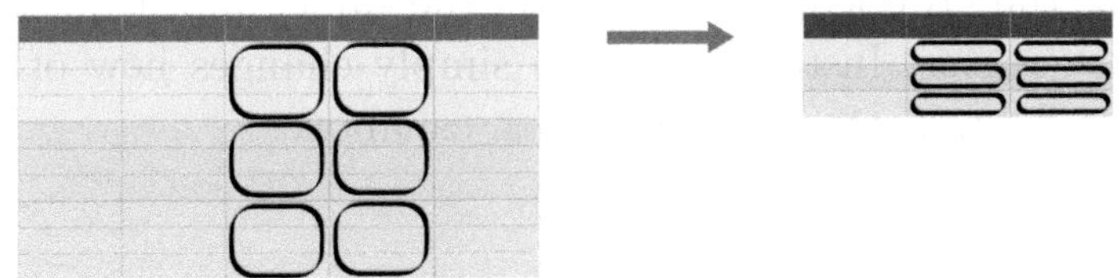

FIGURE 3.16: `summarize` and `group_by` is useful for calculating summary statistics on one or more column(s) for each group. It creates a new data frame—with one row for each group—containing the summary statistic(s) for each column being summarized. It also creates a column listing the value of the grouping variable. The darker, top row of each table represents the column headers. The gray, blue, and green colored rows correspond to the rows that belong to each of the three groups being represented in this cartoon example.

The `group_by` function takes at least two arguments. The first is the data frame that will be grouped, and the second and onwards are columns to use in the grouping. Here we use only one column for grouping (`region`), but more than one can also be used. To do this, list additional columns separated by commas.

```
group_by(region_lang, region) |>
  summarize(
    min_most_at_home = min(most_at_home),
    max_most_at_home = max(most_at_home)
  )
```

```
## # A tibble: 35 x 3
##    region                  min_most_at_home max_most_at_home
##    <chr>                              <dbl>            <dbl>
##  1 Abbotsford - Mission                   0           137445
##  2 Barrie                                 0           182390
##  3 Belleville                             0            97840
##  4 Brantford                              0           124560
##  5 Calgary                                0          1065070
##  6 Edmonton                               0          1050410
##  7 Greater Sudbury                        0           133960
##  8 Guelph                                 0           130950
##  9 Halifax                                0           371215
## 10 Hamilton                               0           630380
## # ... with 25 more rows
```

Notice that `group_by` on its own doesn't change the way the data looks. In the output below, the grouped data set looks the same, and it doesn't *appear* to be grouped by region. Instead, group_by simply changes how other functions work with the data, as we saw with summarize above.

```
group_by(region_lang, region)
```

```
## # A tibble: 7,490 x 7
## # Groups:   region [35]
##    region  category language  mother_tongue most_at_home most_at_work lang_known
##    <chr>   <chr>    <chr>             <dbl>        <dbl>        <dbl>      <dbl>
##  1 St. Jo~ Aborigi~ Aborigin~             5            0            0          0
##  2 Halifax Aborigi~ Aborigin~             5            0            0          0
##  3 Moncton Aborigi~ Aborigin~             0            0            0          0
##  4 Saint ~ Aborigi~ Aborigin~             0            0            0          0
##  5 Saguen~ Aborigi~ Aborigin~             5            5            0          0
##  6 Québec  Aborigi~ Aborigin~             0            5            0         20
##  7 Sherbr~ Aborigi~ Aborigin~             0            0            0          0
##  8 Trois-~ Aborigi~ Aborigin~             0            0            0          0
##  9 Montré~ Aborigi~ Aborigin~            30           15            0         10
## 10 Kingst~ Aborigi~ Aborigin~             0            0            0          0
## # ... with 7,480 more rows
```

3.9.4 Calculating summary statistics on many columns

Sometimes we need to summarize statistics across many columns. An example of this is illustrated in Figure 3.17. In such a case, using summarize alone means that we have to type out the name of each column we want to summarize. In this section we will meet two strategies for performing this task. First we will see how we can do this using summarize + across. Then we will also explore how we can use a more general iteration function, map, to also accomplish this.

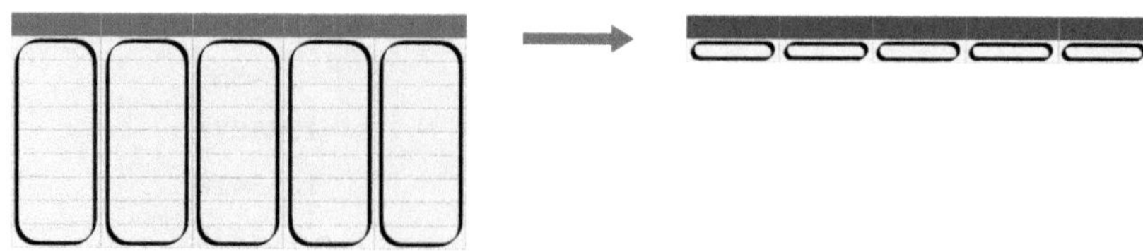

FIGURE 3.17: summarize + across or map is useful for efficiently calculating summary statistics on many columns at once. The darker, top row of each table represents the column headers.

summarize and across for calculating summary statistics on many columns

To summarize statistics across many columns, we can use the summarize function we have just recently learned about. However, in such a case, using summarize alone means that we have to type out the name of each column we want to summarize. To do this more efficiently, we can pair summarize with across and use a colon : to specify a range of columns we would like to perform the statistical summaries on. Here we demonstrate finding the maximum value of each of the numeric columns of the region_lang data set.

```
region_lang |>
  summarize(across(mother_tongue:lang_known, max))
```

```
## # A tibble: 1 x 4
##   mother_tongue most_at_home most_at_work lang_known
##           <dbl>        <dbl>        <dbl>      <dbl>
## 1       3061820      3836770      3218725    5600480
```

> **Note:** Similar to when we use base R statistical summary functions (e.g., max, min, mean, sum, etc) with summarize alone, the use of the summarize + across functions paired with base R statistical summary functions also return NAs when we apply them to columns that contain NAs in the data frame.
>
> To avoid this, again we need to add the argument na.rm = TRUE, but in this case we need to use it a little bit differently. In this case, we need to add a , and then na.rm = TRUE, after specifying the function we want summarize + across to apply, as illustrated below:

```
region_lang_na |>
  summarize(across(mother_tongue:lang_known, max, na.rm = TRUE))
```

```
## # A tibble: 1 x 4
##   mother_tongue most_at_home most_at_work lang_known
##           <dbl>        <dbl>        <dbl>      <dbl>
## 1       3061820      3836770      3218725    5600480
```

---

map for calculating summary statistics on many columns

An alternative to summarize and across for applying a function to many columns is the map family of functions. Let's again find the maximum value of each column of the region_lang data frame, but using map with the max function this time. map takes two arguments: an object (a vector, data frame or list) that you want to apply the function to, and the function that you would like to apply to each column. Note that `map` does not have an argument to specify *which* columns to apply the function to. Therefore, we will use the `select` function before calling `map` to choose the columns for which we want the maximum.

```
region_lang |>
  select(mother_tongue:lang_known) |>
  map(max)
```

```
## $mother_tongue
## [1] 3061820
##
## $most_at_home
## [1] 3836770
##
## $most_at_work
## [1] 3218725
##
## $lang_known
## [1] 5600480
```

Note: The `map` function comes from the `purrr` package. But since `purrr` is part of the tidyverse, once we call `library(tidyverse)` we do not need to load the `purrr` package separately.

The output looks a bit weird... we passed in a data frame, but the output doesn't look like a data frame. As it so happens, it is *not* a data frame, but rather a plain list:

```
region_lang |>
  select(mother_tongue:lang_known) |>
  map(max) |>
  typeof()
```

```
## [1] "list"
```

So what do we do? Should we convert this to a data frame? We could, but a simpler alternative is to just use a different `map` function. There are quite a few to choose from, they all work similarly, but their name reflects the type of output you want from the mapping operation. Table 3.3 lists the commonly used map functions as well as their output type.

TABLE 3.3: The map functions in R.

map function	Output
map	list
map_lgl	logical vector
map_int	integer vector
map_dbl	double vector
map_chr	character vector
map_dfc	data frame, combining column-wise
map_dfr	data frame, combining row-wise

Let's get the columns' maximums again, but this time use the map_dfr function to return the output as a data frame:

```
region_lang |>
  select(mother_tongue:lang_known) |>
  map_dfr(max)
```

```
## # A tibble: 1 x 4
##   mother_tongue most_at_home most_at_work lang_known
##           <dbl>        <dbl>        <dbl>      <dbl>
## 1       3061820      3836770      3218725    5600480
```

Note: Similar to when we use base R statistical summary functions (e.g., max, min, mean, sum, etc.) with summarize, map functions paired with base R statistical summary functions also return NA values when we apply them to columns that contain NA values.

To avoid this, again we need to add the argument na.rm = TRUE. When we use this with map, we do this by adding a , and then na.rm = TRUE after specifying the function, as illustrated below:

```
region_lang_na |>
  select(mother_tongue:lang_known) |>
  map_dfr(max, na.rm = TRUE)
```

```
## # A tibble: 1 x 4
```

```
##   mother_tongue most_at_home most_at_work lang_known
##           <dbl>        <dbl>        <dbl>      <dbl>
## 1       3061820      3836770      3218725    5600480
```

The `map` functions are generally quite useful for solving many problems involving repeatedly applying functions in R. Additionally, their use is not limited to columns of a data frame; `map` family functions can be used to apply functions to elements of a vector, or a list, and even to lists of (nested!) data frames. To learn more about the `map` functions, see the additional resources section at the end of this chapter.

3.10 Apply functions across many columns with `mutate` and `across`

Sometimes we need to apply a function to many columns in a data frame. For example, we would need to do this when converting units of measurements across many columns. We illustrate such a data transformation in Figure 3.18.

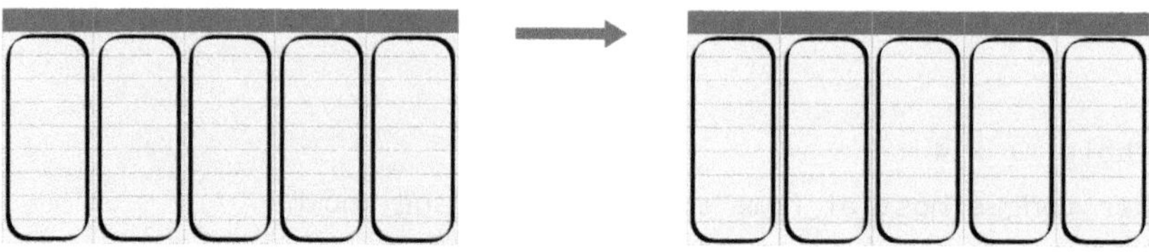

FIGURE 3.18: `mutate` and `across` is useful for applying functions across many columns. The darker, top row of each table represents the column headers.

For example, imagine that we wanted to convert all the numeric columns in the `region_lang` data frame from double type to integer type using the `as.integer` function. When we revisit the `region_lang` data frame, we can see that this would be the columns from `mother_tongue` to `lang_known`.

```
region_lang
```

```
## # A tibble: 7,490 x 7
##   region  category language  mother_tongue most_at_home most_at_work lang_known
##   <chr>   <chr>    <chr>             <dbl>        <dbl>        <dbl>      <dbl>
## 1 St. Jo~ Aborigi~ Aborigin~             5            0            0          0
## 2 Halifax Aborigi~ Aborigin~             5            0            0          0
```

```
## 3 Moncton Aborigi~ Aborigin~              0            0            0          0
## 4 Saint ~ Aborigi~ Aborigin~              0            0            0          0
## 5 Saguen~ Aborigi~ Aborigin~              5            5            0          0
## 6 Québec  Aborigi~ Aborigin~              0            5            0         20
## 7 Sherbr~ Aborigi~ Aborigin~              0            0            0          0
## 8 Trois-~ Aborigi~ Aborigin~              0            0            0          0
## 9 Montré~ Aborigi~ Aborigin~             30           15            0         10
## 10 Kingst~ Aborigi~ Aborigin~             0            0            0          0
## # ... with 7,480 more rows
```

To accomplish such a task, we can use `mutate` paired with `across`. This works in a similar way for column selection, as we saw when we used `summarize` + `across` earlier. As we did above, we again use `across` to specify the columns using `select` syntax as well as the function we want to apply on the specified columns. However, a key difference here is that we are using `mutate`, which means that we get back a data frame with the same number of columns and rows. The only thing that changes is the transformation we applied to the specified columns (here `mother_tongue` to `lang_known`).

```
region_lang |>
  mutate(across(mother_tongue:lang_known, as.integer))
```

```
## # A tibble: 7,490 x 7
##   region  category language  mother_tongue most_at_home most_at_work lang_known
##   <chr>   <chr>    <chr>             <int>        <int>        <int>      <int>
## 1 St. Jo~ Aborigi~ Aborigin~             5            0            0          0
## 2 Halifax Aborigi~ Aborigin~             5            0            0          0
## 3 Moncton Aborigi~ Aborigin~             0            0            0          0
## 4 Saint ~ Aborigi~ Aborigin~             0            0            0          0
## 5 Saguen~ Aborigi~ Aborigin~             5            5            0          0
## 6 Québec  Aborigi~ Aborigin~             0            5            0         20
## 7 Sherbr~ Aborigi~ Aborigin~             0            0            0          0
## 8 Trois-~ Aborigi~ Aborigin~             0            0            0          0
## 9 Montré~ Aborigi~ Aborigin~            30           15            0         10
## 10 Kingst~ Aborigi~ Aborigin~            0            0            0          0
## # ... with 7,480 more rows
```

3.11 Apply functions across columns within one row with `rowwise` and `mutate`

What if you want to apply a function across columns but within one row? We illustrate such a data transformation in Figure 3.19.

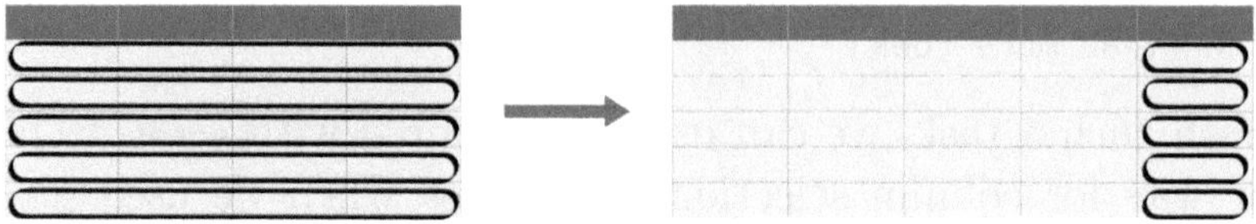

FIGURE 3.19: `rowwise` and `mutate` is useful for applying functions across columns within one row. The darker, top row of each table represents the column headers.

For instance, suppose we want to know the maximum value between `mother_tongue`, `most_at_home`, `most_at_work` and `lang_known` for each language and region in the `region_lang` data set. In other words, we want to apply the `max` function *row-wise.* We will use the (aptly named) `rowwise` function in combination with `mutate` to accomplish this task.

Before we apply `rowwise`, we will `select` only the count columns so we can see all the columns in the data frame's output easily in the book. So for this demonstration, the data set we are operating on looks like this:

```
region_lang |>
  select(mother_tongue:lang_known)
```

```
## # A tibble: 7,490 x 4
##     mother_tongue most_at_home most_at_work lang_known
##             <dbl>        <dbl>        <dbl>      <dbl>
##  1              5            0            0          0
##  2              5            0            0          0
##  3              0            0            0          0
##  4              0            0            0          0
##  5              5            5            0          0
##  6              0            5            0         20
##  7              0            0            0          0
##  8              0            0            0          0
##  9             30           15            0         10
```

```
## 10               0               0               0           0
## # ... with 7,480 more rows
```

Now we apply `rowwise` before `mutate`, to tell R that we would like the mutate function to be applied across, and within, a row, as opposed to being applied on a column (which is the default behavior of `mutate`):

```
region_lang |>
  select(mother_tongue:lang_known) |>
  rowwise() |>
  mutate(maximum = max(c(mother_tongue,
                         most_at_home,
                         most_at_work,
                         lang_known)))
```

```
## # A tibble: 7,490 x 5
## # Rowwise:
##    mother_tongue most_at_home most_at_work lang_known maximum
##            <dbl>        <dbl>        <dbl>      <dbl>   <dbl>
##  1             5            0            0          0       5
##  2             5            0            0          0       5
##  3             0            0            0          0       0
##  4             0            0            0          0       0
##  5             5            5            0          0       5
##  6             0            5            0         20      20
##  7             0            0            0          0       0
##  8             0            0            0          0       0
##  9            30           15            0         10      30
## 10             0            0            0          0       0
## # ... with 7,480 more rows
```

We see that we get an additional column added to the data frame, named `maximum`, which is the maximum value between `mother_tongue`, `most_at_home`, `most_at_work` and `lang_known` for each language and region.

Similar to `group_by`, `rowwise` doesn't appear to do anything when it is called by itself. However, we can apply `rowwise` in combination with other functions to change how these other functions operate on the data. Notice if we used `mutate` without `rowwise`, we would have computed the maximum value across *all* rows rather than the maximum value for *each* row. Below we show what would have happened had we not used `rowwise`. In particular, the same maximum value is reported in every single row; this code does not provide the desired result.

```
region_lang |>
  select(mother_tongue:lang_known) |>
  mutate(maximum = max(c(mother_tongue,
                         most_at_home,
                         most_at_home,
                         lang_known)))
```

```
## # A tibble: 7,490 x 5
##    mother_tongue most_at_home most_at_work lang_known maximum
##            <dbl>        <dbl>        <dbl>      <dbl>   <dbl>
##  1             5            0            0          0 5600480
##  2             5            0            0          0 5600480
##  3             0            0            0          0 5600480
##  4             0            0            0          0 5600480
##  5             5            5            0          0 5600480
##  6             0            5            0         20 5600480
##  7             0            0            0          0 5600480
##  8             0            0            0          0 5600480
##  9            30           15            0         10 5600480
## 10             0            0            0          0 5600480
## # ... with 7,480 more rows
```

3.12 Summary

Cleaning and wrangling data can be a very time-consuming process. However, it is a critical step in any data analysis. We have explored many different functions for cleaning and wrangling data into a tidy format. Table 3.4 summarizes some of the key wrangling functions we learned in this chapter. In the following chapters, you will learn how you can take this tidy data and do so much more with it to answer your burning data science questions!

TABLE 3.4: Summary of wrangling functions

Function	Description
`across`	allows you to apply function(s) to multiple columns
`filter`	subsets rows of a data frame
`group_by`	allows you to apply function(s) to groups of rows
`mutate`	adds or modifies columns in a data frame
`map`	general iteration function
`pivot_longer`	generally makes the data frame longer and narrower
`pivot_wider`	generally makes a data frame wider and decreases the number of rows
`rowwise`	applies functions across columns within one row
`separate`	splits up a character column into multiple columns
`select`	subsets columns of a data frame
`summarize`	calculates summaries of inputs

3.13 Exercises

Practice exercises for the material covered in this chapter can be found in the accompanying worksheets repository[1] in the "Cleaning and wrangling data" row. You can launch an interactive version of the worksheet in your browser by clicking the "launch binder" button. You can also preview a non-interactive version of the worksheet by clicking "view worksheet." If you instead decide to download the worksheet and run it on your own machine, make sure to follow the instructions for computer setup found in Chapter 13. This will ensure that the automated feedback and guidance that the worksheets provide will function as intended.

3.14 Additional resources

- As we mentioned earlier, `tidyverse` is actually an *R meta package*: it installs and loads a collection of R packages that all follow the tidy data philosophy we discussed above. One of the `tidyverse` packages is `dplyr`—a data wrangling workhorse. You have already met many of `dplyr`'s functions (`select`, `filter`, `mutate`, `arrange`, `summarize`, and `group_by`). To learn more about these functions

[1] `https://github.com/UBC-DSCI/data-science-a-first-intro-worksheets#readme`

and meet a few more useful functions, we recommend you check out Chapters 5-9 of the STAT545 online notes[2]. of the data wrangling, exploration, and analysis with R book.

- The `dplyr` R package documentation[3] [Wickham et al., 2021b] is another resource to learn more about the functions in this chapter, the full set of arguments you can use, and other related functions. The site also provides a very nice cheat sheet that summarizes many of the data wrangling functions from this chapter.
- Check out the `tidyselect` R package page[4] [Henry and Wickham, 2021] for a comprehensive list of `select` helpers. These helpers can be used to choose columns in a data frame when paired with the `select` function (and other functions that use the `tidyselect` syntax, such as `pivot_longer`). The documentation for `select` helpers[5] is a useful reference to find the helper you need for your particular problem.
- *R for Data Science* [Wickham and Grolemund, 2016] has a few chapters related to data wrangling that go into more depth than this book. For example, the tidy data chapter[6] covers tidy data, `pivot_longer`/`pivot_wider` and `separate`, but also covers missing values and additional wrangling functions (like `unite`). The data transformation chapter[7] covers `select`, `filter`, `arrange`, `mutate`, and `summarize`. And the `map` functions chapter[8] provides more about the `map` functions.
- You will occasionally encounter a case where you need to iterate over items in a data frame, but none of the above functions are flexible enough to do what you want. In that case, you may consider using a for loop[9].

[2] https://stat545.com/
[3] https://dplyr.tidyverse.org/
[4] https://tidyselect.r-lib.org/index.html
[5] https://tidyselect.r-lib.org/reference/select_helpers.html
[6] https://r4ds.had.co.nz/tidy-data.html
[7] https://r4ds.had.co.nz/transform.html
[8] https://r4ds.had.co.nz/iteration.html#the-map-functions
[9] https://r4ds.had.co.nz/iteration.html#iteration

4

Effective data visualization

4.1 Overview

This chapter will introduce concepts and tools relating to data visualization beyond what we have seen and practiced so far. We will focus on guiding principles for effective data visualization and explaining visualizations independent of any particular tool or programming language. In the process, we will cover some specifics of creating visualizations (scatter plots, bar plots, line plots, and histograms) for data using R.

4.2 Chapter learning objectives

By the end of the chapter, readers will be able to do the following:

- Describe when to use the following kinds of visualizations to answer specific questions using a data set:
  - scatter plots
  - line plots
  - bar plots
  - histogram plots
- Given a data set and a question, select from the above plot types and use R to create a visualization that best answers the question.
- Given a visualization and a question, evaluate the effectiveness of the visualization and suggest improvements to better answer the question.
- Referring to the visualization, communicate the conclusions in non-technical terms.
- Identify rules of thumb for creating effective visualizations.
- Define the three key aspects of ggplot objects:
  - aesthetic mappings
  - geometric objects
  - scales

- Use the `ggplot2` package in R to create and refine the above visualizations using:
  - geometric objects: `geom_point`, `geom_line`, `geom_histogram`, `geom_bar`, `geom_vline`, `geom_hline`
  - scales: `xlim`, `ylim`
  - aesthetic mappings: `x`, `y`, `fill`, `color`, `shape`
  - labeling: `xlab`, `ylab`, `labs`
  - font control and legend positioning: `theme`
  - subplots: `facet_grid`
- Describe the difference in raster and vector output formats.
- Use `ggsave` to save visualizations in `.png` and `.svg` format.

4.3 Choosing the visualization

Ask a question, and answer it

The purpose of a visualization is to answer a question about a data set of interest. So naturally, the first thing to do **before** creating a visualization is to formulate the question about the data you are trying to answer. A good visualization will clearly answer your question without distraction; a *great* visualization will suggest even what the question was itself without additional explanation. Imagine your visualization as part of a poster presentation for a project; even if you aren't standing at the poster explaining things, an effective visualization will convey your message to the audience.

Recall the different data analysis questions from Chapter 1. With the visualizations we will cover in this chapter, we will be able to answer *only descriptive and exploratory* questions. Be careful to not answer any *predictive, inferential, causal or mechanistic* questions with the visualizations presented here, as we have not learned the tools necessary to do that properly just yet.

As with most coding tasks, it is totally fine (and quite common) to make mistakes and iterate a few times before you find the right visualization for your data and question. There are many different kinds of plotting graphics available to use (see Chapter 5 of *Fundamentals of Data Visualization* [Wilke, 2019] for a directory). The types of plot that we introduce in this book are shown in Figure 4.1; which one you should select depends on your data and the question you want to answer. In general, the guiding principles of when to use each type of plot are as follows:

- **scatter plots** visualize the relationship between two quantitative variables
- **line plots** visualize trends with respect to an independent, ordered quantity (e.g., time)
- **bar plots** visualize comparisons of amounts
- **histograms** visualize the distribution of one quantitative variable (i.e., all its possible values and how often they occur)

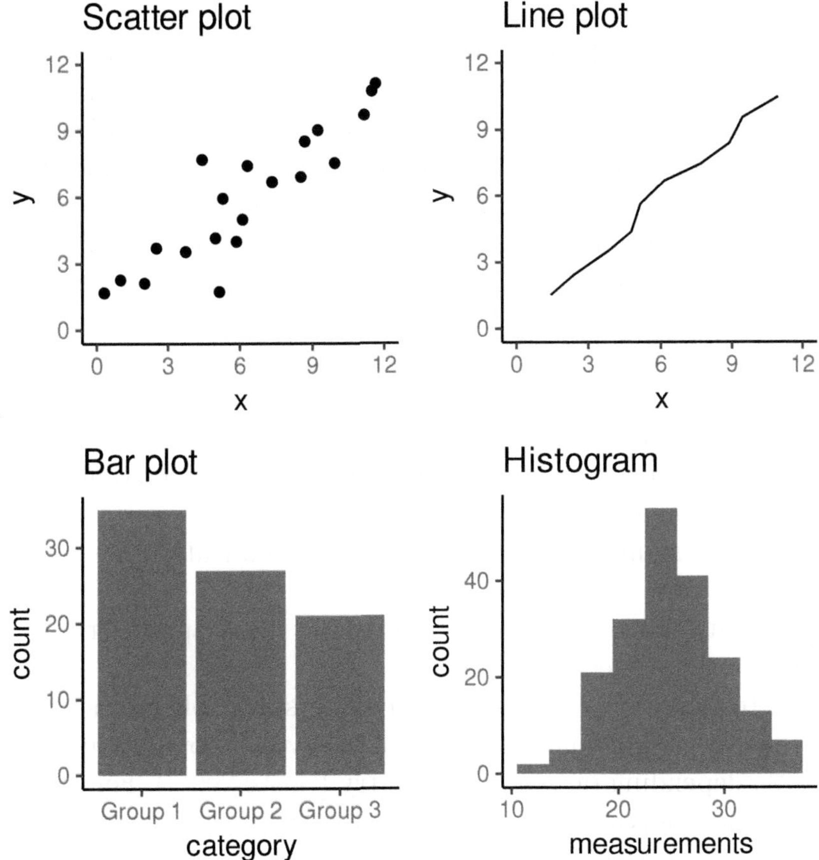

FIGURE 4.1: Examples of scatter, line and bar plots, as well as histograms.

All types of visualization have their (mis)uses, but three kinds are usually hard to understand or are easily replaced with an oft-better alternative. In particular, you should avoid **pie charts**; it is generally better to use bars, as it is easier to compare bar heights than pie slice sizes. You should also not use **3-D visualizations**, as they are typically hard to understand when converted to a static 2-D image format. Finally, do not use tables to make numerical comparisons; humans are much better at quickly processing visual information than text and math. Bar plots are again typically a better alternative.

4.4 Refining the visualization

Convey the message, minimize noise

Just being able to make a visualization in R (or any other language, for that matter) doesn't mean that it effectively communicates your message to others. Once you have selected a broad type of visualization to use, you will have to refine it to suit your particular need. Some rules of thumb for doing this are listed below. They generally fall into two classes: you want to *make your visualization convey your message*, and you want to *reduce visual noise* as much as possible. Humans have limited cognitive ability to process information; both of these types of refinement aim to reduce the mental load on your audience when viewing your visualization, making it easier for them to understand and remember your message quickly.

Convey the message

- Make sure the visualization answers the question you have asked most simply and plainly as possible.
- Use legends and labels so that your visualization is understandable without reading the surrounding text.
- Ensure the text, symbols, lines, etc., on your visualization are big enough to be easily read.
- Ensure the data are clearly visible; don't hide the shape/distribution of the data behind other objects (e.g., a bar).
- Make sure to use color schemes that are understandable by those with colorblindness (a surprisingly large fraction of the overall population—from about 1% to 10%, depending on sex and ancestry [Deeb, 2005]). For example, ColorBrewer[1] and the `RColorBrewer` R package[2] [Neuwirth, 2014] provide the ability to pick such color schemes, and you can check your visualizations after you have created them by uploading to online tools such as a color blindness simulator[3].
- Redundancy can be helpful; sometimes conveying the same message in multiple ways reinforces it for the audience.

Minimize noise

- Use colors sparingly. Too many different colors can be distracting, create false patterns, and detract from the message.
- Be wary of overplotting. Overplotting is when marks that represent the data

[1] https://colorbrewer2.org

[2] https://cran.r-project.org/web/packages/RColorBrewer/index.html

[3] https://www.color-blindness.com/coblis-color-blindness-simulator/

overlap, and is problematic as it prevents you from seeing how many data points are represented in areas of the visualization where this occurs. If your plot has too many dots or lines and starts to look like a mess, you need to do something different.
- Only make the plot area (where the dots, lines, bars are) as big as needed. Simple plots can be made small.
- Don't adjust the axes to zoom in on small differences. If the difference is small, show that it's small!

4.5 Creating visualizations with `ggplot2`

Build the visualization iteratively

This section will cover examples of how to choose and refine a visualization given a data set and a question that you want to answer, and then how to create the visualization in R using the `ggplot2` R package. Given that the `ggplot2`package is loaded by the `tidyverse` metapackage, we still need to load only 'tidyverse':

```
library(tidyverse)
```

4.5.1 Scatter plots and line plots: the Mauna Loa CO_2 data set

The Mauna Loa CO_2 data set[4], curated by Dr. Pieter Tans, NOAA/GML and Dr. Ralph Keeling, Scripps Institution of Oceanography, records the atmospheric concentration of carbon dioxide (CO_2, in parts per million) at the Mauna Loa research station in Hawaii from 1959 onward [Tans and Keeling, 2020]. For this book, we are going to focus on the last 40 years of the data set, 1980-2020.

Question: Does the concentration of atmospheric CO_2 change over time, and are there any interesting patterns to note?

To get started, we will read and inspect the data:

```
# mauna loa carbon dioxide data
co2_df <- read_csv("data/mauna_loa_data.csv")
co2_df
```

```
## # A tibble: 484 x 2
```

[4] https://www.esrl.noaa.gov/gmd/ccgg/trends/data.html

```
##     date_measured   ppm
##     <date>        <dbl>
##   1 1980-02-01     338.
##   2 1980-03-01     340.
##   3 1980-04-01     341.
##   4 1980-05-01     341.
##   5 1980-06-01     341.
##   6 1980-07-01     339.
##   7 1980-08-01     338.
##   8 1980-09-01     336.
##   9 1980-10-01     336.
##  10 1980-11-01     337.
## # ... with 474 more rows
```

We see that there are two columns in the `co2_df` data frame; `date_measured` and `ppm`. The `date_measured` column holds the date the measurement was taken, and is of type `date`. The `ppm` column holds the value of CO_2 in parts per million that was measured on each date, and is type `double`.

Note: `read_csv` was able to parse the `date_measured` column into the `date` vector type because it was entered in the international standard date format, called ISO 8601, which lists dates as `year-month-day`. `date` vectors are `double` vectors with special properties that allow them to handle dates correctly. For example, `date` type vectors allow functions like `ggplot` to treat them as numeric dates and not as character vectors, even though they contain non-numeric characters (e.g., in the `date_measured` column in the `co2_df` data frame). This means R will not accidentally plot the dates in the wrong order (i.e., not alphanumerically as would happen if it was a character vector). An in-depth study of dates and times is beyond the scope of the book, but interested readers may consult the Dates and Times chapter of *R for Data Science* [Wickham and Grolemund, 2016]; see the additional resources at the end of this chapter.

Since we are investigating a relationship between two variables (CO_2 concentration and date), a scatter plot is a good place to start. Scatter plots show the data as individual points with `x` (horizontal axis) and `y` (vertical axis) coordinates. Here, we will use the measurement date as the `x` coordinate and the CO_2 concentration as the `y` coordinate. When using the `ggplot2` package, we create a plot object with the `ggplot` function. There are a few basic aspects of a plot that we need to specify:

- The name of the data frame object to visualize.

  - Here, we specify the `co2_df` data frame.
- The **aesthetic mapping**, which tells `ggplot` how the columns in the data frame map to properties of the visualization.
  - To create an aesthetic mapping, we use the `aes` function.
  - Here, we set the plot `x` axis to the `date_measured` variable, and the plot `y` axis to the `ppm` variable.
- The `+` operator, which tells `ggplot` that we would like to add another layer to the plot.
- The **geometric object**, which specifies how the mapped data should be displayed.
  - To create a geometric object, we use a `geom_*` function (see the ggplot reference[5] for a list of geometric objects).
  - Here, we use the `geom_point` function to visualize our data as a scatter plot.

Figure 4.2 shows how each of these aspects map to code for creating a basic scatter plot of the `co2_df` data. Note that we could pass many other possible arguments to the aesthetic mapping and geometric object to change how the plot looks. For the purposes of quickly testing things out to see what they look like, though, we can just start with the default settings.

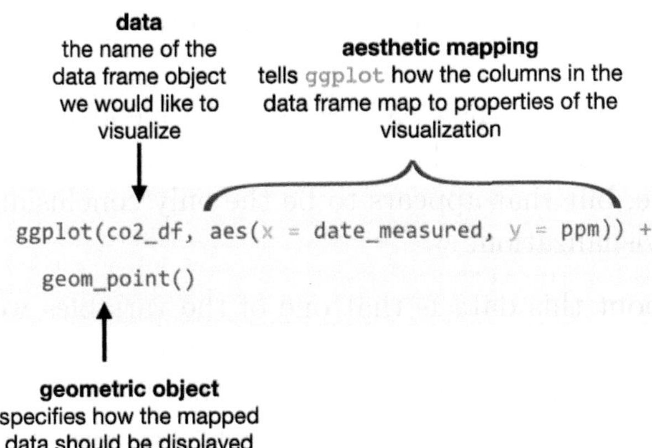

FIGURE 4.2: Creating a scatter plot with the `ggplot` function.

[5]https://ggplot2.tidyverse.org/reference/

```
co2_scatter <- ggplot(co2_df, aes(x = date_measured, y = ppm)) +
  geom_point()

co2_scatter
```

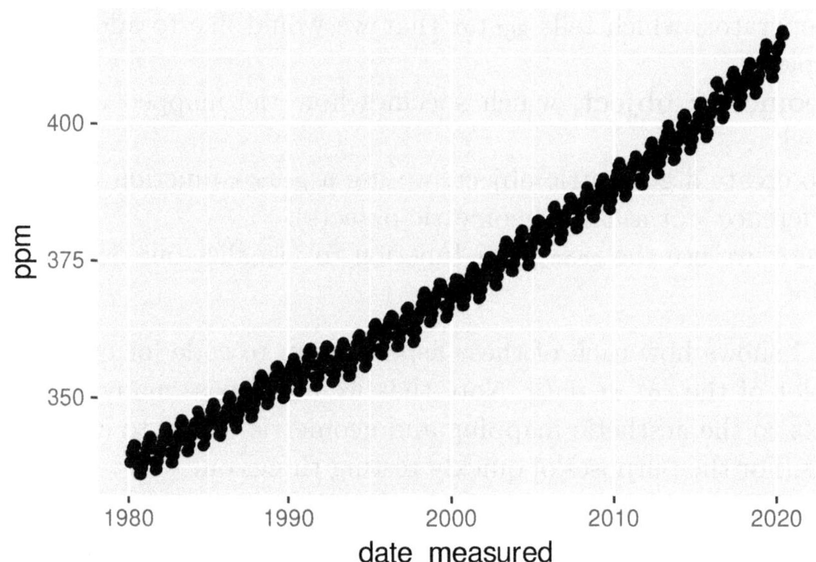

FIGURE 4.3: Scatter plot of atmospheric concentration of CO_2 over time.

Certainly, the visualization in Figure 4.3 shows a clear upward trend in the atmospheric concentration of CO_2 over time. This plot answers the first part of our question in the affirmative, but that appears to be the only conclusion one can make from the scatter visualization.

One important thing to note about this data is that one of the variables we are exploring is time. Time is a special kind of quantitative variable because it forces additional structure on the data—the data points have a natural order. Specifically, each observation in the data set has a predecessor and a successor, and the order of the observations matters; changing their order alters their meaning. In situations like this, we typically use a line plot to visualize the data. Line plots connect the sequence of `x` and `y` coordinates of the observations with line segments, thereby emphasizing their order.

We can create a line plot in `ggplot` using the `geom_line` function. Let's now try to visualize the `co2_df` as a line plot with just the default arguments:

```
co2_line <- ggplot(co2_df, aes(x = date_measured, y = ppm)) +
  geom_line()
```

```
co2_line
```

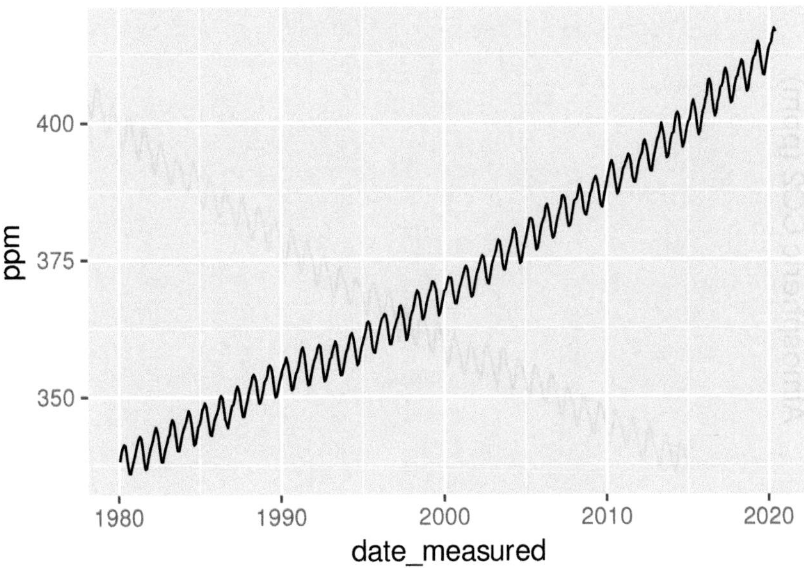

FIGURE 4.4: Line plot of atmospheric concentration of CO_2 over time.

Aha! Figure 4.4 shows us there *is* another interesting phenomenon in the data: in addition to increasing over time, the concentration seems to oscillate as well. Given the visualization as it is now, it is still hard to tell how fast the oscillation is, but nevertheless, the line seems to be a better choice for answering the question than the scatter plot was. The comparison between these two visualizations also illustrates a common issue with scatter plots: often, the points are shown too close together or even on top of one another, muddling information that would otherwise be clear (*overplotting*).

Now that we have settled on the rough details of the visualization, it is time to refine things. This plot is fairly straightforward, and there is not much visual noise to remove. But there are a few things we must do to improve clarity, such as adding informative axis labels and making the font a more readable size. To add axis labels, we use the `xlab` and `ylab` functions. To change the font size, we use the `theme` function with the `text` argument:

```
co2_line <- ggplot(co2_df, aes(x = date_measured, y = ppm)) +
  geom_line() +
  xlab("Year") +
  ylab("Atmospheric CO2 (ppm)") +
  theme(text = element_text(size = 12))
```

```
co2_line
```

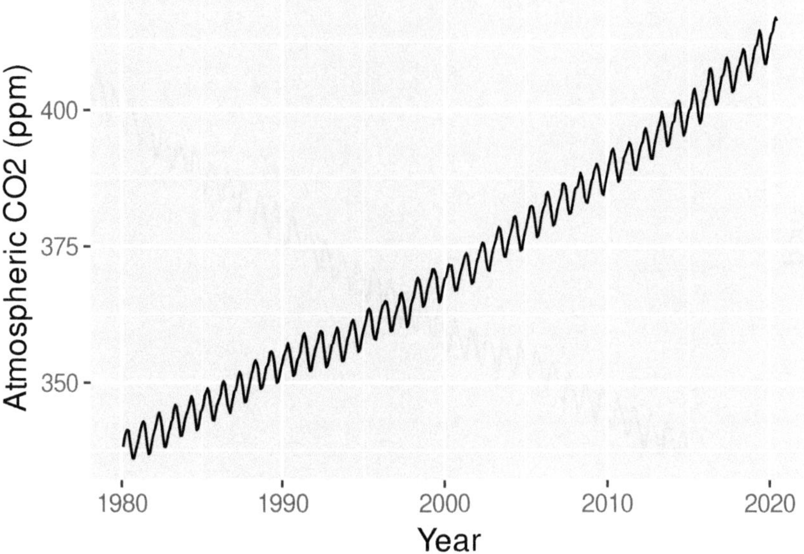

FIGURE 4.5: Line plot of atmospheric concentration of CO_2 over time with clearer axes and labels.

Note: The `theme` function is quite complex and has many arguments that can be specified to control many non-data aspects of a visualization. An in-depth discussion of the `theme` function is beyond the scope of this book. Interested readers may consult the `theme` function documentation; see the additional resources section at the end of this chapter.

Finally, let's see if we can better understand the oscillation by changing the visualization slightly. Note that it is totally fine to use a small number of visualizations to answer different aspects of the question you are trying to answer. We will accomplish this by using *scales*, another important feature of `ggplot2` that easily transforms the different variables and set limits. We scale the horizontal axis using the `xlim` function, and the vertical axis with the `ylim` function. In particular, here, we will use the `xlim` function to zoom in on just five years of data (say, 1990-1994). `xlim` takes a vector of length two to specify the upper and lower bounds to limit the axis. We can create that using the `c` function. Note that it is important that the vector given to `xlim` must be of the same type as the data that is mapped to that axis. Here, we have mapped a date to the x-axis, and so we need to use the `date` function (from the `tidyverse` `lubridate` R package[6] [Spinu et al., 2021, Grolemund and Wickham, 2011]) to convert the character strings we provide to `c` to `date` vectors.

Note: `lubridate` is a package that is installed by the `tidyverse` metapackage, but is not loaded by it. Hence we need to load it separately in the code below.

```
library(lubridate)

co2_line <- ggplot(co2_df, aes(x = date_measured, y = ppm)) +
  geom_line() +
  xlab("Year") +
  ylab("Atmospheric CO2 (ppm)") +
  xlim(c(date("1990-01-01"), date("1993-12-01"))) +
  theme(text = element_text(size = 12))

co2_line
```

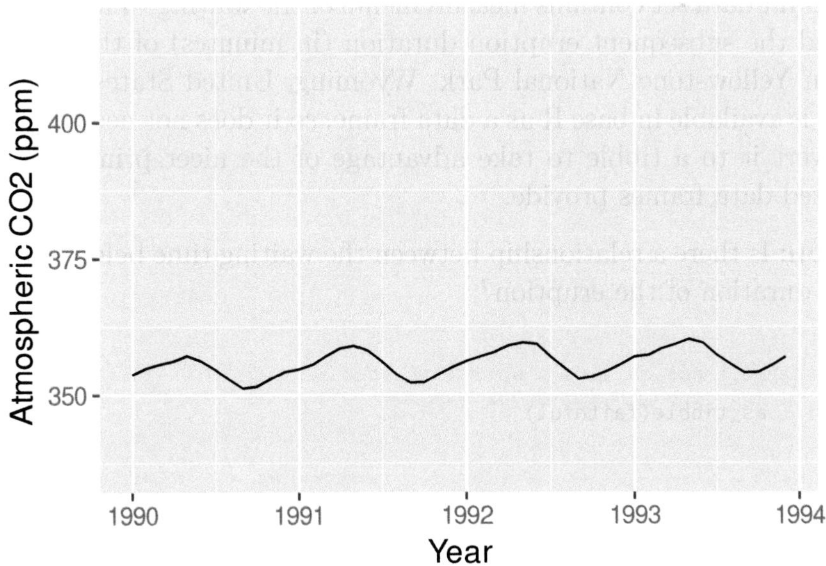

FIGURE 4.6: Line plot of atmospheric concentration of CO_2 from 1990 to 1994.

[6] `https://lubridate.tidyverse.org/`

Interesting! It seems that each year, the atmospheric CO_2 increases until it reaches its peak somewhere around April, decreases until around late September, and finally increases again until the end of the year. In Hawaii, there are two seasons: summer from May through October, and winter from November through April. Therefore, the oscillating pattern in CO_2 matches up fairly closely with the two seasons.

As you might have noticed from the code used to create the final visualization of the `co2_df` data frame, we construct the visualizations in `ggplot` with layers. New layers are added with the `+` operator, and we can really add as many as we would like! A useful analogy to constructing a data visualization is painting a picture. We start with a blank canvas, and the first thing we do is prepare the surface for our painting by adding primer. In our data visualization this is akin to calling `ggplot` and specifying the data set we will be using. Next, we sketch out the background of the painting. In our data visualization, this would be when we map data to the axes in the `aes` function. Then we add our key visual subjects to the painting. In our data visualization, this would be the geometric objects (e.g., `geom_point`, `geom_line`, etc.). And finally, we work on adding details and refinements to the painting. In our data visualization this would be when we fine tune axis labels, change the font, adjust the point size, and do other related things.

4.5.2 Scatter plots: the Old Faithful eruption time data set

The `faithful` data set contains measurements of the waiting time between eruptions and the subsequent eruption duration (in minutes) of the Old Faithful geyser in Yellowstone National Park, Wyoming, United States. The `faithful` data set is available in base R as a data frame, so it does not need to be loaded. We convert it to a tibble to take advantage of the nicer print output these specialized data frames provide.

Question: Is there a relationship between the waiting time before an eruption and the duration of the eruption?

```
# old faithful eruption time / wait time data
faithful <- as_tibble(faithful)
faithful
```

```
## # A tibble: 272 x 2
##    eruptions waiting
##        <dbl>   <dbl>
##  1       3.6      79
##  2       1.8      54
```

```
## 3         3.33        74
## 4         2.28        62
## 5         4.53        85
## 6         2.88        55
## 7         4.7         88
## 8         3.6         85
## 9         1.95        51
## 10        4.35        85
## # ... with 262 more rows
```

Here again, we investigate the relationship between two quantitative variables (waiting time and eruption time). But if you look at the output of the data frame, you'll notice that unlike time in the Mauna Loa CO_2 data set, neither of the variables here have a natural order to them. So a scatter plot is likely to be the most appropriate visualization. Let's create a scatter plot using the `ggplot` function with the `waiting` variable on the horizontal axis, the `eruptions` variable on the vertical axis, and the `geom_point` geometric object. The result is shown in Figure 4.7.

```
faithful_scatter <- ggplot(faithful, aes(x = waiting, y = eruptions)) +
  geom_point()

faithful_scatter
```

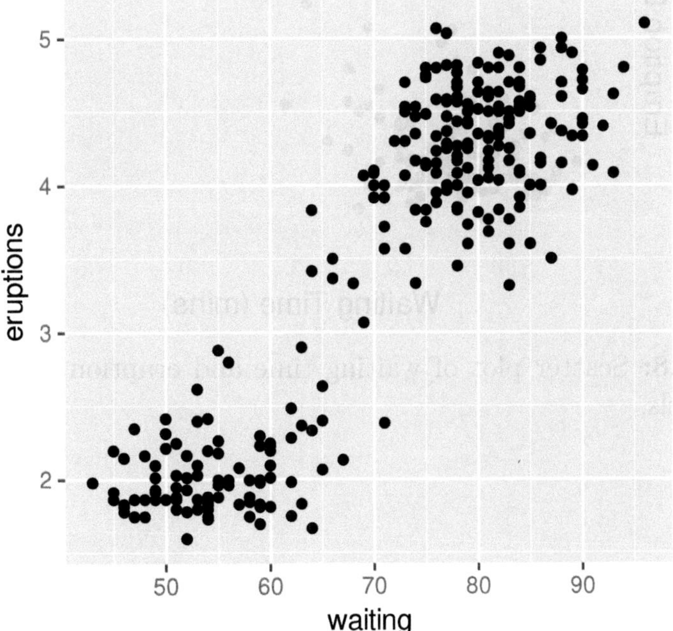

FIGURE 4.7: Scatter plot of waiting time and eruption time.

We can see in Figure 4.7 that the data tend to fall into two groups: one with short waiting and eruption times, and one with long waiting and eruption times. Note that in this case, there is no overplotting: the points are generally nicely visually separated, and the pattern they form is clear. In order to refine the visualization, we need only to add axis labels and make the font more readable:

```
faithful_scatter <- ggplot(faithful, aes(x = waiting, y = eruptions)) +
  geom_point() +
  xlab("Waiting Time (mins)") +
  ylab("Eruption Duration (mins)") +
  theme(text = element_text(size = 12))

faithful_scatter
```

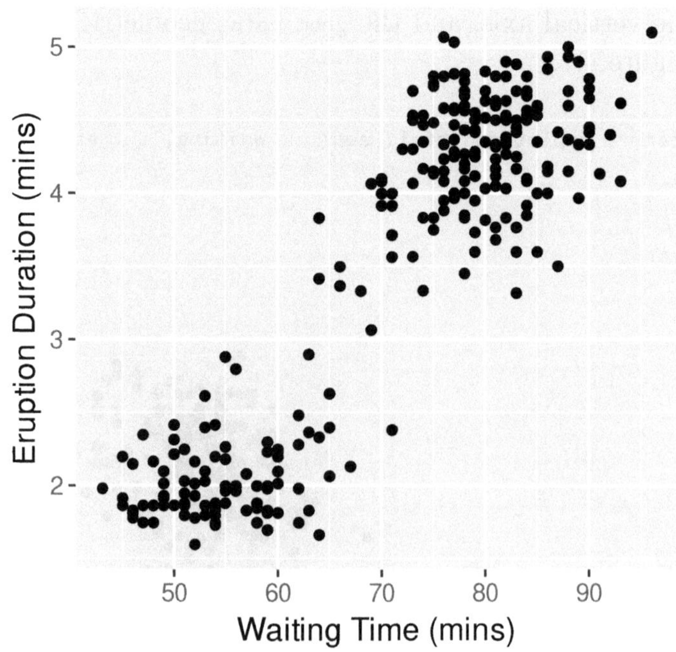

FIGURE 4.8: Scatter plot of waiting time and eruption time with clearer axes and labels.

4.5.3 Axis transformation and colored scatter plots: the Canadian languages data set

Recall the `can_lang` data set [Timbers, 2020] from Chapters 1, 2, and 3, which contains counts of languages from the 2016 Canadian census.

Question: Is there a relationship between the percentage of people who speak a language as their mother tongue and the percentage for whom that is the primary language spoken at home? And is there a pattern in the strength of this relationship in the higher-level language categories (Official languages, Aboriginal languages, or non-official and non-Aboriginal languages)?

To get started, we will read and inspect the data:

```
can_lang <- read_csv("data/can_lang.csv")
can_lang
```

```
## # A tibble: 214 x 6
##    category       language    mother_tongue most_at_home most_at_work lang_known
##    <chr>          <chr>               <dbl>        <dbl>        <dbl>      <dbl>
##  1 Aboriginal la~ Aboriginal~           590          235           30        665
##  2 Non-Official ~ Afrikaans           10260         4785           85      23415
##  3 Non-Official ~ Afro-Asiat~          1150          445           10       2775
##  4 Non-Official ~ Akan (Twi)          13460         5985           25      22150
##  5 Non-Official ~ Albanian            26895        13135          345      31930
##  6 Aboriginal la~ Algonquian~            45           10            0        120
##  7 Aboriginal la~ Algonquin            1260          370           40       2480
##  8 Non-Official ~ American S~          2685         3020         1145      21930
##  9 Non-Official ~ Amharic             22465        12785          200      33670
## 10 Non-Official ~ Arabic             419890       223535         5585     629055
## # ... with 204 more rows
```

We will begin with a scatter plot of the `mother_tongue` and `most_at_home` columns from our data frame. The resulting plot is shown in Figure 4.9.

```
ggplot(can_lang, aes(x = most_at_home, y = mother_tongue)) +
  geom_point()
```

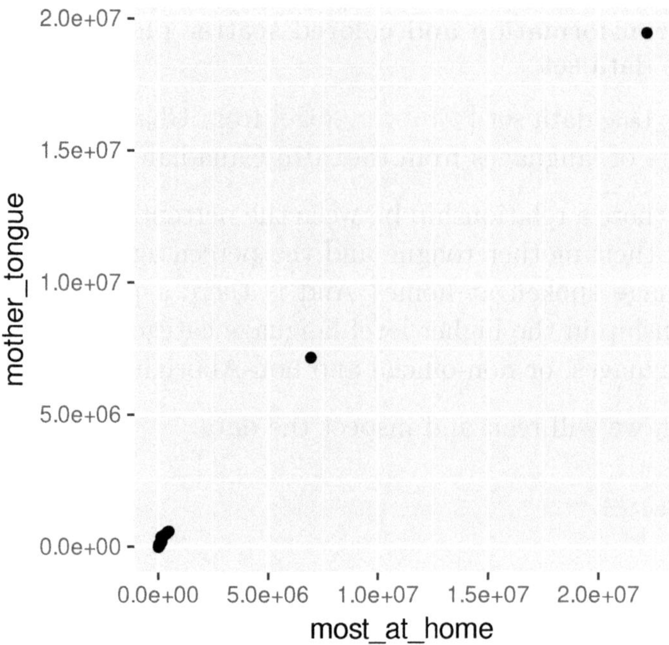

FIGURE 4.9: Scatter plot of number of Canadians reporting a language as their mother tongue vs the primary language at home.

To make an initial improvement in the interpretability of Figure 4.9, we should replace the default axis names with more informative labels. We can use `\n` to create a line break in the axis names so that the words after `\n` are printed on a new line. This will make the axes labels on the plots more readable. We should also increase the font size to further improve readability.

```
ggplot(can_lang, aes(x = most_at_home, y = mother_tongue)) +
  geom_point() +
  xlab("Language spoken most at home \n (number of Canadian residents)") +
  ylab("Mother tongue \n (number of Canadian residents)") +
  theme(text = element_text(size = 12))
```

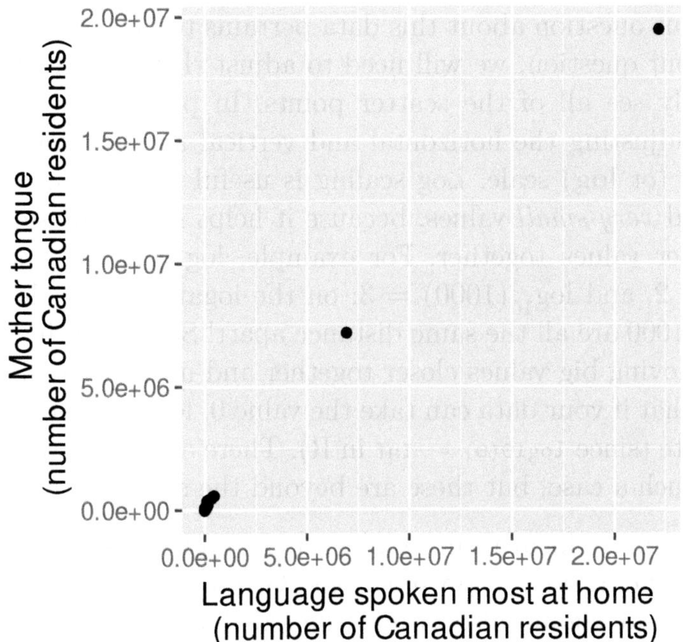

FIGURE 4.10: Scatter plot of number of Canadians reporting a language as their mother tongue vs the primary language at home with x and y labels.

Okay! The axes and labels in Figure 4.10 are much more readable and interpretable now. However, the scatter points themselves could use some work; most of the 214 data points are bunched up in the lower left-hand side of the visualization. The data is clumped because many more people in Canada speak English or French (the two points in the upper right corner) than other languages. In particular, the most common mother tongue language has 19,460,850 speakers, while the least common has only 10. That's a 6-decimal-place difference in the magnitude of these two numbers! We can confirm that the two points in the upper right-hand corner correspond to Canada's two official languages by filtering the data:

```
can_lang |>
  filter(language == "English" | language == "French")
```

```
## # A tibble: 2 x 6
##   category           language mother_tongue most_at_home most_at_work lang_known
##   <chr>              <chr>            <dbl>        <dbl>        <dbl>      <dbl>
## 1 Official languages English       19460850     22162865     15265335   29748265
## 2 Official languages French         7166700      6943800      3825215   10242945
```

Recall that our question about this data pertains to *all* languages; so to properly answer our question, we will need to adjust the scale of the axes so that we can clearly see all of the scatter points. In particular, we will improve the plot by adjusting the horizontal and vertical axes so that they are on a **logarithmic** (or **log**) scale. Log scaling is useful when your data take both *very large* and *very small* values, because it helps space out small values and squishes larger values together. For example, $\log_{10}(1) = 0$, $\log_{10}(10) = 1$, $\log_{10}(100) = 2$, and $\log_{10}(1000) = 3$; on the logarithmic scale, the values 1, 10, 100, and 1000 are all the same distance apart! So we see that applying this function is moving big values closer together and moving small values farther apart. Note that if your data can take the value 0, logarithmic scaling may not be appropriate (since `log10(0) = -Inf` in R). There are other ways to transform the data in such a case, but these are beyond the scope of the book.

We can accomplish logarithmic scaling in a `ggplot` visualization using the `scale_x_log10` and `scale_y_log10` functions. Given that the x and y axes have large numbers, we should also format the axis labels to put commas in these numbers to increase their readability. We can do this in R by passing the `label_comma` function (from the `scales` package) to the `labels` argument of the `scale_x_log10` and `scale_x_log10` functions.

```
library(scales)

ggplot(can_lang, aes(x = most_at_home, y = mother_tongue)) +
  geom_point() +
  xlab("Language spoken most at home \n (number of Canadian residents)") +
  ylab("Mother tongue \n (number of Canadian residents)") +
  theme(text = element_text(size = 12)) +
  scale_x_log10(labels = label_comma()) +
  scale_y_log10(labels = label_comma())
```

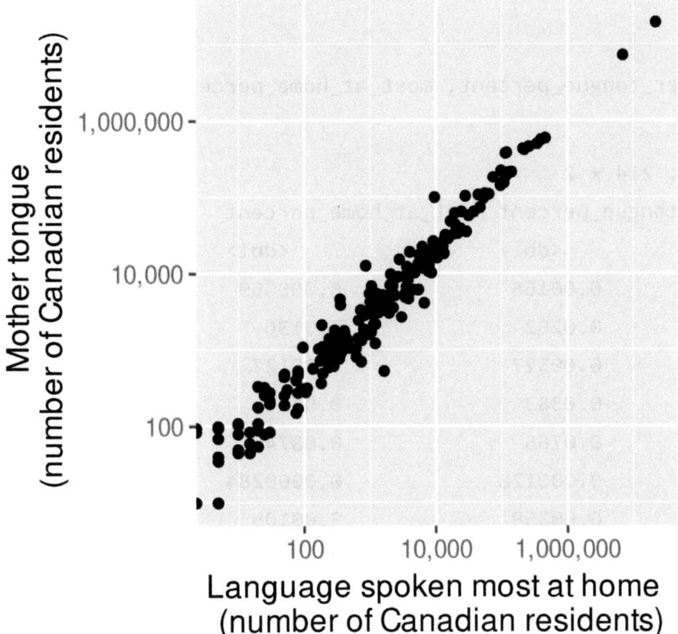

FIGURE 4.11: Scatter plot of number of Canadians reporting a language as their mother tongue vs the primary language at home with log adjusted x and y axes.

Similar to some of the examples in Chapter 3, we can convert the counts to percentages to give them context and make them easier to understand. We can do this by dividing the number of people reporting a given language as their mother tongue or primary language at home by the number of people who live in Canada and multiplying by 100%. For example, the percentage of people who reported that their mother tongue was English in the 2016 Canadian census was 19,460,850 / 35,151,728 $\times$ 100 % = 55.36%.

Below we use `mutate` to calculate the percentage of people reporting a given language as their mother tongue and primary language at home for all the languages in the `can_lang` data set. Since the new columns are appended to the end of the data table, we selected the new columns after the transformation so you can clearly see the mutated output from the table.

```
can_lang <- can_lang |>
  mutate(
    mother_tongue_percent = (mother_tongue / 35151728) * 100,
    most_at_home_percent = (most_at_home / 35151728) * 100
  )
```

```
can_lang |>
  select(mother_tongue_percent, most_at_home_percent)
```

```
## # A tibble: 214 x 2
##    mother_tongue_percent most_at_home_percent
##                    <dbl>                <dbl>
##  1              0.00168             0.000669
##  2              0.0292              0.0136
##  3              0.00327             0.00127
##  4              0.0383              0.0170
##  5              0.0765              0.0374
##  6              0.000128            0.0000284
##  7              0.00358             0.00105
##  8              0.00764             0.00859
##  9              0.0639              0.0364
## 10              1.19                0.636
## # ... with 204 more rows
```

Finally, we will edit the visualization to use the percentages we just computed (and change our axis labels to reflect this change in units). Figure 4.12 displays the final result.

```
ggplot(can_lang, aes(x = most_at_home_percent, y = mother_tongue_percent)) +
  geom_point() +
  xlab("Language spoken most at home \n (percentage of Canadian residents)") +
  ylab("Mother tongue \n (percentage of Canadian residents)") +
  theme(text = element_text(size = 12)) +
  scale_x_log10(labels = comma) +
  scale_y_log10(labels = comma)
```

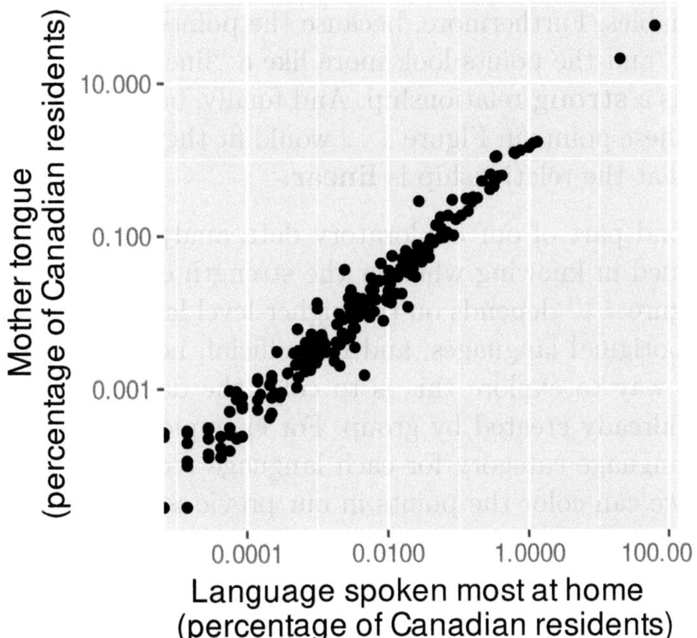

FIGURE 4.12: Scatter plot of percentage of Canadians reporting a language as their mother tongue vs the primary language at home.

Figure 4.12 is the appropriate visualization to use to answer the first question in this section, i.e., whether there is a relationship between the percentage of people who speak a language as their mother tongue and the percentage for whom that is the primary language spoken at home. To fully answer the question, we need to use Figure 4.12 to assess a few key characteristics of the data:

- **Direction:** if the y variable tends to increase when the x variable increases, then y has a **positive** relationship with x. If y tends to decrease when x increases, then y has a **negative** relationship with x. If y does not meaningfully increase or decrease as x increases, then y has **little or no** relationship with x.
- **Strength:** if the y variable *reliably* increases, decreases, or stays flat as x increases, then the relationship is **strong**. Otherwise, the relationship is **weak**. Intuitively, the relationship is strong when the scatter points are close together and look more like a "line" or "curve" than a "cloud."
- **Shape:** if you can draw a straight line roughly through the data points, the relationship is **linear**. Otherwise, it is **nonlinear**.

In Figure 4.12, we see that as the percentage of people who have a language as their mother tongue increases, so does the percentage of people who speak that language at home. Therefore, there is a **positive** relationship between these two variables. Furthermore, because the points in Figure 4.12 are fairly close together, and the points look more like a "line" than a "cloud", we can say that this is a **strong** relationship. And finally, because drawing a straight line through these points in Figure 4.12 would fit the pattern we observe quite well, we say that the relationship is **linear**.

Onto the second part of our exploratory data analysis question! Recall that we are interested in knowing whether the strength of the relationship we uncovered in Figure 4.12 depends on the higher-level language category (Official languages, Aboriginal languages, and non-official, non-Aboriginal languages). One common way to explore this is to color the data points on the scatter plot we have already created by group. For example, given that we have the higher-level language category for each language recorded in the 2016 Canadian census, we can color the points in our previous scatter plot to represent each language's higher-level language category.

Here we want to distinguish the values according to the `category` group with which they belong. We can add an argument to the `aes` function, specifying that the `category` column should color the points. Adding this argument will color the points according to their group and add a legend at the side of the plot.

```
ggplot(can_lang, aes(x = most_at_home_percent,
                     y = mother_tongue_percent,
                     color = category)) +
  geom_point() +
  xlab("Language spoken most at home \n (percentage of Canadian residents)") +
  ylab("Mother tongue \n (percentage of Canadian residents)") +
  theme(text = element_text(size = 12)) +
  scale_x_log10(labels = comma) +
  scale_y_log10(labels = comma)
```

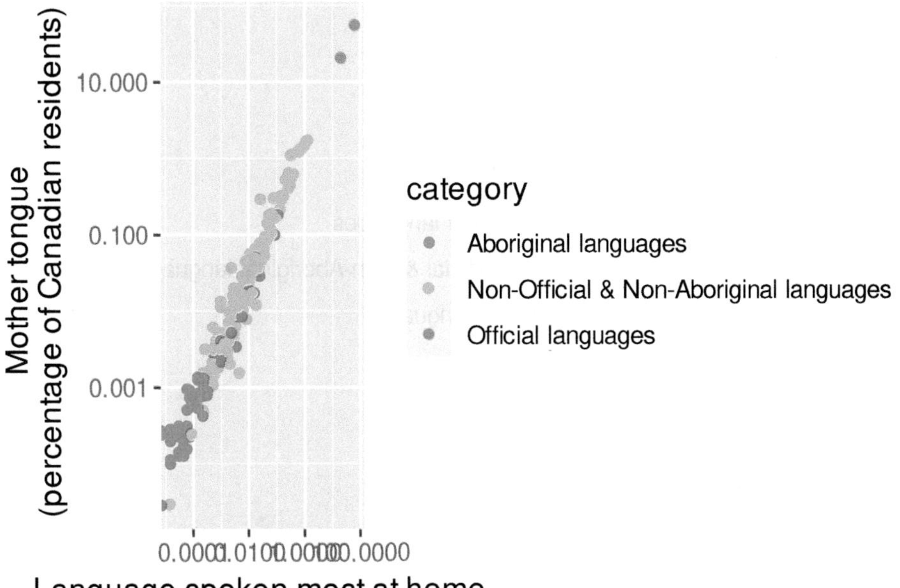

FIGURE 4.13: Scatter plot of percentage of Canadians reporting a language as their mother tongue vs the primary language at home colored by language category.

The legend in Figure 4.13 takes up valuable plot area. We can improve this by moving the legend title using the `legend.position` and `legend.direction` arguments of the `theme` function. Here we set `legend.position` to `"top"` to put the legend above the plot and `legend.direction` to `"vertical"` so that the legend items remain vertically stacked on top of each other. When the `legend.position` is set to either `"top"` or `"bottom"` the default direction is to stack the legend items horizontally. However, that will not work well for this particular visualization because the legend labels are quite long and would run off the page if displayed this way.

```
ggplot(can_lang, aes(x = most_at_home_percent,
                     y = mother_tongue_percent,
                     color = category)) +
  geom_point() +
  xlab("Language spoken most at home \n (percentage of Canadian residents)") +
  ylab("Mother tongue \n (percentage of Canadian residents)") +
  theme(text = element_text(size = 12),
        legend.position = "top",
        legend.direction = "vertical") +
```

```
  scale_x_log10(labels = comma) +
  scale_y_log10(labels = comma)
```

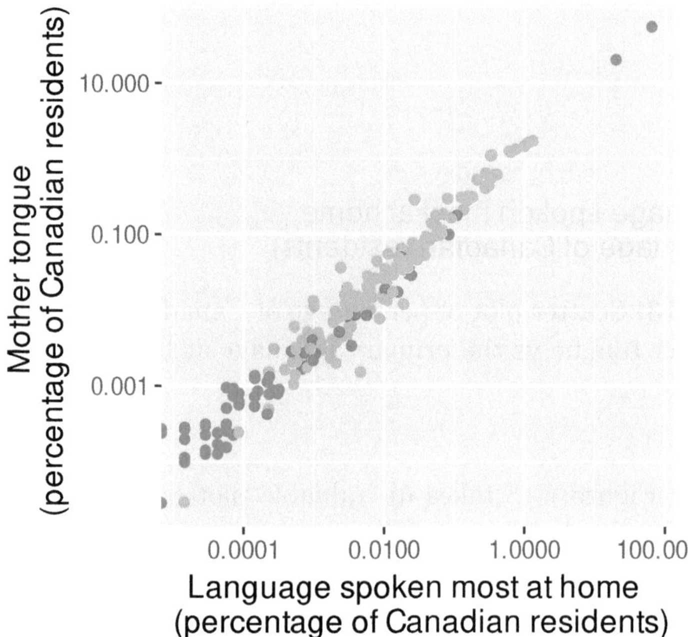

FIGURE 4.14: Scatter plot of percentage of Canadians reporting a language as their mother tongue vs the primary language at home colored by language category with the legend edited.

In Figure 4.14, the points are colored with the default `ggplot2` color palette. But what if you want to use different colors? In R, two packages that provide alternative color palettes are `RColorBrewer` [Neuwirth, 2014] and `ggthemes` [Arnold, 2019]; in this book we will cover how to use `RColorBrewer`. You can visualize the list of color palettes that `RColorBrewer` has to offer with the `display.brewer.all` function. You can also print a list of color-blind friendly palettes by adding `colorblindFriendly = TRUE` to the function.

```
library(RColorBrewer)
display.brewer.all(colorblindFriendly = TRUE)
```

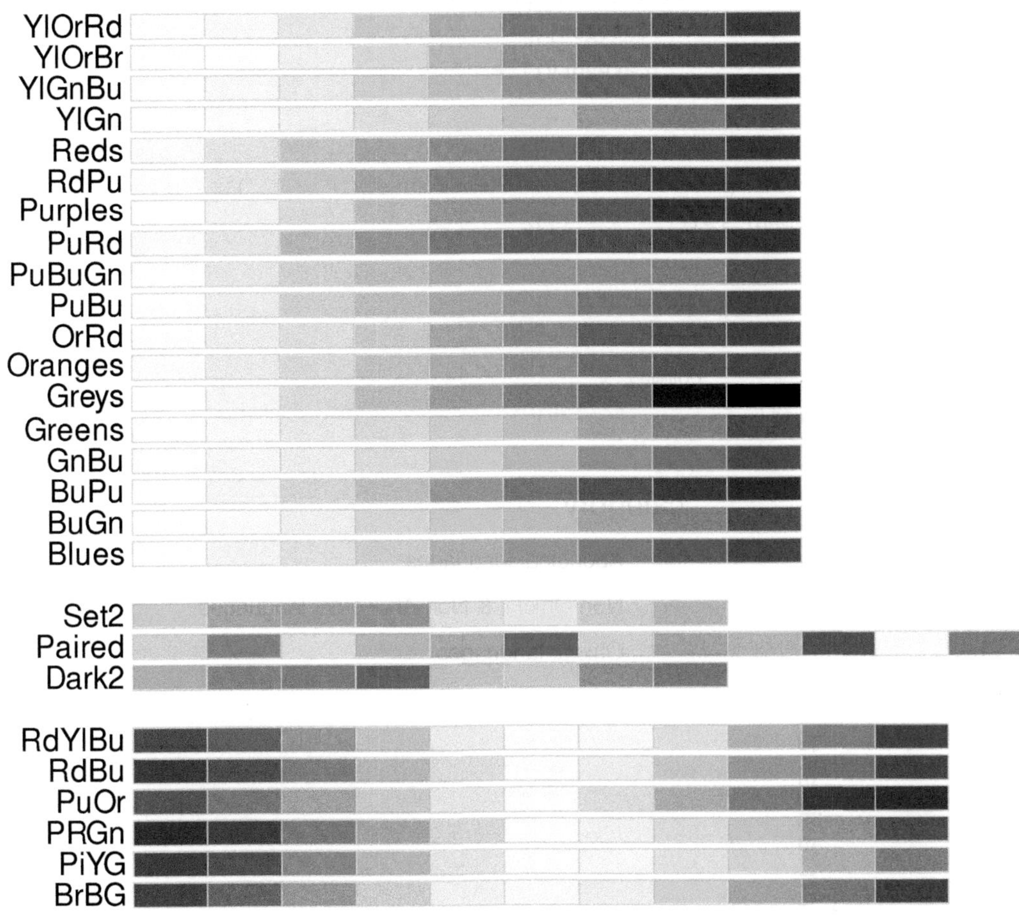

FIGURE 4.15: Color palettes available from the `RColorBrewer` R package.

From Figure 4.15, we can choose the color palette we want to use in our plot. To change the color palette, we add the `scale_color_brewer` layer indicating the palette we want to use. You can use this color blindness simulator[7] to check if your visualizations are color-blind friendly. Below we pick the `"Set2"` palette, with the result shown in Figure 4.16. We also set the `shape` aesthetic mapping to the `category` variable as well; this makes the scatter point shapes different for each category. This kind of visual redundancy—i.e., conveying the same information with both scatter point color and shape—can further improve the clarity and accessibility of your visualization.

[7] https://www.color-blindness.com/coblis-color-blindness-simulator/

```
ggplot(can_lang, aes(x = most_at_home_percent,
                     y = mother_tongue_percent,
                     color = category,
                     shape = category)) +
  geom_point() +
  xlab("Language spoken most at home \n (percentage of Canadian residents)") +
  ylab("Mother tongue \n (percentage of Canadian residents)") +
  theme(text = element_text(size = 12),
        legend.position = "top",
        legend.direction = "vertical") +
  scale_x_log10(labels = comma) +
  scale_y_log10(labels = comma) +
  scale_color_brewer(palette = "Set2")
```

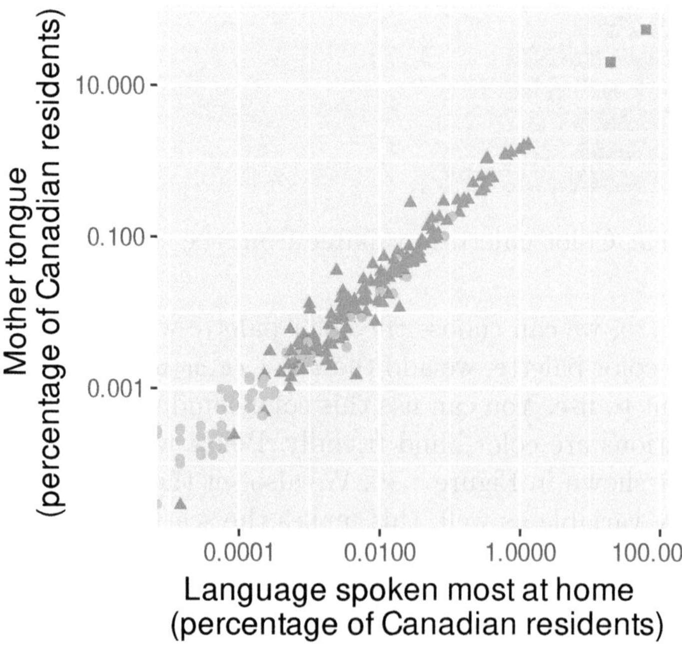

FIGURE 4.16: Scatter plot of percentage of Canadians reporting a language as their mother tongue vs the primary language at home colored by language category with color-blind friendly colors.

From the visualization in Figure 4.16, we can now clearly see that the vast majority of Canadians reported one of the official languages as their mother tongue and as the language they speak most often at home. What do we see when considering the second part of our exploratory question? Do we see a difference in the relationship between languages spoken as a mother tongue and as a primary language at home across the higher-level language categories? Based on Figure 4.16, there does not appear to be much of a difference. For each higher-level language category, there appears to be a strong, positive, and linear relationship between the percentage of people who speak a language as their mother tongue and the percentage who speak it as their primary language at home. The relationship looks similar regardless of the category.

Does this mean that this relationship is positive for all languages in the world? And further, can we use this data visualization on its own to predict how many people have a given language as their mother tongue if we know how many people speak it as their primary language at home? The answer to both these questions is "no!" However, with exploratory data analysis, we can create new hypotheses, ideas, and questions (like the ones at the beginning of this paragraph). Answering those questions often involves doing more complex analyses, and sometimes even gathering additional data. We will see more of such complex analyses later on in this book.

4.5.4 Bar plots: the island landmass data set

The `islands.csv` data set contains a list of Earth's landmasses as well as their area (in thousands of square miles) [McNeil, 1977].

Question: Are the continents (North / South America, Africa, Europe, Asia, Australia, Antarctica) Earth's seven largest landmasses? If so, what are the next few largest landmasses after those?

To get started, we will read and inspect the data:

```
# islands data
islands_df <- read_csv("data/islands.csv")
islands_df
```

```
## # A tibble: 48 x 3
##    landmass      size landmass_type
##    <chr>        <dbl> <chr>
##  1 Africa       11506 Continent
##  2 Antarctica    5500 Continent
##  3 Asia         16988 Continent
##  4 Australia     2968 Continent
```

```
## 5 Axel Heiberg     16 Other
## 6 Baffin          184 Other
## 7 Banks            23 Other
## 8 Borneo          280 Other
## 9 Britain          84 Other
## 10 Celebes         73 Other
## # ... with 38 more rows
```

Here, we have a data frame of Earth's landmasses, and are trying to compare their sizes. The right type of visualization to answer this question is a bar plot. In a bar plot, the height of the bar represents the value of a summary statistic (usually a size, count, proportion or percentage). They are particularly useful for comparing summary statistics between different groups of a categorical variable.

We specify that we would like to use a bar plot via the `geom_bar` function in `ggplot2`. However, by default, `geom_bar` sets the heights of bars to the number of times a value appears in a data frame (its *count*); here, we want to plot exactly the values in the data frame, i.e., the landmass sizes. So we have to pass the `stat = "identity"` argument to `geom_bar`. The result is shown in Figure 4.17.

```
islands_bar <- ggplot(islands_df, aes(x = landmass, y = size)) +
  geom_bar(stat = "identity")

islands_bar
```

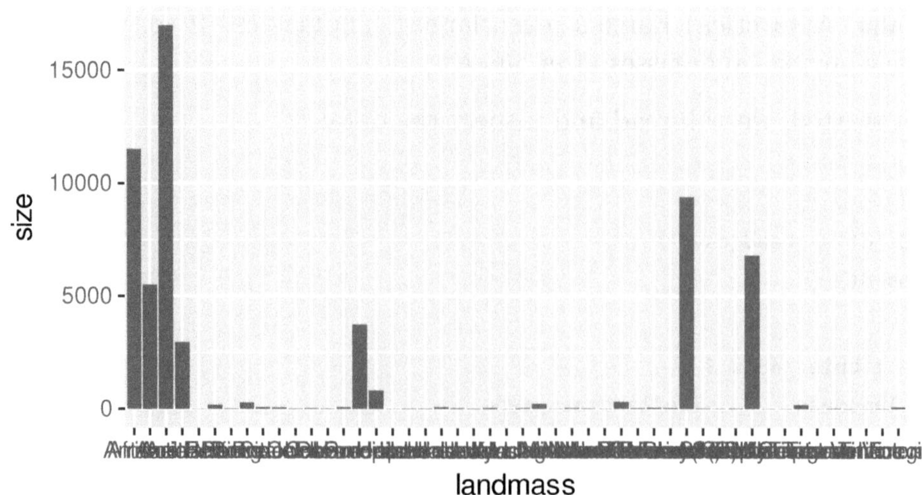

FIGURE 4.17: Bar plot of all Earth's landmasses' size with squished labels.

Alright, not bad! The plot in Figure 4.17 is definitely the right kind of visualization, as we can clearly see and compare sizes of landmasses. The major issues are that the smaller landmasses' sizes are hard to distinguish, and the names of the landmasses are obscuring each other as they have been squished into too little space. But remember that the question we asked was only about the largest landmasses; let's make the plot a little bit clearer by keeping only the largest 12 landmasses. We do this using the `slice_max` function. Then to help us make sure the labels have enough space, we'll use horizontal bars instead of vertical ones. We do this by swapping the `x` and `y` variables:

```
islands_top12 <- slice_max(islands_df, order_by = size, n = 12)
islands_bar <- ggplot(islands_top12, aes(x = size, y = landmass)) +
  geom_bar(stat = "identity")

islands_bar
```

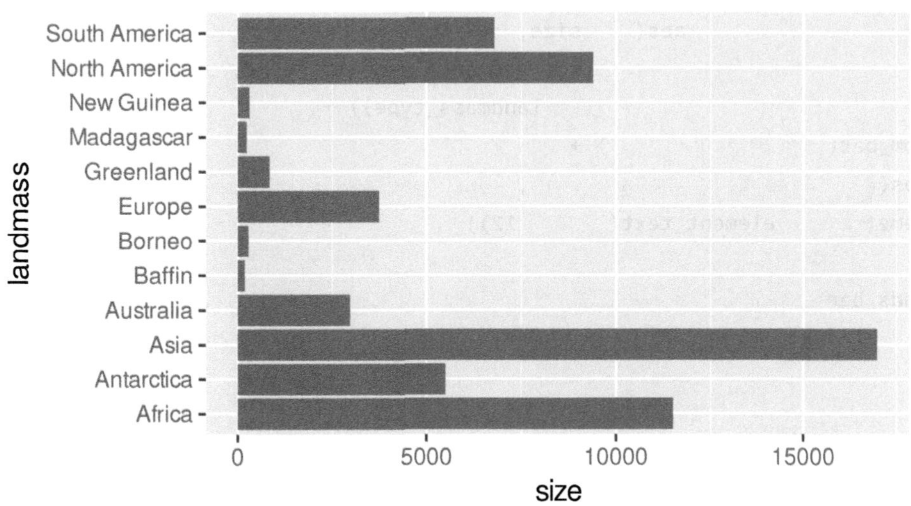

FIGURE 4.18: Bar plot of size for Earth's largest 12 landmasses.

The plot in Figure 4.18 is definitely clearer now, and allows us to answer our question ("are the top 7 largest landmasses continents?") in the affirmative. But the question could be made clearer from the plot by organizing the bars not by alphabetical order but by size, and to color them based on whether they are a continent. The data for this is stored in the `landmass_type` column. To use this to color the bars, we add the `fill` argument to the aesthetic mapping and set it to `landmass_type`.

To organize the landmasses by their `size` variable, we will use the `tidyverse` `fct_reorder` function in the aesthetic mapping to organize the landmasses by their `size` variable. The first argument passed to `fct_reorder` is the name of the factor column whose levels we would like to reorder (here, `landmass`). The second argument is the column name that holds the values we would like to use to do the ordering (here, `size`). The `fct_reorder` function uses ascending order by default, but this can be changed to descending order by setting `.desc = TRUE`. We do this here so that the largest bar will be closest to the axis line, which is more visually appealing.

To label the x and y axes, we will use the `labs` function instead of the `xlab` and `ylab` functions from earlier in this chapter. The `labs` function is more general; we are using it in this case because we would also like to change the legend label. The default label is the name of the column being mapped to `fill`. Here that would be `landmass_type`; however `landmass_type` is not proper English (and so is less readable). Thus we use the `fill` argument inside `labs` to change that to "Type." Finally, we again use the `theme` function to change the font size.

```
islands_bar <- ggplot(islands_top12,
                      aes(x = size,
                          y = fct_reorder(landmass, size, .desc = TRUE),
                          fill = landmass_type)) +
  geom_bar(stat = "identity") +
  labs(x = "Size (1000 square mi)", y = "Landmass",  fill = "Type") +
  theme(text = element_text(size = 12))

islands_bar
```

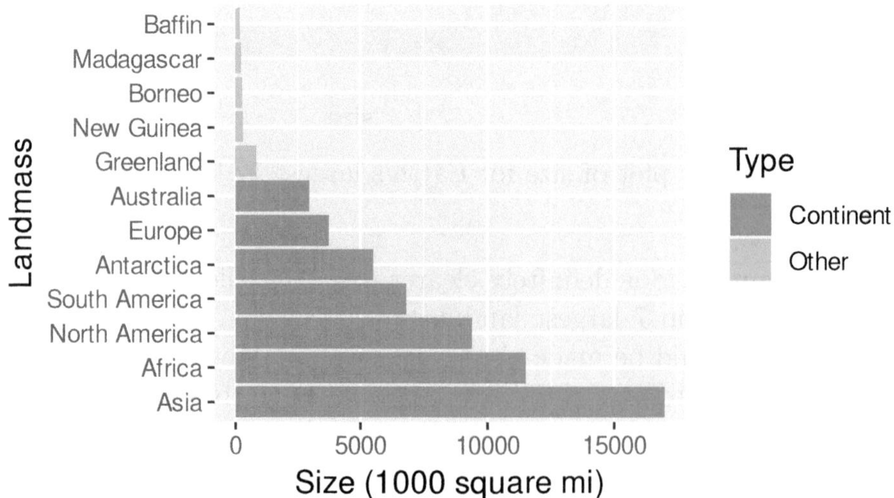

FIGURE 4.19: Bar plot of size for Earth's largest 12 landmasses colored by whether its a continent with clearer axes and labels.

The plot in Figure 4.19 is now a very effective visualization for answering our original questions. Landmasses are organized by their size, and continents are colored differently than other landmasses, making it quite clear that continents are the largest seven landmasses.

4.5.5 Histograms: the Michelson speed of light data set

The `morley` data set contains measurements of the speed of light collected in experiments performed in 1879. Five experiments were performed, and in each experiment, 20 runs were performed—meaning that 20 measurements of the speed of light were collected in each experiment [Michelson, 1882]. The `morley` data set is available in base R as a data frame, so it does not need to be loaded. Because the speed of light is a very large number (the true value is 299,792.458 km/sec), the data is coded to be the measured speed of light minus 299,000. This coding allows us to focus on the variations in the measurements, which are generally much smaller than 299,000. If we used the full large speed measurements, the variations in the measurements would not be noticeable, making it difficult to study the differences between the experiments. Note that we convert the `morley` data to a tibble to take advantage of the nicer print output these specialized data frames provide.

Question: Given what we know now about the speed of light (299,792.458 kilometres per second), how accurate were each of the experiments?

```
# michelson morley experimental data
morley <- as_tibble(morley)
morley
```

```
## # A tibble: 100 x 3
##     Expt   Run Speed
##    <int> <int> <int>
##  1     1     1   850
##  2     1     2   740
##  3     1     3   900
##  4     1     4  1070
##  5     1     5   930
##  6     1     6   850
##  7     1     7   950
##  8     1     8   980
##  9     1     9   980
## 10     1    10   880
## # ... with 90 more rows
```

In this experimental data, Michelson was trying to measure just a single quantitative number (the speed of light). The data set contains many measurements of this single quantity. To tell how accurate the experiments were, we need to visualize the distribution of the measurements (i.e., all their possible values and how often each occurs). We can do this using a *histogram*. A histogram helps us visualize how a particular variable is distributed in a data set by separating the data into bins, and then using vertical bars to show how many data points fell in each bin.

To create a histogram in `ggplot2` we will use the `geom_histogram` geometric object, setting the `x` axis to the `Speed` measurement variable. As usual, let's use the default arguments just to see how things look.

```
morley_hist <- ggplot(morley, aes(x = Speed)) +
  geom_histogram()

morley_hist
```

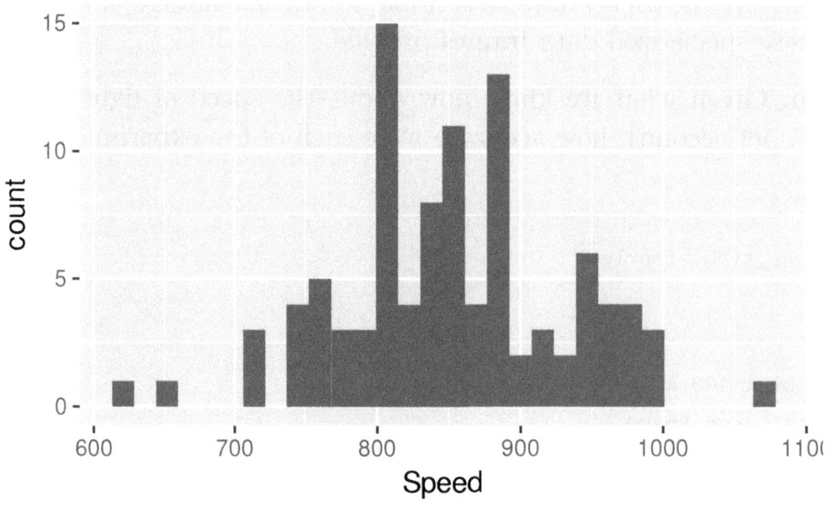

FIGURE 4.20: Histogram of Michelson's speed of light data.

Figure 4.20 is a great start. However, we cannot tell how accurate the measurements are using this visualization unless we can see the true value. In order to visualize the true speed of light, we will add a vertical line with the `geom_vline` function. To draw a vertical line with `geom_vline`, we need to specify where on the x-axis the line should be drawn. We can do this by setting the `xintercept` argument. Here we set it to 792.458, which is the true value of light speed minus 299,000; this ensures it is coded the same way as the measurements in the `morley` data frame. We would also like to fine tune this vertical line, styling it so that it is dashed and 1 point in thickness. A point is a measurement unit commonly used with fonts, and 1 point is about 0.353 mm. We do this by setting `linetype = "dashed"` and `size = 1`, respectively. There is a similar function, `geom_hline`, that is used for plotting horizontal lines. Note that *vertical lines* are used to denote quantities on the *horizontal axis*, while *horizontal lines* are used to denote quantities on the *vertical axis*.

```
morley_hist <- ggplot(morley, aes(x = Speed)) +
  geom_histogram() +
  geom_vline(xintercept = 792.458, linetype = "dashed", size = 1)

morley_hist
```

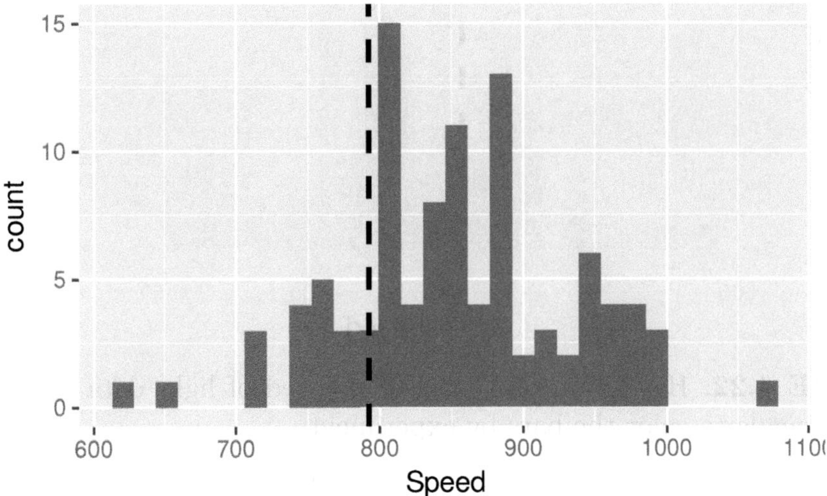

FIGURE 4.21: Histogram of Michelson's speed of light data with vertical line indicating true speed of light.

In Figure 4.21, we still cannot tell which experiments (denoted in the `Expt` column) led to which measurements; perhaps some experiments were more accurate than others. To fully answer our question, we need to separate the measurements from each other visually. We can try to do this using a *colored* histogram, where counts from different experiments are stacked on top of each other in different colors. We can create a histogram colored by the `Expt` variable by adding it to the `fill` aesthetic mapping. We make sure the different colors can be seen (despite them all sitting on top of each other) by setting the `alpha` argument in `geom_histogram` to `0.5` to make the bars slightly translucent. We also specify `position = "identity"` in `geom_histogram` to ensure the histograms for each experiment will be overlaid side-by-side, instead of stacked bars (which is the default for bar plots or histograms when they are colored by another categorical variable).

```
morley_hist <- ggplot(morley, aes(x = Speed, fill = Expt)) +
  geom_histogram(alpha = 0.5, position = "identity") +
  geom_vline(xintercept = 792.458, linetype = "dashed", size = 1.0)

morley_hist
```

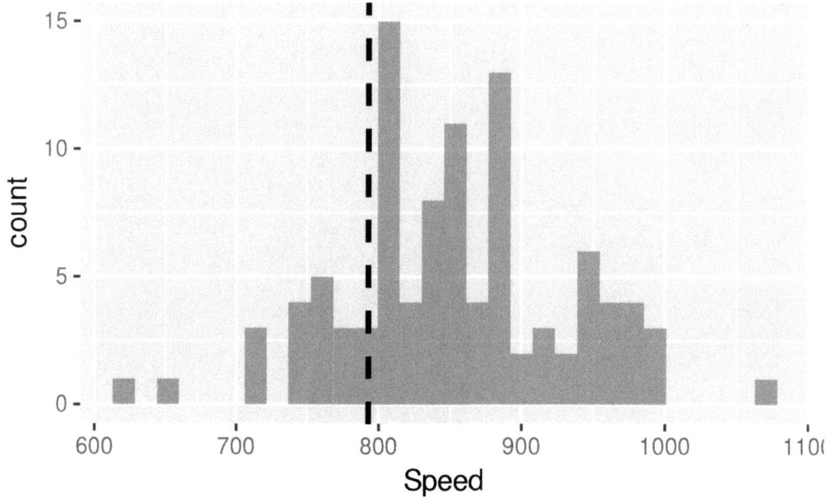

FIGURE 4.22: Histogram of Michelson's speed of light data where an attempt is made to color the bars by experiment.

Alright great, Figure 4.22 looks...wait a second! The histogram is still all the same color! What is going on here? Well, if you recall from Chapter 3, the *data type* you use for each variable can influence how R and `tidyverse` treats it. Here, we indeed have an issue with the data types in the `morley` data frame. In particular, the `Expt` column is currently an *integer* (you can see the label `<int>` underneath the `Expt` column in the printed data frame at the start of this section). But we want to treat it as a *category*, i.e., there should be one category per type of experiment.

To fix this issue we can convert the `Expt` variable into a *factor* by passing it to `as_factor` in the `fill` aesthetic mapping. Recall that factor is a data type in R that is often used to represent categories. By writing `as_factor(Expt)` we are ensuring that R will treat this variable as a factor, and the color will be mapped discretely.

```
morley_hist <- ggplot(morley, aes(x = Speed, fill = as_factor(Expt))) +
  geom_histogram(alpha = 0.5, position = "identity") +
  geom_vline(xintercept = 792.458, linetype = "dashed", size = 1.0)

morley_hist
```

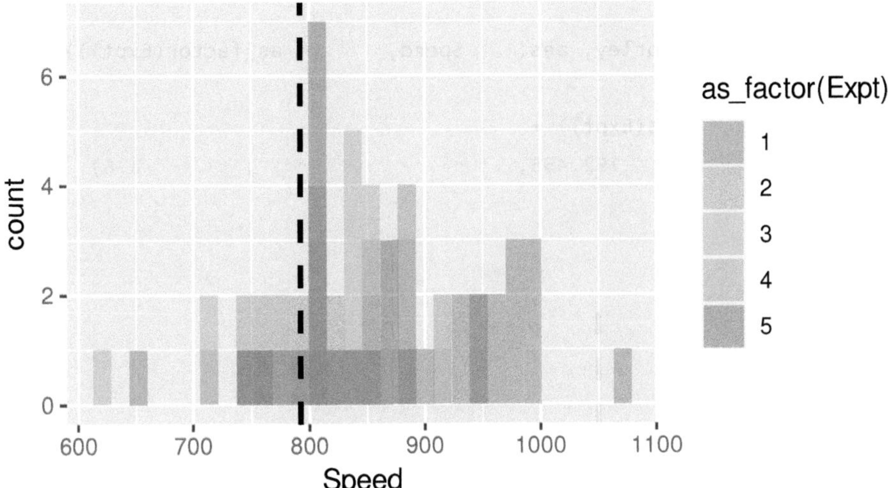

FIGURE 4.23: Histogram of Michelson's speed of light data colored by experiment as factor.

Note: Factors impact plots in two ways: (1) ensuring a color is mapped as discretely where appropriate (as in this example) and (2) the ordering of levels in a plot. `ggplot` takes into account the order of the factor levels as opposed to the order of data in your data frame. Learning how to reorder your factor levels will help you with reordering the labels of a factor on a plot.

Unfortunately, the attempt to separate out the experiment number visually has created a bit of a mess. All of the colors in Figure 4.23 are blending together, and although it is possible to derive *some* insight from this (e.g., experiments 1 and 3 had some of the most incorrect measurements), it isn't the clearest way to convey our message and answer the question. Let's try a different strategy of creating grid of separate histogram plots.

We use the `facet_grid` function to create a plot that has multiple subplots arranged in a grid. The argument to `facet_grid` specifies the variable(s) used to split the plot into subplots, and how to split them (i.e., into rows or columns). If the plot is to be split horizontally, into rows, then the `rows` argument is used. If the plot is to be split vertically, into columns, then the `columns` argument is used. Both the `rows` and `columns` arguments take the column names on which to split the data when creating the subplots. Note that the column names must be surrounded by the `vars` function. This function allows the column names to be correctly evaluated in the context of the data frame.

```
morley_hist <- ggplot(morley, aes(x = Speed, fill = as_factor(Expt))) +
  geom_histogram() +
  facet_grid(rows = vars(Expt)) +
  geom_vline(xintercept = 792.458, linetype = "dashed", size = 1.0)

morley_hist
```

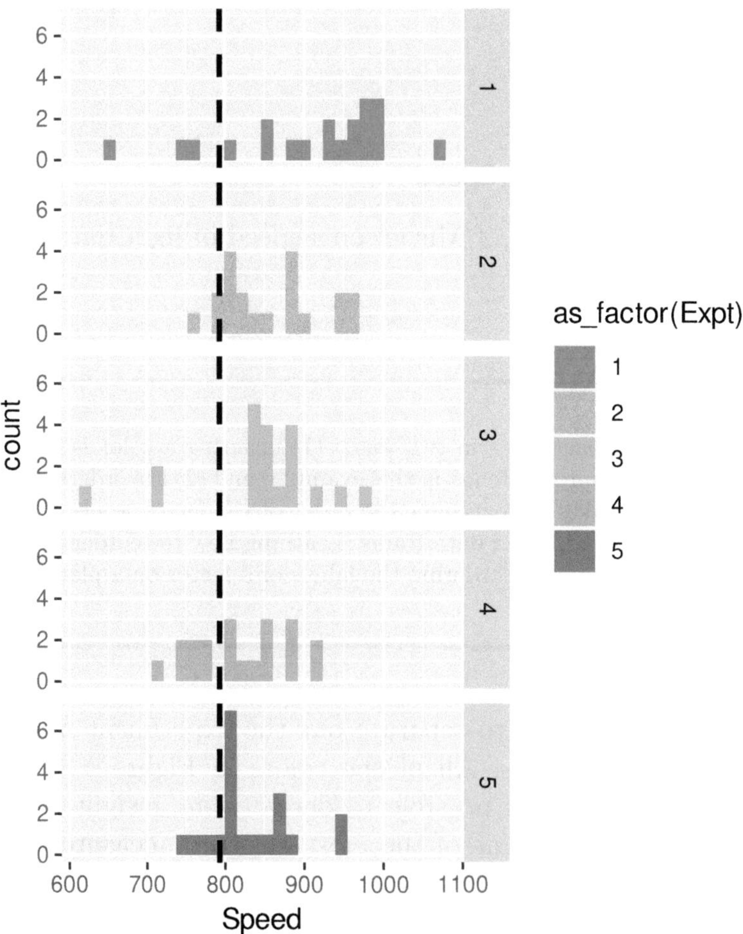

FIGURE 4.24: Histogram of Michelson's speed of light data split vertically by experiment.

The visualization in Figure 4.24 now makes it quite clear how accurate the different experiments were with respect to one another. The most variable measurements came from Experiment 1. There the measurements ranged from about 650–1050 km/sec. The least variable measurements came from Experiment 2. There, the measurements ranged from about 750–950 km/sec. The most different experiments still obtained quite similar results!

There are two finishing touches to make this visualization even clearer. First and foremost, we need to add informative axis labels using the `labs` function, and increase the font size to make it readable using the `theme` function. Second, and perhaps more subtly, even though it is easy to compare the experiments on this plot to one another, it is hard to get a sense of just how accurate all the experiments were overall. For example, how accurate is the value 800 on the plot, relative to the true speed of light? To answer this question, we'll use the `mutate` function to transform our data into a relative measure of accuracy rather than absolute measurements:

```
morley_rel <- mutate(morley,
                     relative_accuracy = 100 *
                       ((299000 + Speed) - 299792.458) / (299792.458))

morley_hist <- ggplot(morley_rel,
                      aes(x = relative_accuracy,
                          fill = as_factor(Expt))) +
  geom_histogram() +
  facet_grid(rows = vars(Expt)) +
  geom_vline(xintercept = 0, linetype = "dashed", size = 1.0) +
  labs(x = "Relative Accuracy (%)",
       y = "# Measurements",
       fill = "Experiment ID") +
  theme(text = element_text(size = 12))

morley_hist
```

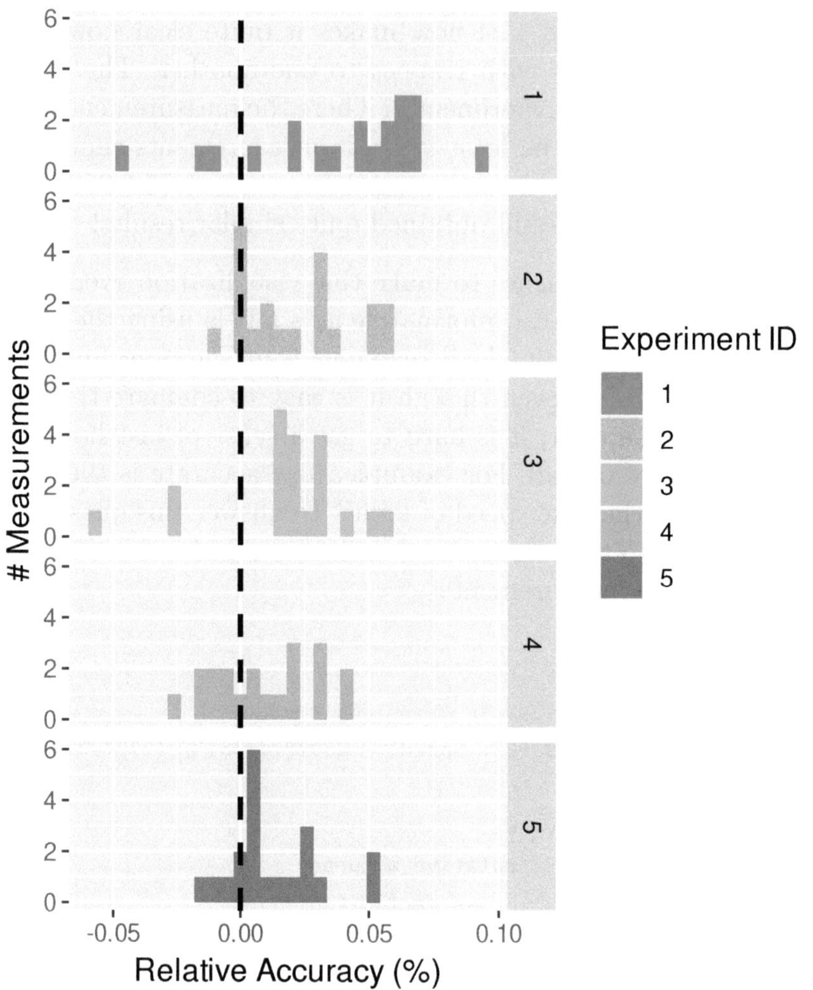

FIGURE 4.25: Histogram of relative accuracy split vertically by experiment with clearer axes and labels.

Wow, impressive! These measurements of the speed of light from 1879 had errors around *0.05%* of the true speed. Figure 4.25 shows you that even though experiments 2 and 5 were perhaps the most accurate, all of the experiments did quite an admirable job given the technology available at the time.

Choosing a binwidth for histograms

When you create a histogram in R, the default number of bins used is 30. Naturally, this is not always the right number to use. You can set the number of bins yourself by using the `bins` argument in the `geom_histogram` geometric object. You can also set the *width* of the bins using the `binwidth` argument in the `geom_histogram` geometric object. But what number of bins, or bin width, is the right one to use?

Unfortunately there is no hard rule for what the right bin number or width is. It depends entirely on your problem; the *right* number of bins or bin width is the one that *helps you answer the question* you asked. Choosing the correct setting for your problem is something that commonly takes iteration. We recommend setting the *bin width* (not the *number of bins*) because it often more directly corresponds to values in your problem of interest. For example, if you are looking at a histogram of human heights, a bin width of 1 inch would likely be reasonable, while the number of bins to use is not immediately clear. It's usually a good idea to try out several bin widths to see which one most clearly captures your data in the context of the question you want to answer.

To get a sense for how different bin widths affect visualizations, let's experiment with the histogram that we have been working on in this section. In Figure 4.26, we compare the default setting with three other histograms where we set the `binwidth` to 0.001, 0.01 and 0.1. In this case, we can see that both the default number of bins and the binwidth of 0.01 are effective for helping answer our question. On the other hand, the bin widths of 0.001 and 0.1 are too small and too big, respectively.

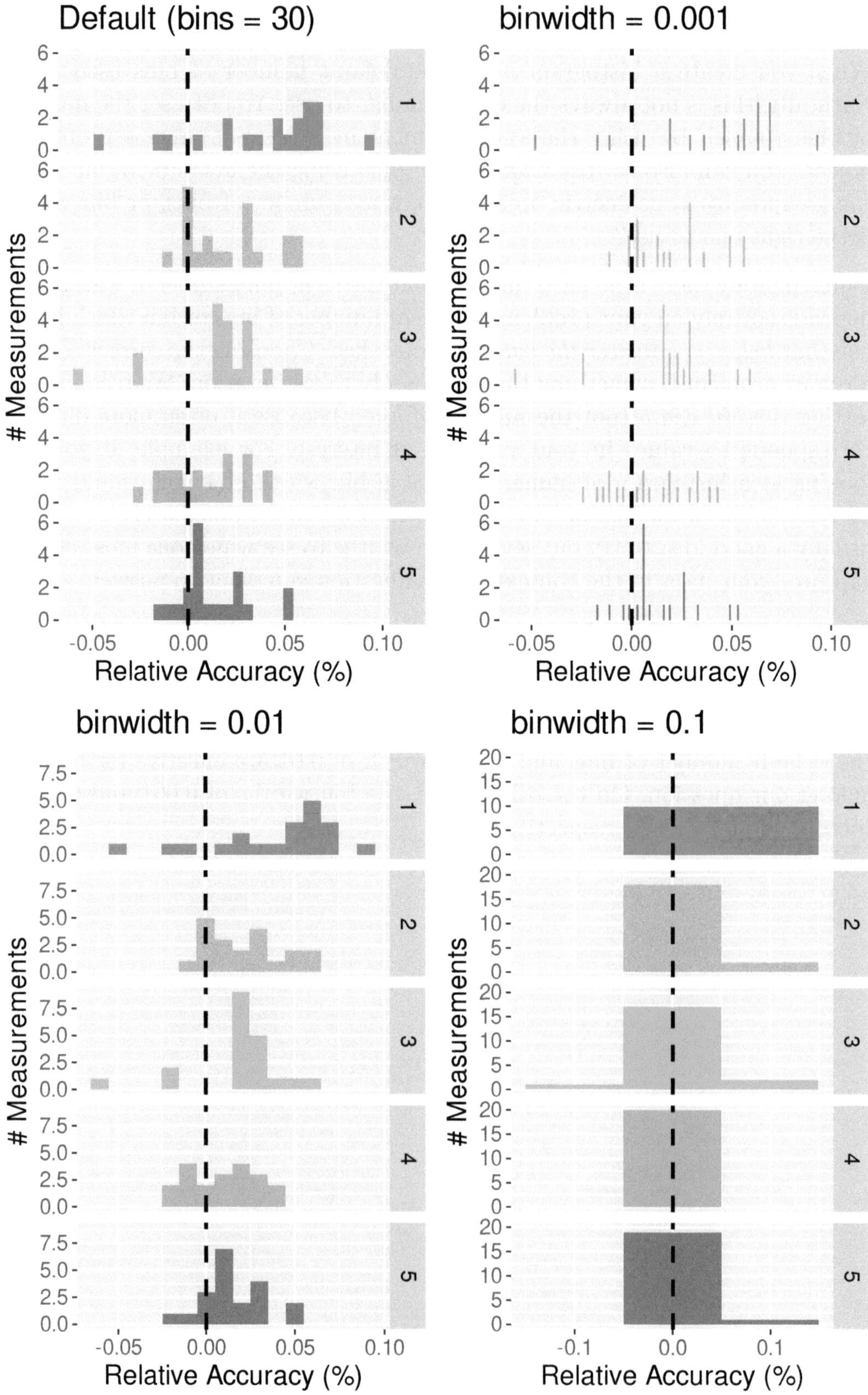

FIGURE 4.26: Effect of varying bin width on histograms.

Adding layers to a `ggplot` plot object

One of the powerful features of `ggplot` is that you can continue to iterate on a single plot object, adding and refining one layer at a time. If you stored your plot as a named object using the assignment symbol (`<-`), you can add to it using the `+` operator. For example, if we wanted to add a title to the last plot we created (`morley_hist`), we can use the `+` operator to add a title layer with the `ggtitle` function. The result is shown in Figure 4.27.

```
morley_hist_title <- morley_hist +
  ggtitle("Speed of light experiments \n were accurate to about 0.05%")

morley_hist_title
```

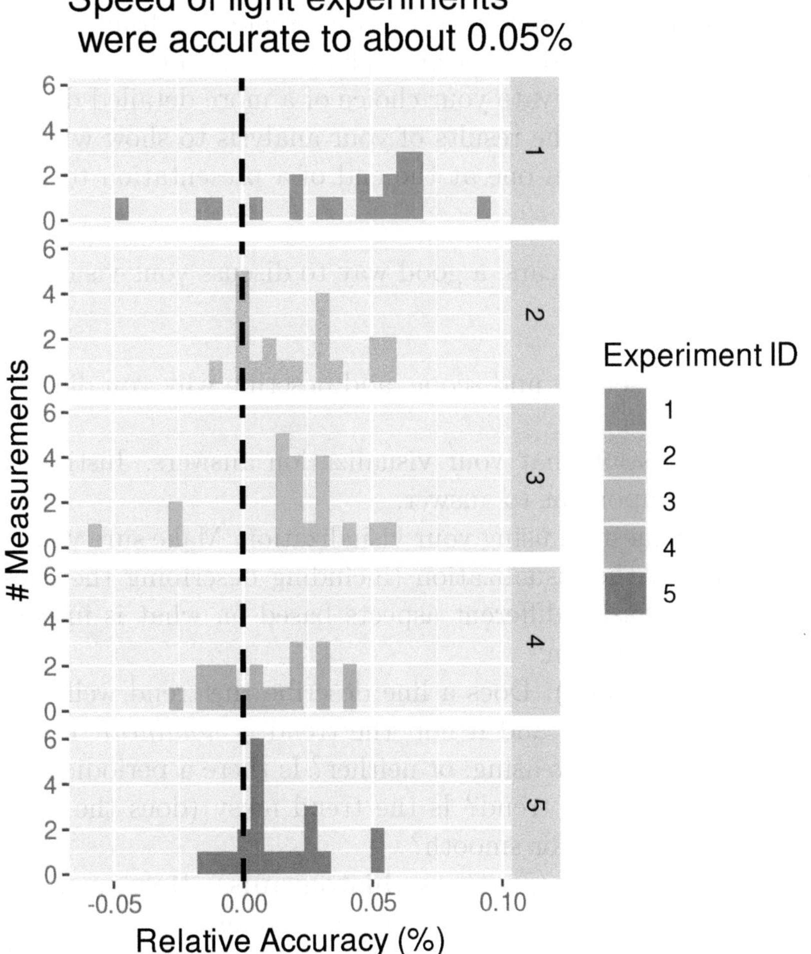

FIGURE 4.27: Histogram of relative accuracy split vertically by experiment with a descriptive title highlighting the take home message of the visualization.

Note: Good visualization titles clearly communicate the take home message to the audience. Typically, that is the answer to the question you posed before making the visualization.

4.6 Explaining the visualization

Tell a story

Typically, your visualization will not be shown entirely on its own, but rather it will be part of a larger presentation. Further, visualizations can provide supporting information for any aspect of a presentation, from opening to conclusion. For example, you could use an exploratory visualization in the opening of the presentation to motivate your choice of a more detailed data analysis / model, a visualization of the results of your analysis to show what your analysis has uncovered, or even one at the end of a presentation to help suggest directions for future work.

Regardless of where it appears, a good way to discuss your visualization is as a story:

1) Establish the setting and scope, and describe why you did what you did.
2) Pose the question that your visualization answers. Justify why the question is important to answer.
3) Answer the question using your visualization. Make sure you describe *all* aspects of the visualization (including describing the axes). But you can emphasize different aspects based on what is important to answer your question:
    - **trends (lines):** Does a line describe the trend well? If so, the trend is *linear*, and if not, the trend is *nonlinear*. Is the trend increasing, decreasing, or neither? Is there a periodic oscillation (wiggle) in the trend? Is the trend noisy (does the line "jump around" a lot) or smooth?
    - **distributions (scatters, histograms):** How spread out are the data? Where are they centered, roughly? Are there any obvious "clusters" or "subgroups", which would be visible as multiple bumps in the histogram?
    - **distributions of two variables (scatters):** Is there a clear /

strong relationship between the variables (points fall in a distinct pattern), a weak one (points fall in a pattern but there is some noise), or no discernible relationship (the data are too noisy to make any conclusion)?

- **amounts (bars):** How large are the bars relative to one another? Are there patterns in different groups of bars?

4) Summarize your findings, and use them to motivate whatever you will discuss next.

Below are two examples of how one might take these four steps in describing the example visualizations that appeared earlier in this chapter. Each of the steps is denoted by its numeral in parentheses, e.g. (3).

Mauna Loa Atmospheric CO_2 Measurements: (1) Many current forms of energy generation and conversion—from automotive engines to natural gas power plants—rely on burning fossil fuels and produce greenhouse gases, typically primarily carbon dioxide (CO_2), as a byproduct. Too much of these gases in the Earth's atmosphere will cause it to trap more heat from the sun, leading to global warming. (2) In order to assess how quickly the atmospheric concentration of CO_2 is increasing over time, we (3) used a data set from the Mauna Loa observatory in Hawaii, consisting of CO_2 measurements from 1980 to 2020. We plotted the measured concentration of CO_2 (on the vertical axis) over time (on the horizontal axis). From this plot, you can see a clear, increasing, and generally linear trend over time. There is also a periodic oscillation that occurs once per year and aligns with Hawaii's seasons, with an amplitude that is small relative to the growth in the overall trend. This shows that atmospheric CO_2 is clearly increasing over time, and (4) it is perhaps worth investigating more into the causes.

Michelson Light Speed Experiments: (1) Our modern understanding of the physics of light has advanced significantly from the late 1800s when Michelson and Morley's experiments first demonstrated that it had a finite speed. We now know, based on modern experiments, that it moves at roughly 299,792.458 kilometers per second. (2) But how accurately were we first able to measure this fundamental physical constant, and did certain experiments produce more accurate results than others? (3) To better understand this, we plotted data from 5 experiments by Michelson in 1879, each with 20 trials, as histograms stacked on top of one another. The horizontal axis shows the accuracy of the measurements relative to the true speed of light as we know it today, expressed as a percentage. From this visualization, you can see that most results had relative errors of at most 0.05%. You can also see that experiments 1 and 3 had measurements that were the farthest from the true value, and experiment 5 tended to provide the most consistently accurate result. (4) It would be worth further investigating the differences between these experiments to see why they produced different results.

4.7 Saving the visualization

Choose the right output format for your needs

Just as there are many ways to store data sets, there are many ways to store visualizations and images. Which one you choose can depend on several factors, such as file size/type limitations (e.g., if you are submitting your visualization as part of a conference paper or to a poster printing shop) and where it will be displayed (e.g., online, in a paper, on a poster, on a billboard, in talk slides). Generally speaking, images come in two flavors: *raster* formats and *vector* formats.

Raster images are represented as a 2-D grid of square pixels, each with its own color. Raster images are often *compressed* before storing so they take up less space. A compressed format is *lossy* if the image cannot be perfectly re-created when loading and displaying, with the hope that the change is not noticeable. *Lossless* formats, on the other hand, allow a perfect display of the original image.

- *Common file types:*
  - JPEG[8] (`.jpg`, `.jpeg`): lossy, usually used for photographs
  - PNG[9] (`.png`): lossless, usually used for plots / line drawings
  - BMP[10] (`.bmp`): lossless, raw image data, no compression (rarely used)
  - TIFF[11] (`.tif`, `.tiff`): typically lossless, no compression, used mostly in graphic arts, publishing
- *Open-source software:* GIMP[12]

Vector images are represented as a collection of mathematical objects (lines, surfaces, shapes, curves). When the computer displays the image, it redraws all of the elements using their mathematical formulas.

- *Common file types:*
  - SVG[13] (`.svg`): general-purpose use

---

[8] https://en.wikipedia.org/wiki/JPEG

[9] https://en.wikipedia.org/wiki/Portable_Network_Graphics

[10] https://en.wikipedia.org/wiki/BMP_file_format

[11] https://en.wikipedia.org/wiki/TIFF

[12] https://www.gimp.org/

[13] https://en.wikipedia.org/wiki/Scalable_Vector_Graphics

  - EPS[14] (`.eps`), general-purpose use (rarely used)
- *Open-source software:* Inkscape[15]

Raster and vector images have opposing advantages and disadvantages. A raster image of a fixed width / height takes the same amount of space and time to load regardless of what the image shows (the one caveat is that the compression algorithms may shrink the image more or run faster for certain images). A vector image takes space and time to load corresponding to how complex the image is, since the computer has to draw all the elements each time it is displayed. For example, if you have a scatter plot with 1 million points stored as an SVG file, it may take your computer some time to open the image. On the other hand, you can zoom into / scale up vector graphics as much as you like without the image looking bad, while raster images eventually start to look "pixelated."

Note: The portable document format PDF[16] (`.pdf`) is commonly used to store *both* raster and vector formats. If you try to open a PDF and it's taking a long time to load, it may be because there is a complicated vector graphics image that your computer is rendering.

Let's learn how to save plot images to these different file formats using a scatter plot of the Old Faithful data set[17] [Hardle, 1991], shown in Figure 4.28.

```
library(svglite) # we need this to save SVG files
faithful_plot <- ggplot(data = faithful, aes(x = waiting, y = eruptions)) +
  geom_point() +
  labs(x = "Waiting time to next eruption \n (minutes)",
       y = "Eruption time \n (minutes)") +
  theme(text = element_text(size = 12))

faithful_plot
```

[14]https://en.wikipedia.org/wiki/Encapsulated_PostScript
[15]https://inkscape.org/
[16]https://en.wikipedia.org/wiki/PDF
[17]https://www.stat.cmu.edu/~larry/all-of-statistics/=data/faithful.dat

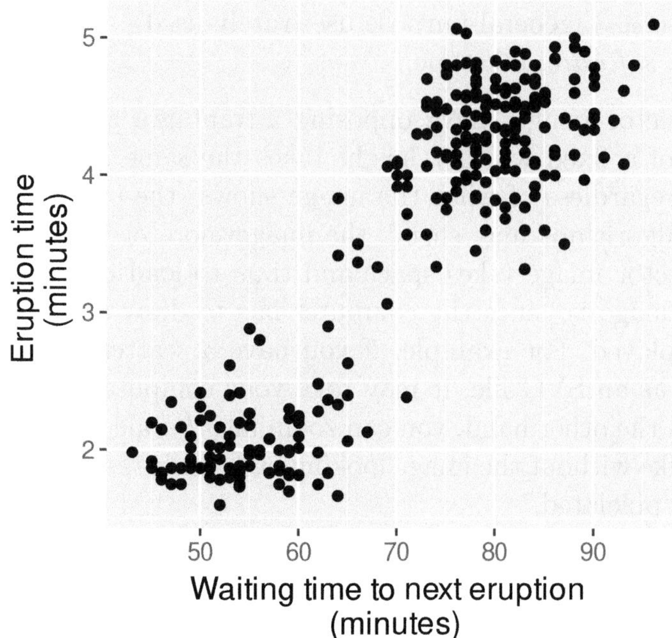

FIGURE 4.28: Scatter plot of waiting time and eruption time.

Now that we have a named `ggplot` plot object, we can use the `ggsave` function to save a file containing this image. `ggsave` works by taking a file name to create for the image as its first argument. This can include the path to the directory where you would like to save the file (e.g., `img/filename.png` to save a file named `filename` to the `img` directory), and the name of the plot object to save as its second argument. The kind of image to save is specified by the file extension. For example, to create a PNG image file, we specify that the file extension is `.png`. Below we demonstrate how to save PNG, JPG, BMP, TIFF and SVG file types for the `faithful_plot`:

```
ggsave("img/faithful_plot.png", faithful_plot)
ggsave("img/faithful_plot.jpg", faithful_plot)
ggsave("img/faithful_plot.bmp", faithful_plot)
ggsave("img/faithful_plot.tiff", faithful_plot)
ggsave("img/faithful_plot.svg", faithful_plot)
```

Take a look at the file sizes in Table 4.1. Wow, that's quite a difference! Notice that for such a simple plot with few graphical elements (points), the vector graphics format (SVG) is over 100 times smaller than the uncompressed raster images (BMP, TIFF). Also, note that the JPG format is twice as large as the PNG format since the JPG compression algorithm is designed for natural images (not plots).

TABLE 4.1: File sizes of the scatter plot of the Old Faithful data set when saved as different file formats.

Image type	File type	Image size
Raster	PNG	0.14 MB
Raster	JPG	0.38 MB
Raster	BMP	2.63 MB
Raster	TIFF	7.89 MB
Vector	SVG	0.03 MB

In Figure 4.29, we also show what the images look like when we zoom in to a rectangle with only 2 data points. You can see why vector graphics formats are so useful: because they're just based on mathematical formulas, vector graphics can be scaled up to arbitrary sizes. This makes them great for presentation media of all sizes, from papers to posters to billboards.

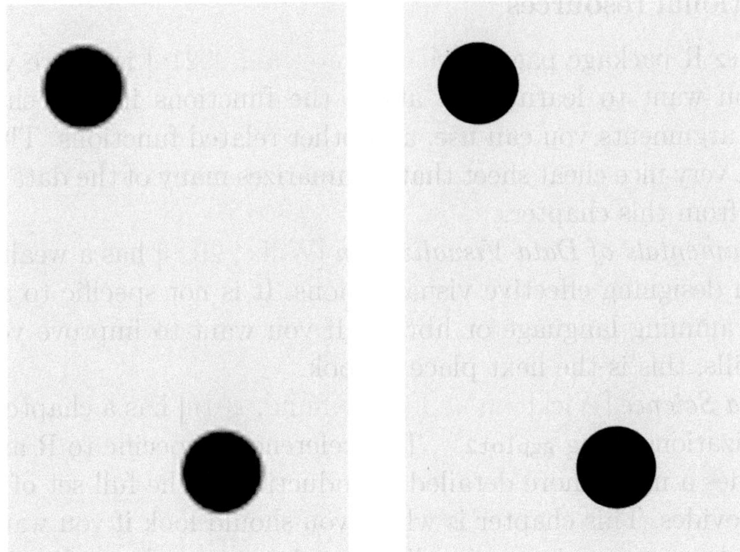

FIGURE 4.29: Zoomed in `faithful`, raster (PNG, left) and vector (SVG, right) formats.

4.8 Exercises

Practice exercises for the material covered in this chapter can be found in the accompanying worksheets repository[18] in the "Effective data visualization" row. You can launch an interactive version of the worksheet in your browser by clicking the "launch binder" button. You can also preview a non-interactive version of the worksheet by clicking "view worksheet." If you instead decide to download the worksheet and run it on your own machine, make sure to follow the instructions for computer setup found in Chapter 13. This will ensure that the automated feedback and guidance that the worksheets provide will function as intended.

4.9 Additional resources

- The `ggplot2` R package page[19] [Wickham et al., 2021a] is where you should look if you want to learn more about the functions in this chapter, the full set of arguments you can use, and other related functions. The site also provides a very nice cheat sheet that summarizes many of the data wrangling functions from this chapter.
- The *Fundamentals of Data Visualization* [Wilke, 2019] has a wealth of information on designing effective visualizations. It is not specific to any particular programming language or library. If you want to improve your visualization skills, this is the next place to look.
- *R for Data Science* [Wickham and Grolemund, 2016] has a chapter on creating visualizations using `ggplot2`[20]. This reference is specific to R and `ggplot2`, but provides a much more detailed introduction to the full set of tools that `ggplot2` provides. This chapter is where you should look if you want to learn how to make more intricate visualizations in `ggplot2` than what is included in this chapter.
- The `theme` function documentation[21] is an excellent reference to see how you can fine tune the non-data aspects of your visualization.

[18] https://github.com/UBC-DSCI/data-science-a-first-intro-worksheets#readme

[19] https://ggplot2.tidyverse.org

[20] https://r4ds.had.co.nz/data-visualisation.html

[21] https://ggplot2.tidyverse.org/reference/theme.html

- *R for Data Science* [Wickham and Grolemund, 2016] has a chapter on dates and times[22]. This chapter is where you should look if you want to learn about `date` vectors, including how to create them, and how to use them to effectively handle durations, periods and intervals using the `lubridate` package.

[22] https://r4ds.had.co.nz/dates-and-times.html

5

Classification I: training & predicting

5.1 Overview

In previous chapters, we focused solely on descriptive and exploratory data analysis questions. This chapter and the next together serve as our first foray into answering *predictive* questions about data. In particular, we will focus on *classification*, i.e., using one or more variables to predict the value of a categorical variable of interest. This chapter will cover the basics of classification, how to preprocess data to make it suitable for use in a classifier, and how to use our observed data to make predictions. The next chapter will focus on how to evaluate how accurate the predictions from our classifier are, as well as how to improve our classifier (where possible) to maximize its accuracy.

5.2 Chapter learning objectives

By the end of the chapter, readers will be able to do the following:

- Recognize situations where a classifier would be appropriate for making predictions.
- Describe what a training data set is and how it is used in classification.
- Interpret the output of a classifier.
- Compute, by hand, the straight-line (Euclidean) distance between points on a graph when there are two predictor variables.
- Explain the K-nearest neighbor classification algorithm.
- Perform K-nearest neighbor classification in R using `tidymodels`.
- Use a `recipe` to preprocess data to be centered, scaled, and balanced.
- Combine preprocessing and model training using a `workflow`.

5.3 The classification problem

In many situations, we want to make predictions based on the current situation as well as past experiences. For instance, a doctor may want to diagnose a patient as either diseased or healthy based on their symptoms and the doctor's past experience with patients; an email provider might want to tag a given email as "spam" or "not spam" based on the email's text and past email text data; or a credit card company may want to predict whether a purchase is fraudulent based on the current purchase item, amount, and location as well as past purchases. These tasks are all examples of **classification**, i.e., predicting a categorical class (sometimes called a *label*) for an observation given its other variables (sometimes called *features*).

Generally, a classifier assigns an observation without a known class (e.g., a new patient) to a class (e.g., diseased or healthy) on the basis of how similar it is to other observations for which we do know the class (e.g., previous patients with known diseases and symptoms). These observations with known classes that we use as a basis for prediction are called a **training set**; this name comes from the fact that we use these data to train, or teach, our classifier. Once taught, we can use the classifier to make predictions on new data for which we do not know the class.

There are many possible methods that we could use to predict a categorical class/label for an observation. In this book, we will focus on the widely used **K-nearest neighbors** algorithm [Fix and Hodges, 1951, Cover and Hart, 1967]. In your future studies, you might encounter decision trees, support vector machines (SVMs), logistic regression, neural networks, and more; see the additional resources section at the end of the next chapter for where to begin learning more about these other methods. It is also worth mentioning that there are many variations on the basic classification problem. For example, we focus on the setting of **binary classification** where only two classes are involved (e.g., a diagnosis of either healthy or diseased), but you may also run into multiclass classification problems with more than two categories (e.g., a diagnosis of healthy, bronchitis, pneumonia, or a common cold).

5.4 Exploring a data set

In this chapter and the next, we will study a data set of digitized breast cancer image features[1], created by Dr. William H. Wolberg, W. Nick Street, and Olvi L. Mangasarian [Street et al., 1993]. Each row in the data set represents an image of a tumor sample, including the diagnosis (benign or malignant) and several other measurements (nucleus texture, perimeter, area, and more). Diagnosis for each image was conducted by physicians.

As with all data analyses, we first need to formulate a precise question that we want to answer. Here, the question is *predictive*: can we use the tumor image measurements available to us to predict whether a future tumor image (with unknown diagnosis) shows a benign or malignant tumor? Answering this question is important because traditional, non-data-driven methods for tumor diagnosis are quite subjective and dependent upon how skilled and experienced the diagnosing physician is. Furthermore, benign tumors are not normally dangerous; the cells stay in the same place, and the tumor stops growing before it gets very large. By contrast, in malignant tumors, the cells invade the surrounding tissue and spread into nearby organs, where they can cause serious damage [Stanford Health Care, 2021]. Thus, it is important to quickly and accurately diagnose the tumor type to guide patient treatment.

5.4.1 Loading the cancer data

Our first step is to load, wrangle, and explore the data using visualizations in order to better understand the data we are working with. We start by loading the `tidyverse` package needed for our analysis.

```
library(tidyverse)
```

In this case, the file containing the breast cancer data set is a `.csv` file with headers. We'll use the `read_csv` function with no additional arguments, and then inspect its contents:

```
cancer <- read_csv("data/wdbc.csv")
cancer
```

```
## # A tibble: 569 x 12
##       ID Class Radius Texture Perimeter  Area Smoothness Compactness Concavity
```

[1] https://archive.ics.uci.edu/ml/datasets/Breast+Cancer+Wisconsin+%28Diagnostic%29

```
##     <dbl> <chr>  <dbl>   <dbl>     <dbl>  <dbl>      <dbl>       <dbl>     <dbl>
##  1 8.42e5 M       1.10   -2.07      1.27    0.984      1.57        3.28      2.65
##  2 8.43e5 M      1.83   -0.353    1.68    1.91     -0.826     -0.487   -0.0238
##  3 8.43e7 M       1.58    0.456     1.57    1.56       0.941       1.05      1.36
##  4 8.43e7 M     -0.768   0.254   -0.592  -0.764      3.28        3.40      1.91
##  5 8.44e7 M       1.75   -1.15      1.78    1.82       0.280       0.539     1.37
##  6 8.44e5 M     -0.476  -0.835   -0.387  -0.505     2.24        1.24      0.866
##  7 8.44e5 M       1.17    0.161    1.14    1.09     -0.123      0.0882    0.300
##  8 8.45e7 M     -0.118   0.358   -0.0728 -0.219     1.60       1.14      0.0610
##  9 8.45e5 M     -0.320   0.588   -0.184  -0.384      2.20        1.68      1.22
## 10 8.45e7 M     -0.473   1.10    -0.329  -0.509      1.58        2.56      1.74
## # ... with 559 more rows, and 3 more variables: Concave_Points <dbl>,
## #   Symmetry <dbl>, Fractal_Dimension <dbl>
```

5.4.2 Describing the variables in the cancer data set

Breast tumors can be diagnosed by performing a *biopsy*, a process where tissue is removed from the body and examined for the presence of disease. Traditionally these procedures were quite invasive; modern methods such as fine needle aspiration, used to collect the present data set, extract only a small amount of tissue and are less invasive. Based on a digital image of each breast tissue sample collected for this data set, ten different variables were measured for each cell nucleus in the image (items 3–12 of the list of variables below), and then the mean for each variable across the nuclei was recorded. As part of the data preparation, these values have been *standardized (centered and scaled)*; we will discuss what this means and why we do it later in this chapter. Each image additionally was given a unique ID and a diagnosis by a physician. Therefore, the total set of variables per image in this data set is:

1. ID: identification number
2. Class: the diagnosis (M = malignant or B = benign)
3. Radius: the mean of distances from center to points on the perimeter
4. Texture: the standard deviation of gray-scale values
5. Perimeter: the length of the surrounding contour
6. Area: the area inside the contour
7. Smoothness: the local variation in radius lengths
8. Compactness: the ratio of squared perimeter and area
9. Concavity: severity of concave portions of the contour
10. Concave Points: the number of concave portions of the contour
11. Symmetry: how similar the nucleus is when mirrored
12. Fractal Dimension: a measurement of how "rough" the perimeter is

Below we use `glimpse` to preview the data frame. This function can make it easier to inspect the data when we have a lot of columns, as it prints the data such that the columns go down the page (instead of across).

```
glimpse(cancer)
```

```
## Rows: 569
## Columns: 12
## $ ID                <dbl> 842302, 842517, 84300903, 84348301, 84358402, 843786~
## $ Class             <chr> "M", "M", "M", "M", "M", "M", "M", "M", "M", "M", "M~
## $ Radius            <dbl> 1.0960995, 1.8282120, 1.5784992, -0.7682333, 1.74875~
## $ Texture           <dbl> -2.0715123, -0.3533215, 0.4557859, 0.2535091, -1.150~
## $ Perimeter         <dbl> 1.26881726, 1.68447255, 1.56512598, -0.59216612, 1.7~
## $ Area              <dbl> 0.98350952, 1.90703027, 1.55751319, -0.76379174, 1.8~
## $ Smoothness        <dbl> 1.56708746, -0.82623545, 0.94138212, 3.28066684, 0.2~
## $ Compactness       <dbl> 3.28062806, -0.48664348, 1.05199990, 3.39991742, 0.5~
## $ Concavity         <dbl> 2.65054179, -0.02382489, 1.36227979, 1.91421287, 1.3~
## $ Concave_Points    <dbl> 2.53024886, 0.54766227, 2.03543978, 1.45043113, 1.42~
## $ Symmetry          <dbl> 2.215565542, 0.001391139, 0.938858720, 2.864862154, ~
## $ Fractal_Dimension <dbl> 2.25376381, -0.86788881, -0.39765801, 4.90660199, -
0~
```

From the summary of the data above, we can see that `Class` is of type character (denoted by `<chr>`). Since we will be working with `Class` as a categorical statistical variable, we will convert it to a factor using the function `as_factor`.

```
cancer <- cancer |>
  mutate(Class = as_factor(Class))
glimpse(cancer)
```

```
## Rows: 569
## Columns: 12
## $ ID                <dbl> 842302, 842517, 84300903, 84348301, 84358402, 843786~
## $ Class             <fct> M, M, M, M, M, M, M, M, M, M, M, M, M, M, M, M, M, M~
## $ Radius            <dbl> 1.0960995, 1.8282120, 1.5784992, -0.7682333, 1.74875~
## $ Texture           <dbl> -2.0715123, -0.3533215, 0.4557859, 0.2535091, -1.150~
## $ Perimeter         <dbl> 1.26881726, 1.68447255, 1.56512598, -0.59216612, 1.7~
## $ Area              <dbl> 0.98350952, 1.90703027, 1.55751319, -0.76379174, 1.8~
## $ Smoothness        <dbl> 1.56708746, -0.82623545, 0.94138212, 3.28066684, 0.2~
## $ Compactness       <dbl> 3.28062806, -0.48664348, 1.05199990, 3.39991742, 0.5~
## $ Concavity         <dbl> 2.65054179, -0.02382489, 1.36227979, 1.91421287, 1.3~
## $ Concave_Points    <dbl> 2.53024886, 0.54766227, 2.03543978, 1.45043113, 1.42~
```

```
## $ Symmetry          <dbl> 2.215565542, 0.001391139, 0.938858720, 2.864862154, ~
## $ Fractal_Dimension <dbl> 2.25376381, -0.86788881, -0.39765801, 4.90660199, -
0~
```

Recall that factors have what are called "levels", which you can think of as categories. We can verify the levels of the `Class` column by using the `levels` function. This function should return the name of each category in that column. Given that we only have two different values in our `Class` column (B for benign and M for malignant), we only expect to get two names back. Note that the `levels` function requires a *vector* argument; so we use the `pull` function to extract a single column (`Class`) and pass that into the `levels` function to see the categories in the `Class` column.

```
cancer |>
  pull(Class) |>
  levels()
```

```
## [1] "M" "B"
```

5.4.3 Exploring the cancer data

Before we start doing any modeling, let's explore our data set. Below we use the `group_by`, `summarize` and `n` functions to find the number and percentage of benign and malignant tumor observations in our data set. The `n` function within `summarize`, when paired with `group_by`, counts the number of observations in each `Class` group. Then we calculate the percentage in each group by dividing by the total number of observations. We have 357 (63%) benign and 212 (37%) malignant tumor observations.

```
num_obs <- nrow(cancer)
cancer |>
  group_by(Class) |>
  summarize(
    count = n(),
    percentage = n() / num_obs * 100
  )
```

```
## # A tibble: 2 x 3
##   Class count percentage
##   <fct> <int>      <dbl>
## 1 M       212       37.3
## 2 B       357       62.7
```

Next, let's draw a scatter plot to visualize the relationship between the perimeter and concavity variables. Rather than use `ggplot's` default palette, we select our own colorblind-friendly colors—`"orange2"` for light orange and `"steelblue2"` for light blue—and pass them as the `values` argument to the `scale_color_manual` function. We also make the category labels ("B" and "M") more readable by changing them to "Benign" and "Malignant" using the `labels` argument.

```
perim_concav <- cancer |>
  ggplot(aes(x = Perimeter, y = Concavity, color = Class)) +
  geom_point(alpha = 0.6) +
  labs(x = "Perimeter (standardized)",
       y = "Concavity (standardized)",
       color = "Diagnosis") +
  scale_color_manual(labels = c("Malignant", "Benign"),
                     values = c("orange2", "steelblue2")) +
  theme(text = element_text(size = 12))
perim_concav
```

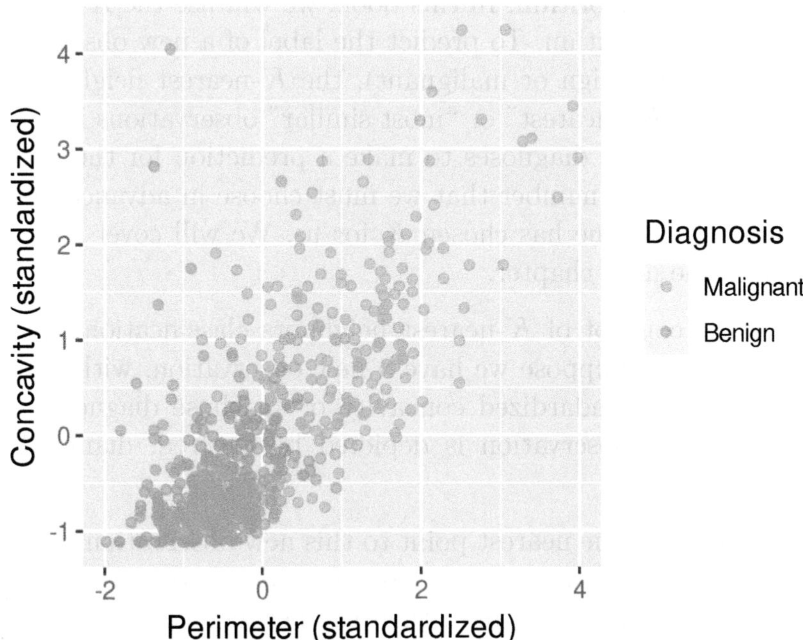

FIGURE 5.1: Scatter plot of concavity versus perimeter colored by diagnosis label.

In Figure 5.1, we can see that malignant observations typically fall in the upper right-hand corner of the plot area. By contrast, benign observations typically fall in the lower left-hand corner of the plot. In other words, benign observations tend to have lower concavity and perimeter values, and malignant ones tend to have larger values. Suppose we obtain a new observation not in the current data set that has all the variables measured *except* the label (i.e., an image without the physician's diagnosis for the tumor class). We could compute the standardized perimeter and concavity values, resulting in values of, say, 1 and 1. Could we use this information to classify that observation as benign or malignant? Based on the scatter plot, how might you classify that new observation? If the standardized concavity and perimeter values are 1 and 1 respectively, the point would lie in the middle of the orange cloud of malignant points and thus we could probably classify it as malignant. Based on our visualization, it seems like the *prediction of an unobserved label* might be possible.

5.5 Classification with K-nearest neighbors

In order to actually make predictions for new observations in practice, we will need a classification algorithm. In this book, we will use the K-nearest neighbors classification algorithm. To predict the label of a new observation (here, classify it as either benign or malignant), the K-nearest neighbors classifier generally finds the K "nearest" or "most similar" observations in our training set, and then uses their diagnoses to make a prediction for the new observation's diagnosis. K is a number that we must choose in advance; for now, we will assume that someone has chosen K for us. We will cover how to choose K ourselves in the next chapter.

To illustrate the concept of K-nearest neighbors classification, we will walk through an example. Suppose we have a new observation, with standardized perimeter of 2 and standardized concavity of 4, whose diagnosis "Class" is unknown. This new observation is depicted by the red, diamond point in Figure 5.2.

Figure 5.3 shows that the nearest point to this new observation is **malignant** and located at the coordinates (2.1, 3.6). The idea here is that if a point is close to another in the scatter plot, then the perimeter and concavity values are similar, and so we may expect that they would have the same diagnosis.

Suppose we have another new observation with standardized perimeter 0.2 and concavity of 3.3. Looking at the scatter plot in Figure 5.4, how would you classify this red, diamond observation? The nearest neighbor to this new point is a **benign** observation at (0.2, 2.7). Does this seem like the right prediction to make for this observation? Probably not, if you consider the other nearby points.

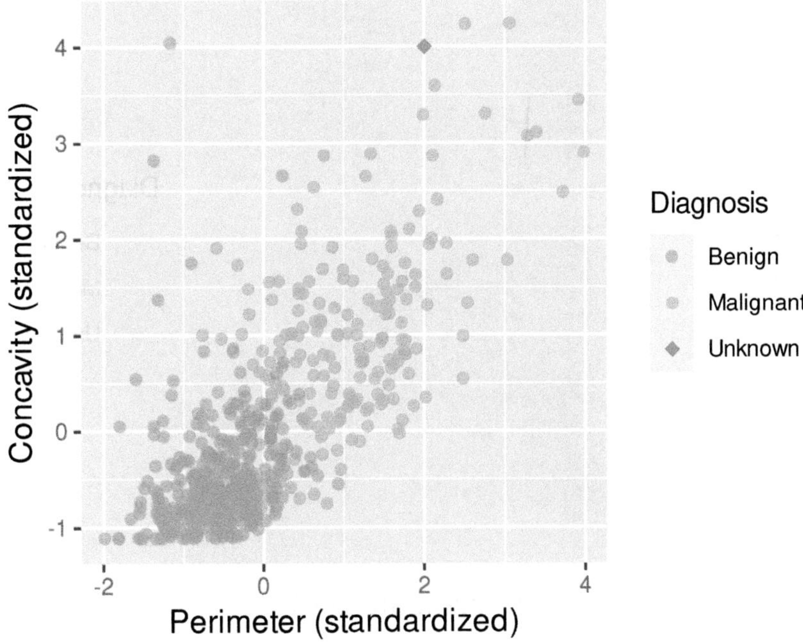

FIGURE 5.2: Scatter plot of concavity versus perimeter with new observation represented as a red diamond.

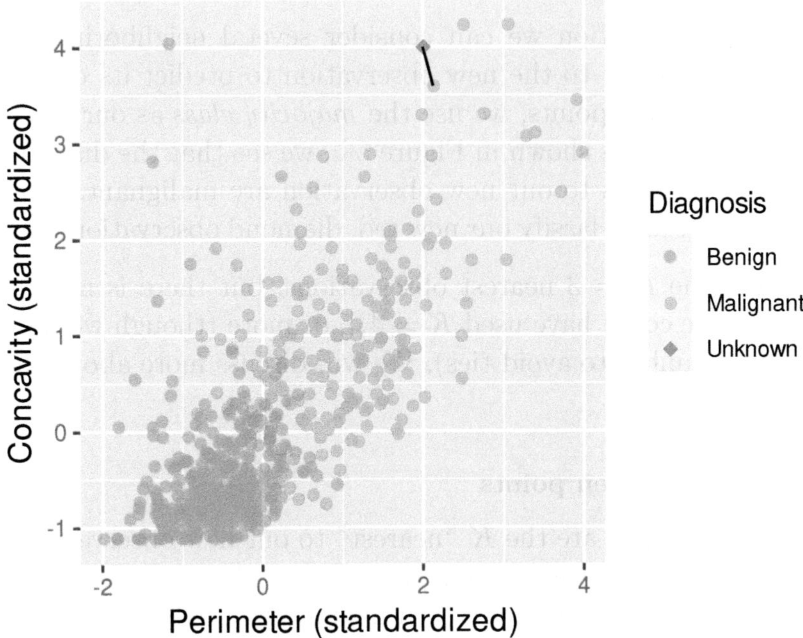

FIGURE 5.3: Scatter plot of concavity versus perimeter. The new observation is represented as a red diamond with a line to the one nearest neighbor, which has a malignant label.

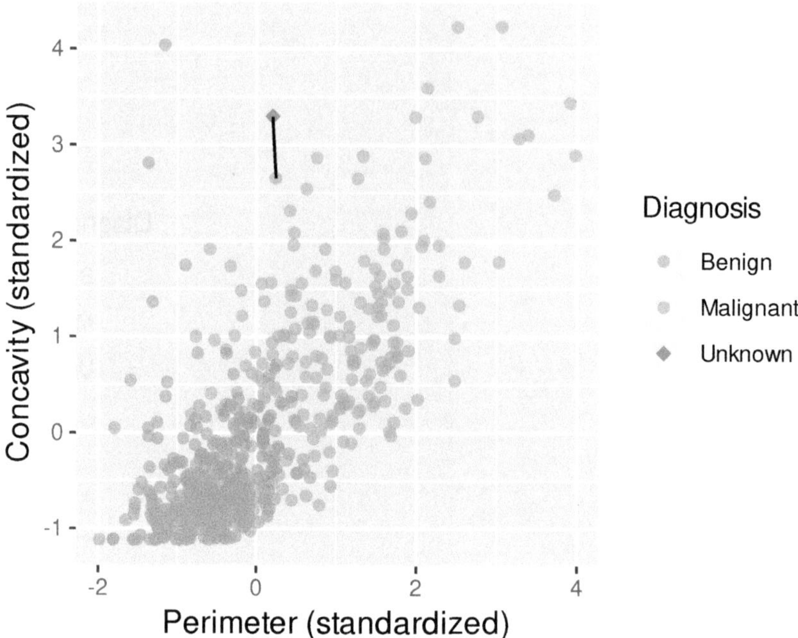

FIGURE 5.4: Scatter plot of concavity versus perimeter. The new observation is represented as a red diamond with a line to the one nearest neighbor, which has a benign label.

To improve the prediction we can consider several neighboring points, say $K = 3$, that are closest to the new observation to predict its diagnosis class. Among those 3 closest points, we use the *majority class* as our prediction for the new observation. As shown in Figure 5.5, we see that the diagnoses of 2 of the 3 nearest neighbors to our new observation are malignant. Therefore we take majority vote and classify our new red, diamond observation as malignant.

Here we chose the $K = 3$ nearest observations, but there is nothing special about $K = 3$. We could have used $K = 4, 5$ or more (though we may want to choose an odd number to avoid ties). We will discuss more about choosing K in the next chapter.

5.5.1 Distance between points

We decide which points are the K "nearest" to our new observation using the *straight-line distance* (we will often just refer to this as *distance*). Suppose we have two observations a and b, each having two predictor variables, x and y. Denote a_x and a_y to be the values of variables x and y for observation a; b_x and b_y have similar definitions for observation b. Then the straight-line distance between observation a and b on the x-y plane can be computed using the following formula:

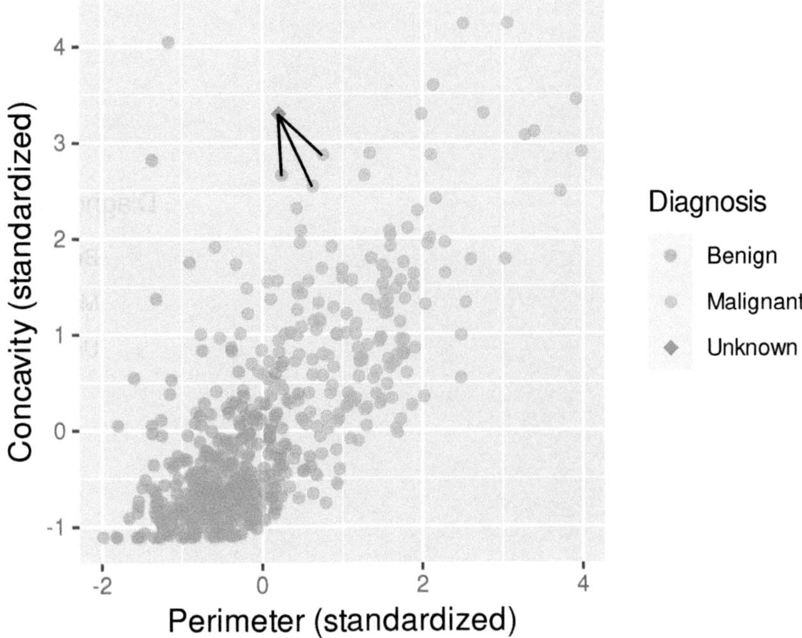

FIGURE 5.5: Scatter plot of concavity versus perimeter with three nearest neighbors.

$$\text{Distance} = \sqrt{(a_x - b_x)^2 + (a_y - b_y)^2}$$

To find the K nearest neighbors to our new observation, we compute the distance from that new observation to each observation in our training data, and select the K observations corresponding to the K *smallest* distance values. For example, suppose we want to use $K = 5$ neighbors to classify a new observation with perimeter of 0 and concavity of 3.5, shown as a red diamond in Figure 5.6. Let's calculate the distances between our new point and each of the observations in the training set to find the $K = 5$ neighbors that are nearest to our new point. You will see in the `mutate` step below, we compute the straight-line distance using the formula above: we square the differences between the two observations' perimeter and concavity coordinates, add the squared differences, and then take the square root.

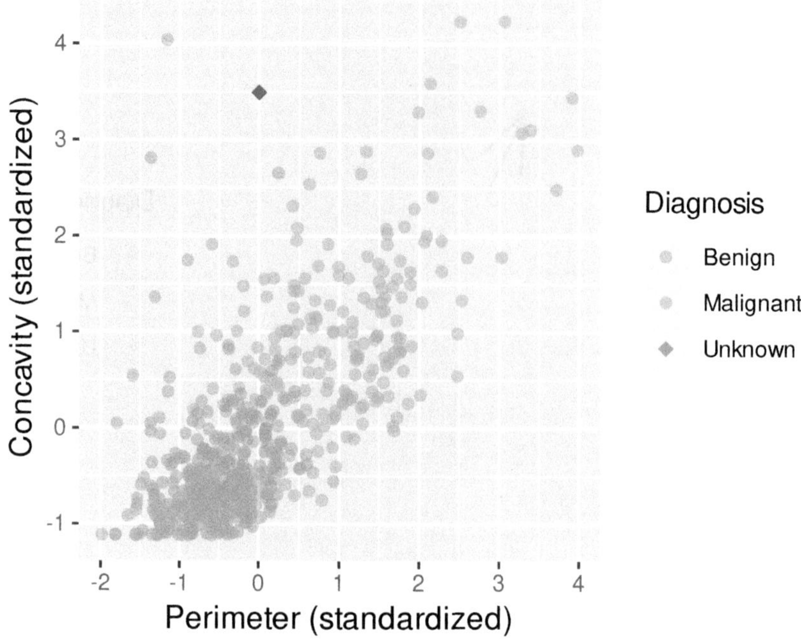

FIGURE 5.6: Scatter plot of concavity versus perimeter with new observation represented as a red diamond.

```
new_obs_Perimeter <- 0
new_obs_Concavity <- 3.5
cancer |>
  select(ID, Perimeter, Concavity, Class) |>
  mutate(dist_from_new = sqrt((Perimeter - new_obs_Perimeter)^2 +
                              (Concavity - new_obs_Concavity)^2)) |>
  arrange(dist_from_new) |>
  slice(1:5) # take the first 5 rows
```

```
## # A tibble: 5 x 5
##        ID Perimeter Concavity Class dist_from_new
##     <dbl>     <dbl>     <dbl> <fct>         <dbl>
## 1   86409     0.241      2.65 B             0.881
## 2  887181     0.750      2.87 M             0.980
## 3  899667     0.623      2.54 M             1.14
## 4  907914     0.417      2.31 M             1.26
## 5 8710441    -1.16       4.04 B             1.28
```

In Table 5.1 we show in mathematical detail how the `mutate` step was used to compute the `dist_from_new` variable (the distance to the new observation) for each of the 5 nearest neighbors in the training data.

TABLE 5.1: Evaluating the distances from the new observation to each of its 5 nearest neighbors

Perimeter	Concavity	Distance	Class
0.24	2.65	$\sqrt{(0-0.24)^2+(3.5-2.65)^2}=0.88$	B
0.75	2.87	$\sqrt{(0-0.75)^2+(3.5-2.87)^2}=0.98$	M
0.62	2.54	$\sqrt{(0-0.62)^2+(3.5-2.54)^2}=1.14$	M
0.42	2.31	$\sqrt{(0-0.42)^2+(3.5-2.31)^2}=1.26$	M
-1.16	4.04	$\sqrt{(0-(-1.16))^2+(3.5-4.04)^2}=1.28$	B

The result of this computation shows that 3 of the 5 nearest neighbors to our new observation are malignant (`M`); since this is the majority, we classify our new observation as malignant. These 5 neighbors are circled in Figure 5.7.

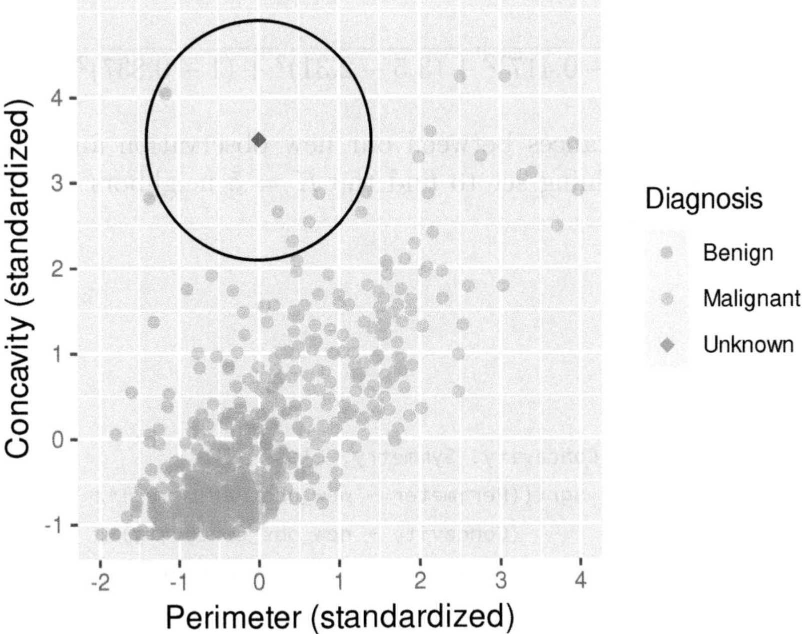

FIGURE 5.7: Scatter plot of concavity versus perimeter with 5 nearest neighbors circled.

5.5.2 More than two explanatory variables

Although the above description is directed toward two predictor variables, exactly the same K-nearest neighbors algorithm applies when you have a higher number of predictor variables. Each predictor variable may give us new information to help create our classifier. The only difference is the formula for the distance between points. Suppose we have m predictor variables for two observations a and b, i.e., $a = (a_1, a_2, \dots, a_m)$ and $b = (b_1, b_2, \dots, b_m)$.

The distance formula becomes

$$\mathrm{Distance} = \sqrt{(a_1 - b_1)^2 + (a_2 - b_2)^2 + \dots + (a_m - b_m)^2}.$$

This formula still corresponds to a straight-line distance, just in a space with more dimensions. Suppose we want to calculate the distance between a new observation with a perimeter of 0, concavity of 3.5, and symmetry of 1, and another observation with a perimeter, concavity, and symmetry of 0.417, 2.31, and 0.837 respectively. We have two observations with three predictor variables: perimeter, concavity, and symmetry. Previously, when we had two variables, we added up the squared difference between each of our (two) variables, and then took the square root. Now we will do the same, except for our three variables. We calculate the distance as follows

$$\mathrm{Distance} = \sqrt{(0 - 0.417)^2 + (3.5 - 2.31)^2 + (1 - 0.837)^2} = 1.27.$$

Let's calculate the distances between our new observation and each of the observations in the training set to find the $K = 5$ neighbors when we have these three predictors.

```
new_obs_Perimeter <- 0
new_obs_Concavity <- 3.5
new_obs_Symmetry <- 1
cancer |>
  select(ID, Perimeter, Concavity, Symmetry, Class) |>
  mutate(dist_from_new = sqrt((Perimeter - new_obs_Perimeter)^2 +
                                (Concavity - new_obs_Concavity)^2 +
                                  (Symmetry - new_obs_Symmetry)^2)) |>
  arrange(dist_from_new) |>
  slice(1:5) # take the first 5 rows
```

```
## # A tibble: 5 x 6
##         ID Perimeter Concavity Symmetry Class dist_from_new
##      <dbl>     <dbl>     <dbl>    <dbl> <fct>         <dbl>
## 1   907914     0.417      2.31    0.837 M              1.27
## 2 90439701     1.33       2.89    1.10  M              1.47
## 3   925622     0.470      2.08    1.15  M              1.50
## 4   859471    -1.37       2.81    1.09  B              1.53
## 5   899667     0.623      2.54    2.06  M              1.56
```

Based on $K = 5$ nearest neighbors with these three predictors we would classify the new observation as malignant since 4 out of 5 of the nearest neighbors are malignant class. Figure 5.8 shows what the data look like when we visualize them as a 3-dimensional scatter with lines from the new observation to its five nearest neighbors.

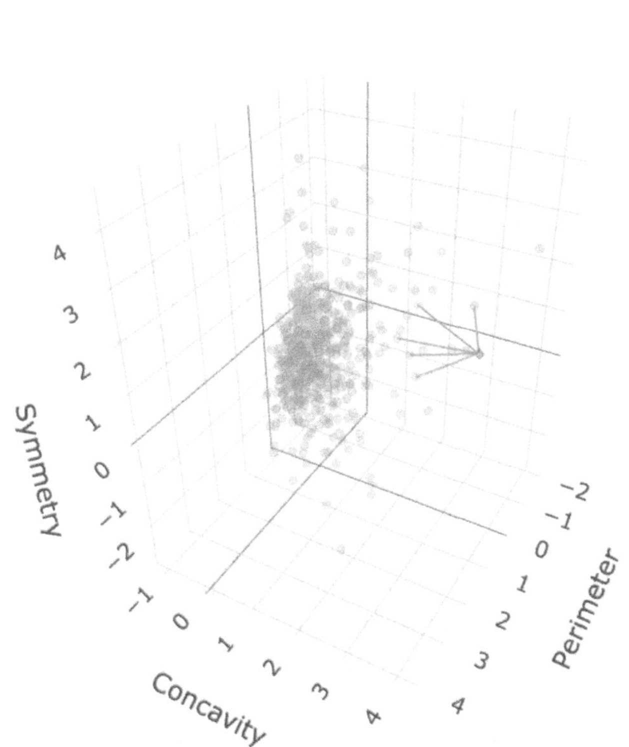

FIGURE 5.8: 3D scatter plot of the standardized symmetry, concavity, and perimeter variables. Note that in general we recommend against using 3D visualizations; here we show the data in 3D only to illustrate what higher dimensions and nearest neighbors look like, for learning purposes.

5.5.3 Summary of K-nearest neighbors algorithm

In order to classify a new observation using a K-nearest neighbor classifier, we have to do the following:

1. Compute the distance between the new observation and each observation in the training set.
2. Sort the data table in ascending order according to the distances.
3. Choose the top K rows of the sorted table.

4. Classify the new observation based on a majority vote of the neighbor classes.

5.6 K-nearest neighbors with `tidymodels`

Coding the K-nearest neighbors algorithm in R ourselves can get complicated, especially if we want to handle multiple classes, more than two variables, or predict the class for multiple new observations. Thankfully, in R, the K-nearest neighbors algorithm is implemented in the `parsnip` R package[2] [Kuhn and Vaughan, 2021] included in `tidymodels`, along with many other models[3] that you will encounter in this and future chapters of the book. The `tidymodels` collection provides tools to help make and use models, such as classifiers. Using the packages in this collection will help keep our code simple, readable and accurate; the less we have to code ourselves, the fewer mistakes we will likely make. We start by loading `tidymodels`.

```
library(tidymodels)
```

Let's walk through how to use `tidymodels` to perform K-nearest neighbors classification. We will use the `cancer` data set from above, with perimeter and concavity as predictors and $K = 5$ neighbors to build our classifier. Then we will use the classifier to predict the diagnosis label for a new observation with perimeter 0, concavity 3.5, and an unknown diagnosis label. Let's pick out our two desired predictor variables and class label and store them as a new data set named `cancer_train`:

```
cancer_train <- cancer |>
  select(Class, Perimeter, Concavity)
cancer_train
```

```
## # A tibble: 569 x 3
##    Class Perimeter Concavity
##    <fct>     <dbl>     <dbl>
##  1 M         1.27     2.65
##  2 M         1.68    -0.0238
##  3 M         1.57     1.36
##  4 M        -0.592    1.91
```

[2] `https://parsnip.tidymodels.org/`

[3] `https://www.tidymodels.org/find/parsnip/`

```
## 5 M 1.78 1.37
## 6 M -0.387 0.866
## 7 M 1.14 0.300
## 8 M -0.0728 0.0610
## 9 M -0.184 1.22
## 10 M -0.329 1.74
## # ... with 559 more rows
```

Next, we create a *model specification* for K-nearest neighbors classification by calling the `nearest_neighbor` function, specifying that we want to use $K = 5$ neighbors (we will discuss how to choose K in the next chapter) and the straight-line distance (`weight_func = "rectangular"`). The `weight_func` argument controls how neighbors vote when classifying a new observation; by setting it to `"rectangular"`, each of the K nearest neighbors gets exactly 1 vote as described above. Other choices, which weigh each neighbor's vote differently, can be found on the `parsnip` website[4]. In the `set_engine` argument, we specify which package or system will be used for training the model. Here `kknn` is the R package we will use for performing K-nearest neighbors classification. Finally, we specify that this is a classification problem with the `set_mode` function.

```
knn_spec <- nearest_neighbor(weight_func = "rectangular", neighbors = 5) |>
  set_engine("kknn") |>
  set_mode("classification")
knn_spec
```

```
## K-Nearest Neighbor Model Specification (classification)
##
## Main Arguments:
##   neighbors = 5
##   weight_func = rectangular
##
## Computational engine: kknn
```

In order to fit the model on the breast cancer data, we need to pass the model specification and the data set to the `fit` function. We also need to specify what variables to use as predictors and what variable to use as the target. Below, the `Class ~ Perimeter + Concavity` argument specifies that `Class` is the target variable (the one we want to predict), and both `Perimeter` and `Concavity` are to be used as the predictors.

---

[4] https://parsnip.tidymodels.org/reference/nearest_neighbor.html

```
knn_fit <- knn_spec |>
  fit(Class ~ Perimeter + Concavity, data = cancer_train)
```

We can also use a convenient shorthand syntax using a period, `Class ~ .`, to indicate that we want to use every variable *except* `Class` as a predictor in the model. In this particular setup, since `Concavity` and `Perimeter` are the only two predictors in the `cancer_train` data frame, `Class ~ Perimeter + Concavity` and `Class ~ .` are equivalent. In general, you can choose individual predictors using the `+` symbol, or you can specify to use *all* predictors using the `.` symbol.

```
knn_fit <- knn_spec |>
  fit(Class ~ ., data = cancer_train)
knn_fit
```

```
## parsnip model object
##
## Fit time:  17ms
##
## Call:
## kknn::train.kknn(formula = Class ~ ., data = data, ks = min_rows(5,    data, 5)
## , kernel = ~"rectangular")
##
## Type of response variable: nominal
## Minimal misclassification: 0.07557118
## Best kernel: rectangular
## Best k: 5
```

Here you can see the final trained model summary. It confirms that the computational engine used to train the model was `kknn::train.kknn`. It also shows the fraction of errors made by the nearest neighbor model, but we will ignore this for now and discuss it in more detail in the next chapter. Finally, it shows (somewhat confusingly) that the "best" weight function was "rectangular" and "best" setting of K was 5; but since we specified these earlier, R is just repeating those settings to us here. In the next chapter, we will actually let R find the value of K for us.

Finally, we make the prediction on the new observation by calling the `predict` function, passing both the fit object we just created and the new observation itself. As above, when we ran the K-nearest neighbors classification algorithm manually, the `knn_fit` object classifies the new observation as malignant ("M"). Note that the `predict` function outputs a data frame with a single variable named `.pred_class`.

```
new_obs <- tibble(Perimeter = 0, Concavity = 3.5)
predict(knn_fit, new_obs)
```

```
## # A tibble: 1 x 1
##   .pred_class
##   <fct>
## 1 M
```

Is this predicted malignant label the true class for this observation? Well, we don't know because we do not have this observation's diagnosis— that is what we were trying to predict! The classifier's prediction is not necessarily correct, but in the next chapter, we will learn ways to quantify how accurate we think our predictions are.

5.7 Data preprocessing with tidymodels

5.7.1 Centering and scaling

When using K-nearest neighbor classification, the *scale* of each variable (i.e., its size and range of values) matters. Since the classifier predicts classes by identifying observations nearest to it, any variables with a large scale will have a much larger effect than variables with a small scale. But just because a variable has a large scale *doesn't mean* that it is more important for making accurate predictions. For example, suppose you have a data set with two features, salary (in dollars) and years of education, and you want to predict the corresponding type of job. When we compute the neighbor distances, a difference of $1000 is huge compared to a difference of 10 years of education. But for our conceptual understanding and answering of the problem, it's the opposite; 10 years of education is huge compared to a difference of $1000 in yearly salary!

In many other predictive models, the *center* of each variable (e.g., its mean) matters as well. For example, if we had a data set with a temperature variable measured in degrees Kelvin, and the same data set with temperature measured in degrees Celsius, the two variables would differ by a constant shift of 273 (even though they contain exactly the same information). Likewise, in our hypothetical job classification example, we would likely see that the center of the salary variable is in the tens of thousands, while the center of the years of education variable is in the single digits. Although this doesn't affect the K-nearest neighbor classification algorithm, this large shift can change the outcome of using many other predictive models.

To scale and center our data, we need to find our variables' *mean* (the average, which quantifies the "central" value of a set of numbers) and *standard deviation* (a number quantifying how spread out values are). For each observed value of the variable, we subtract the mean (i.e., center the variable) and divide by the standard deviation (i.e., scale the variable). When we do this, the data is said to be *standardized*, and all variables in a data set will have a mean of 0 and a standard deviation of 1. To illustrate the effect that standardization can have on the K-nearest neighbor algorithm, we will read in the original, unstandardized Wisconsin breast cancer data set; we have been using a standardized version of the data set up until now. To keep things simple, we will just use the `Area`, `Smoothness`, and `Class` variables:

```
unscaled_cancer <- read_csv("data/unscaled_wdbc.csv") |>
  mutate(Class = as_factor(Class)) |>
  select(Class, Area, Smoothness)
unscaled_cancer
```

```
## # A tibble: 569 x 3
##    Class  Area Smoothness
##    <fct> <dbl>      <dbl>
##  1 M      1001     0.118
##  2 M      1326     0.0847
##  3 M      1203     0.110
##  4 M       386.    0.142
##  5 M      1297     0.100
##  6 M       477.    0.128
##  7 M      1040     0.0946
##  8 M       578.    0.119
##  9 M       520.    0.127
## 10 M       476.    0.119
## # ... with 559 more rows
```

Looking at the unscaled and uncentered data above, you can see that the differences between the values for area measurements are much larger than those for smoothness. Will this affect predictions? In order to find out, we will create a scatter plot of these two predictors (colored by diagnosis) for both the unstandardized data we just loaded, and the standardized version of that same data. But first, we need to standardize the `unscaled_cancer` data set with `tidymodels`.

In the `tidymodels` framework, all data preprocessing happens using a `recipe` from the recipes R package[5] [Kuhn and Wickham, 2021]. Here we will initialize a recipe for the unscaled_cancer data above, specifying that the Class variable is the target, and all other variables are predictors:

```
uc_recipe <- recipe(Class ~ ., data = unscaled_cancer)
print(uc_recipe)
```

```
## Recipe
##
## Inputs:
##
##       role #variables
##    outcome          1
##  predictor          2
```

So far, there is not much in the recipe; just a statement about the number of targets and predictors. Let's add scaling (step_scale) and centering (step_center) steps for all of the predictors so that they each have a mean of 0 and standard deviation of 1. Note that tidyverse actually provides step_normalize, which does both centering and scaling in a single recipe step; in this book we will keep step_scale and step_center separate to emphasize conceptually that there are two steps happening. The prep function finalizes the recipe by using the data (here, unscaled_cancer) to compute anything necessary to run the recipe (in this case, the column means and standard deviations):

```
uc_recipe <- uc_recipe |>
  step_scale(all_predictors()) |>
  step_center(all_predictors()) |>
  prep()
uc_recipe
```

```
## Recipe
##
## Inputs:
##
##       role #variables
##    outcome          1
##  predictor          2
##
## Training data contained 569 data points and no missing data.
```

[5] https://recipes.tidymodels.org/

```
##
## Operations:
##
## Scaling for Area, Smoothness [trained]
## Centering for Area, Smoothness [trained]
```

You can now see that the recipe includes a scaling and centering step for all predictor variables. Note that when you add a step to a recipe, you must specify what columns to apply the step to. Here we used the `all_predictors()` function to specify that each step should be applied to all predictor variables. However, there are a number of different arguments one could use here, as well as naming particular columns with the same syntax as the `select` function. For example:

- `all_nominal()` and `all_numeric()`: specify all categorical or all numeric variables
- `all_predictors()` and `all_outcomes()`: specify all predictor or all target variables
- `Area, Smoothness`: specify both the `Area` and `Smoothness` variable
- `-Class`: specify everything except the `Class` variable

You can find a full set of all the steps and variable selection functions on the `recipes` reference page[6].

At this point, we have calculated the required statistics based on the data input into the recipe, but the data are not yet scaled and centered. To actually scale and center the data, we need to apply the `bake` function to the unscaled data.

```
scaled_cancer <- bake(uc_recipe, unscaled_cancer)
scaled_cancer
```

```
## # A tibble: 569 x 3
##      Area Smoothness Class
##     <dbl>      <dbl> <fct>
##  1  0.984      1.57  M
##  2  1.91      -0.826 M
##  3  1.56       0.941 M
##  4 -0.764      3.28  M
##  5  1.82       0.280 M
##  6 -0.505      2.24  M
##  7  1.09      -0.123 M
##  8 -0.219      1.60  M
```

[6] https://recipes.tidymodels.org/reference/index.html

```
##  9 -0.384       2.20  M
## 10 -0.509       1.58  M
## # ... with 559 more rows
```

It may seem redundant that we had to both bake *and* prep to scale and center the data. However, we do this in two steps so we can specify a different data set in the bake step if we want. For example, we may want to specify new data that were not part of the training set.

You may wonder why we are doing so much work just to center and scale our variables. Can't we just manually scale and center the Area and Smoothness variables ourselves before building our K-nearest neighbor model? Well, technically *yes*; but doing so is error-prone. In particular, we might accidentally forget to apply the same centering / scaling when making predictions, or accidentally apply a *different* centering / scaling than what we used while training. Proper use of a recipe helps keep our code simple, readable, and error-free. Furthermore, note that using prep and bake is required only when you want to inspect the result of the preprocessing steps yourself. You will see further on in Section 5.8 that tidymodels provides tools to automatically apply prep and bake as necessary without additional coding effort.

Figure 5.9 shows the two scatter plots side-by-side—one for unscaled_cancer and one for scaled_cancer. Each has the same new observation annotated with its $K = 3$ nearest neighbors. In the original unstandardized data plot, you can see some odd choices for the three nearest neighbors. In particular, the "neighbors" are visually well within the cloud of benign observations, and the neighbors are all nearly vertically aligned with the new observation (which is why it looks like there is only one black line on this plot). Figure 5.10 shows a close-up of that region on the unstandardized plot. Here the computation of nearest neighbors is dominated by the much larger-scale area variable. The plot for standardized data on the right in Figure 5.9 shows a much more intuitively reasonable selection of nearest neighbors. Thus, standardizing the data can change things in an important way when we are using predictive algorithms. Standardizing your data should be a part of the preprocessing you do before predictive modeling and you should always think carefully about your problem domain and whether you need to standardize your data.

5.7.2 Balancing

Another potential issue in a data set for a classifier is *class imbalance*, i.e., when one label is much more common than another. Since classifiers like the K-nearest neighbor algorithm use the labels of nearby points to predict the label of a new point, if there are many more data points with one label overall, the algorithm is more likely to pick that label in general (even if the "pattern"

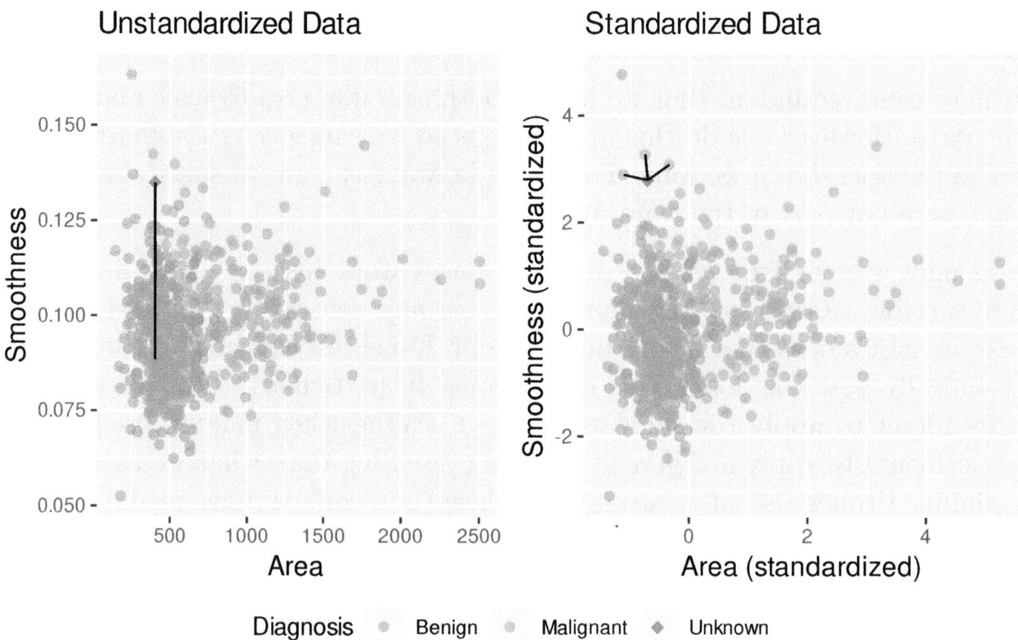

FIGURE 5.9: Comparison of K = 3 nearest neighbors with standardized and unstandardized data.

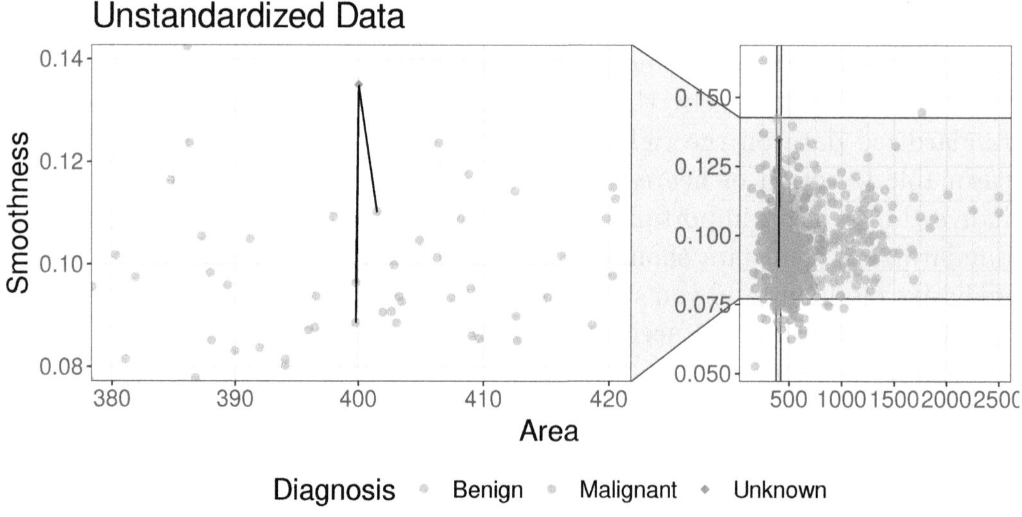

FIGURE 5.10: Close-up of three nearest neighbors for unstandardized data.

of data suggests otherwise). Class imbalance is actually quite a common and important problem: from rare disease diagnosis to malicious email detection, there are many cases in which the "important" class to identify (presence of disease, malicious email) is much rarer than the "unimportant" class (no disease, normal email).

To better illustrate the problem, let's revisit the scaled breast cancer data, `cancer`; except now we will remove many of the observations of malignant tumors, simulating what the data would look like if the cancer was rare. We will do this by picking only 3 observations from the malignant group, and keeping all of the benign observations. We choose these 3 observations using the `slice_head` function, which takes two arguments: a data frame-like object, and the number of rows to select from the top (`n`). The new imbalanced data is shown in Figure 5.11.

```
rare_cancer <- bind_rows(
      filter(cancer, Class == "B"),
      cancer |> filter(Class == "M") |> slice_head(n = 3)
    ) |>
    select(Class, Perimeter, Concavity)

rare_plot <- rare_cancer |>
  ggplot(aes(x = Perimeter, y = Concavity, color = Class)) +
  geom_point(alpha = 0.5) +
  labs(x = "Perimeter (standardized)",
       y = "Concavity (standardized)",
       color = "Diagnosis") +
  scale_color_manual(labels = c("Malignant", "Benign"),
                     values = c("orange2", "steelblue2")) +
  theme(text = element_text(size = 12))

rare_plot
```

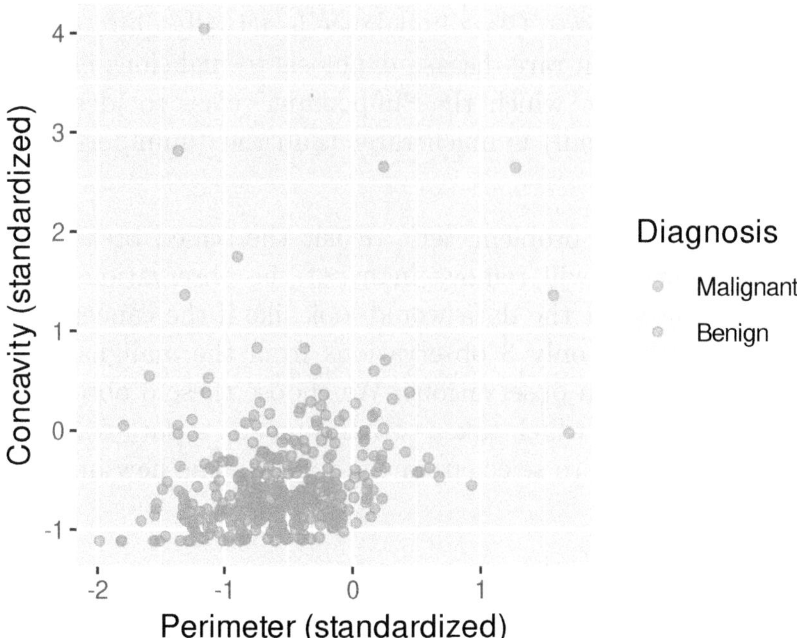

FIGURE 5.11: Imbalanced data.

Suppose we now decided to use $K = 7$ in K-nearest neighbor classification. With only 3 observations of malignant tumors, the classifier will *always predict that the tumor is benign, no matter what its concavity and perimeter are!* This is because in a majority vote of 7 observations, at most 3 will be malignant (we only have 3 total malignant observations), so at least 4 must be benign, and the benign vote will always win. For example, Figure 5.12 shows what happens for a new tumor observation that is quite close to three observations in the training data that were tagged as malignant.

Figure 5.13 shows what happens if we set the background color of each area of the plot to the predictions the K-nearest neighbor classifier would make. We can see that the decision is always "benign," corresponding to the blue color.

Despite the simplicity of the problem, solving it in a statistically sound manner is actually fairly nuanced, and a careful treatment would require a lot more detail and mathematics than we will cover in this textbook. For the present purposes, it will suffice to rebalance the data by *oversampling* the rare class. In other words, we will replicate rare observations multiple times in our data set to give them more voting power in the K-nearest neighbor algorithm. In order to do this, we will add an oversampling step to the earlier `uc_recipe` recipe with the `step_upsample` function from the `themis` R package. We show below how to do this, and also use the `group_by` and `summarize` functions to see that our classes are now balanced:

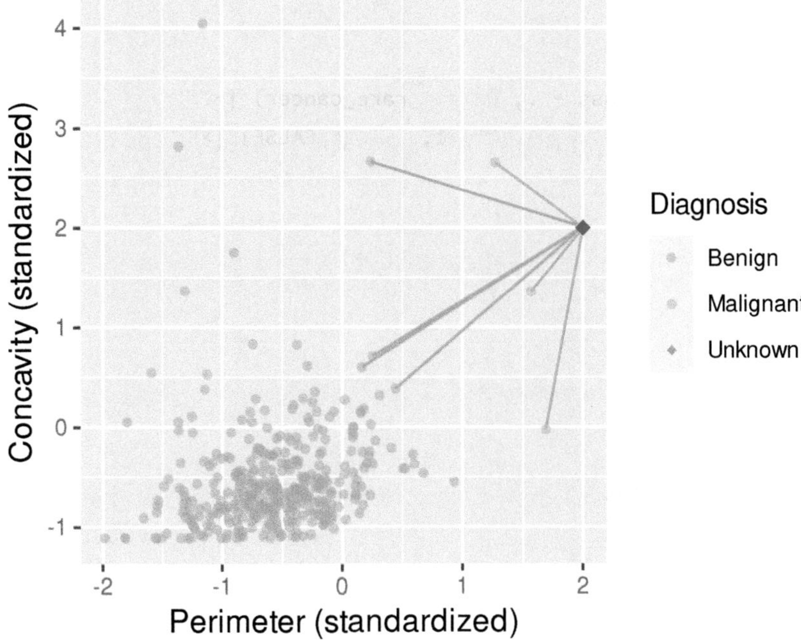

FIGURE 5.12: Imbalanced data with 7 nearest neighbors to a new observation highlighted.

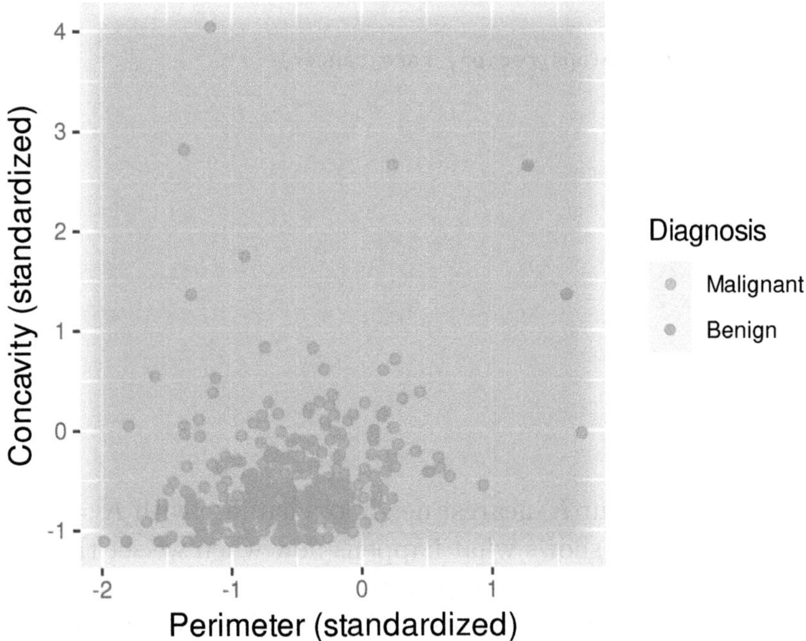

FIGURE 5.13: Imbalanced data with background color indicating the decision of the classifier and the points represent the labeled data.

```r
library(themis)

ups_recipe <- recipe(Class ~ ., data = rare_cancer) |>
  step_upsample(Class, over_ratio = 1, skip = FALSE) |>
  prep()

ups_recipe
```

```
## Recipe
##
## Inputs:
##
##       role #variables
##    outcome          1
##  predictor          2
##
## Training data contained 360 data points and no missing data.
##
## Operations:
##
## Up-sampling based on Class [trained]
```

```r
upsampled_cancer <- bake(ups_recipe, rare_cancer)

upsampled_cancer |>
  group_by(Class) |>
  summarize(n = n())
```

```
## # A tibble: 2 x 2
##   Class     n
##   <fct> <int>
## 1 M       357
## 2 B       357
```

Now suppose we train our K-nearest neighbor classifier with $K = 7$ on this *balanced* data. Figure 5.14 shows what happens now when we set the background color of each area of our scatter plot to the decision the K-nearest neighbor classifier would make. We can see that the decision is more reasonable; when the points are close to those labeled malignant, the classifier predicts a malignant tumor, and vice versa when they are closer to the benign tumor observations.

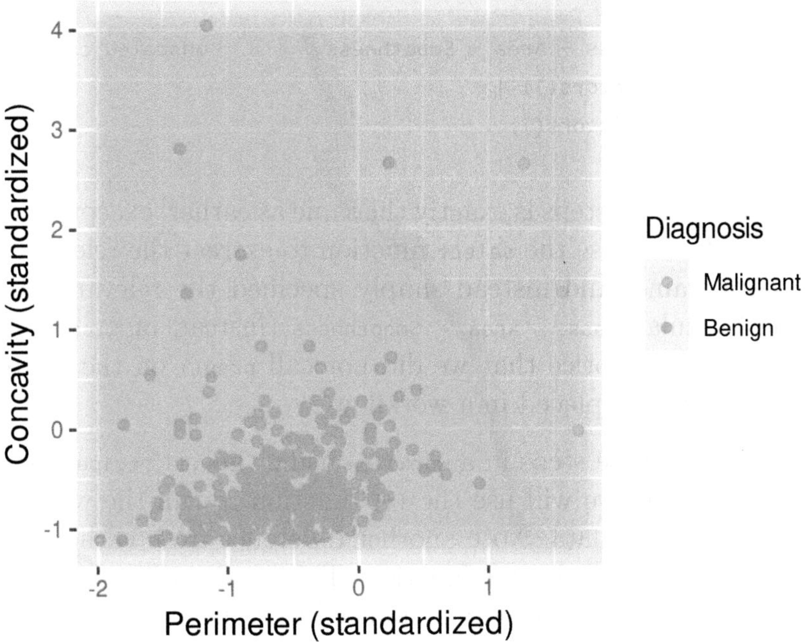

FIGURE 5.14: Upsampled data with background color indicating the decision of the classifier.

5.8 Putting it together in a workflow

The `tidymodels` package collection also provides the `workflow`, a way to chain together multiple data analysis steps without a lot of otherwise necessary code for intermediate steps. To illustrate the whole pipeline, let's start from scratch with the `unscaled_wdbc.csv` data. First we will load the data, create a model, and specify a recipe for how the data should be preprocessed:

```
# load the unscaled cancer data
# and make sure the target Class variable is a factor
unscaled_cancer <- read_csv("data/unscaled_wdbc.csv") |>
  mutate(Class = as_factor(Class))

# create the KNN model
knn_spec <- nearest_neighbor(weight_func = "rectangular", neighbors = 7) |>
  set_engine("kknn") |>
  set_mode("classification")

# create the centering / scaling recipe
```

```
uc_recipe <- recipe(Class ~ Area + Smoothness, data = unscaled_cancer) |>
  step_scale(all_predictors()) |>
  step_center(all_predictors())
```

Note that each of these steps is exactly the same as earlier, except for one major difference: we did not use the `select` function to extract the relevant variables from the data frame, and instead simply specified the relevant variables to use via the formula `Class ~ Area + Smoothness` (instead of `Class ~ .`) in the recipe. You will also notice that we did not call `prep()` on the recipe; this is unnecessary when it is placed in a workflow.

We will now place these steps in a `workflow` using the `add_recipe` and `add_model` functions, and finally we will use the `fit` function to run the whole workflow on the `unscaled_cancer` data. Note another difference from earlier here: we do not include a formula in the `fit` function. This is again because we included the formula in the recipe, so there is no need to respecify it:

```
knn_fit <- workflow() |>
  add_recipe(uc_recipe) |>
  add_model(knn_spec) |>
  fit(data = unscaled_cancer)

knn_fit
```

```
## == Workflow [trained] ==========================================================
## Preprocessor: Recipe
## Model: nearest_neighbor()
##
## -- Preprocessor ------------------------------------------------------------
-----
## 2 Recipe Steps
##
## * step_scale()
## * step_center()
##
## -- Model -------------------------------------------------------------------
-----
##
## Call:
## kknn::train.kknn(formula = ..y ~ ., data = data, ks = min_rows(7,    data, 5),
## kernel = ~"rectangular")
```

```
##
## Type of response variable: nominal
## Minimal misclassification: 0.112478
## Best kernel: rectangular
## Best k: 7
```

As before, the fit object lists the function that trains the model as well as the "best" settings for the number of neighbors and weight function (for now, these are just the values we chose manually when we created `knn_spec` above). But now the fit object also includes information about the overall workflow, including the centering and scaling preprocessing steps. In other words, when we use the `predict` function with the `knn_fit` object to make a prediction for a new observation, it will first apply the same recipe steps to the new observation. As an example, we will predict the class label of two new observations: one with `Area = 500` and `Smoothness = 0.075`, and one with `Area = 1500` and `Smoothness = 0.1`.

```
new_observation <- tibble(Area = c(500, 1500), Smoothness = c(0.075, 0.1))
prediction <- predict(knn_fit, new_observation)

prediction
```

```
## # A tibble: 2 x 1
##   .pred_class
##   <fct>
## 1 B
## 2 M
```

The classifier predicts that the first observation is benign ("B"), while the second is malignant ("M"). Figure 5.15 visualizes the predictions that this trained K-nearest neighbor model will make on a large range of new observations. Although you have seen colored prediction map visualizations like this a few times now, we have not included the code to generate them, as it is a little bit complicated. For the interested reader who wants a learning challenge, we now include it below. The basic idea is to create a grid of synthetic new observations using the `expand.grid` function, predict the label of each, and visualize the predictions with a colored scatter having a very high transparency (low `alpha` value) and large point radius. See if you can figure out what each line is doing!

Note: Understanding this code is not required for the remainder of the textbook. It is included for those readers who would like to use similar visualizations in their own data analyses.

```
# create the grid of area/smoothness vals, and arrange in a data frame
are_grid <- seq(min(unscaled_cancer$Area),
                max(unscaled_cancer$Area),
                length.out = 100)
smo_grid <- seq(min(unscaled_cancer$Smoothness),
                max(unscaled_cancer$Smoothness),
                length.out = 100)
asgrid <- as_tibble(expand.grid(Area = are_grid,
                                Smoothness = smo_grid))

# use the fit workflow to make predictions at the grid points
knnPredGrid <- predict(knn_fit, asgrid)

# bind the predictions as a new column with the grid points
prediction_table <- bind_cols(knnPredGrid, asgrid) |>
  rename(Class = .pred_class)

# plot:
# 1. the colored scatter of the original data
# 2. the faded colored scatter for the grid points
wkflw_plot <-
  ggplot() +
  geom_point(data = unscaled_cancer,
             mapping = aes(x = Area,
                           y = Smoothness,
                           color = Class),
             alpha = 0.75) +
  geom_point(data = prediction_table,
             mapping = aes(x = Area,
                           y = Smoothness,
                           color = Class),
             alpha = 0.02,
             size = 5) +
```

```
  labs(color = "Diagnosis",
       x = "Area (standardized)",
       y = "Smoothness (standardized)") +
  scale_color_manual(labels = c("Malignant", "Benign"),
                     values = c("orange2", "steelblue2")) +
  theme(text = element_text(size = 12))

wkflw_plot
```

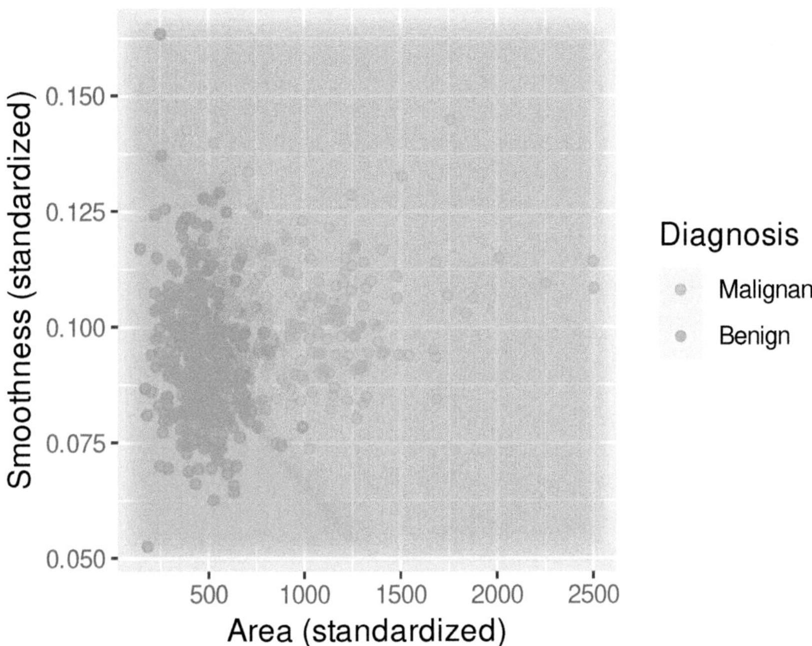

FIGURE 5.15: Scatter plot of smoothness versus area where background color indicates the decision of the classifier.

5.9 Exercises

Practice exercises for the material covered in this chapter can be found in the accompanying worksheets repository[7] in the "Classification I: training and predicting" row. You can launch an interactive version of the worksheet in your browser by clicking the "launch binder" button. You can also preview a non-interactive version of the worksheet by clicking "view worksheet." If you instead

[7] https://github.com/UBC-DSCI/data-science-a-first-intro-worksheets#readme

decide to download the worksheet and run it on your own machine, make sure to follow the instructions for computer setup found in Chapter 13. This will ensure that the automated feedback and guidance that the worksheets provide will function as intended.

6

Classification II: evaluation & tuning

6.1 Overview

This chapter continues the introduction to predictive modeling through classification. While the previous chapter covered training and data preprocessing, this chapter focuses on how to evaluate the accuracy of a classifier, as well as how to improve the classifier (where possible) to maximize its accuracy.

6.2 Chapter learning objectives

By the end of the chapter, readers will be able to do the following:

- Describe what training, validation, and test data sets are and how they are used in classification.
- Split data into training, validation, and test data sets.
- Describe what a random seed is and its importance in reproducible data analysis.
- Set the random seed in R using the `set.seed` function.
- Evaluate classification accuracy in R using a validation data set and appropriate metrics.
- Execute cross-validation in R to choose the number of neighbors in a K-nearest neighbors classifier.
- Describe the advantages and disadvantages of the K-nearest neighbors classification algorithm.

6.3 Evaluating accuracy

Sometimes our classifier might make the wrong prediction. A classifier does not need to be right 100% of the time to be useful, though we don't want the classifier to make too many wrong predictions. How do we measure how "good" our classifier is? Let's revisit the breast cancer images data[1] [Street et al., 1993] and think about how our classifier will be used in practice. A biopsy will be performed on a *new* patient's tumor, the resulting image will be analyzed, and the classifier will be asked to decide whether the tumor is benign or malignant. The key word here is *new*: our classifier is "good" if it provides accurate predictions on data *not seen during training*. But then, how can we evaluate our classifier without visiting the hospital to collect more tumor images?

The trick is to split the data into a **training set** and **test set** (Figure 6.1) and use only the **training set** when building the classifier. Then, to evaluate the accuracy of the classifier, we first set aside the true labels from the **test set**, and then use the classifier to predict the labels in the **test set**. If our predictions match the true labels for the observations in the **test set**, then we have some confidence that our classifier might also accurately predict the class labels for new observations without known class labels.

Note: If there were a golden rule of machine learning, it might be this: *you cannot use the test data to build the model!* If you do, the model gets to "see" the test data in advance, making it look more accurate than it really is. Imagine how bad it would be to overestimate your classifier's accuracy when predicting whether a patient's tumor is malignant or benign!

How exactly can we assess how well our predictions match the true labels for the observations in the test set? One way we can do this is to calculate the **prediction accuracy**. This is the fraction of examples for which the classifier made the correct prediction. To calculate this, we divide the number of correct predictions by the number of predictions made.

$$\text{prediction accuracy} = \frac{\text{number of correct predictions}}{\text{total number of predictions}}$$

The process for assessing if our predictions match the true labels in the test set is illustrated in Figure 6.2. Note that there are other measures for how well classifiers perform, such as *precision* and *recall*; these will not be discussed here, but you will likely encounter them in other more advanced books on this topic.

[1] `https://archive.ics.uci.edu/ml/datasets/Breast+Cancer+Wisconsin+%28Diagnostic%29`

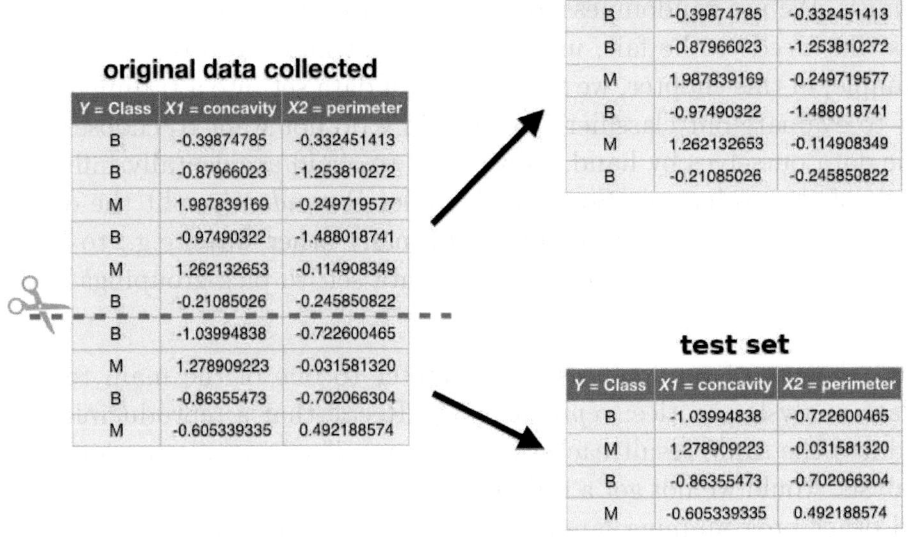

Y = Class	X1 = concavity	X2 = perimeter
B	-0.39874785	-0.332451413
B	-0.87966023	-1.253810272
M	1.987839169	-0.249719577
B	-0.97490322	-1.488018741
M	1.262132653	-0.114908349
B	-0.21085026	-0.245850822
B	-1.03994838	-0.722600465
M	1.278909223	-0.031581320
B	-0.86355473	-0.702066304
M	-0.605339335	0.492188574

FIGURE 6.1: Splitting the data into training and testing sets.

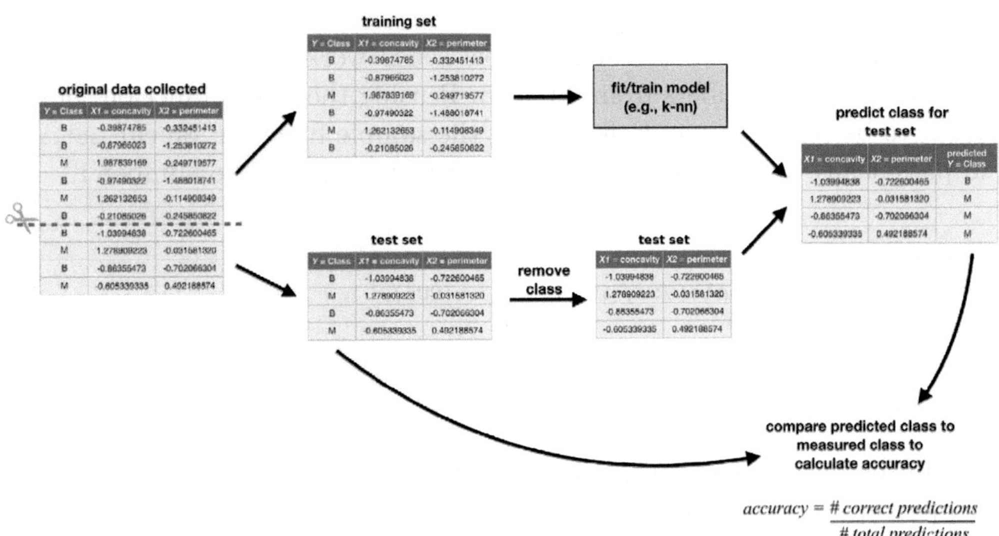

FIGURE 6.2: Process for splitting the data and finding the prediction accuracy.

6.4 Randomness and seeds

Beginning in this chapter, our data analyses will often involve the use of *randomness*. We use randomness any time we need to make a decision in our analysis that needs to be fair, unbiased, and not influenced by human input. For example, in this chapter, we need to split a data set into a training set and test set to evaluate our classifier. We certainly do not want to choose how to split the data ourselves by hand, as we want to avoid accidentally influencing the result of the evaluation. So instead, we let R *randomly* split the data. In future chapters we will use randomness in many other ways, e.g., to help us select a small subset of data from a larger data set, to pick groupings of data, and more.

However, the use of randomness runs counter to one of the main tenets of good data analysis practice: *reproducibility*. Recall that a reproducible analysis produces the same result each time it is run; if we include randomness in the analysis, would we not get a different result each time? The trick is that in R—and other programming languages—randomness is not actually random! Instead, R uses a *random number generator* that produces a sequence of numbers that are completely determined by a *seed value*. Once you set the seed value using the `set.seed` function, everything after that point may *look* random, but is actually totally reproducible. As long as you pick the same seed value, you get the same result!

Let's use an example to investigate how seeds work in R. Say we want to randomly pick 10 numbers from 0 to 9 in R using the `sample` function, but we want it to be reproducible. Before using the sample function, we call `set.seed`, and pass it any integer as an argument. Here, we pass in the number `1`.

```
set.seed(1)
random_numbers1 <- sample(0:9, 10, replace=TRUE)
random_numbers1
```

```
##  [1] 8 3 6 0 1 6 1 2 0 4
```

You can see that `random_numbers1` is a list of 10 numbers from 0 to 9 that, from all appearances, looks random. If we run the `sample` function again, we will get a fresh batch of 10 numbers that also look random.

```
random_numbers2 <- sample(0:9, 10, replace=TRUE)
random_numbers2
```

```
## [1] 4 9 5 9 6 8 4 4 8 8
```

If we want to force R to produce the same sequences of random numbers, we can simply call the `set.seed` function again with the same argument value.

```
set.seed(1)
random_numbers1_again <- sample(0:9, 10, replace=TRUE)
random_numbers1_again
```

```
## [1] 8 3 6 0 1 6 1 2 0 4
```

```
random_numbers2_again <- sample(0:9, 10, replace=TRUE)
random_numbers2_again
```

```
## [1] 4 9 5 9 6 8 4 4 8 8
```

Notice that after setting the seed, we get the same two sequences of numbers in the same order. `random_numbers1` and `random_numbers1_again` produce the same sequence of numbers, and the same can be said about `random_numbers2` and `random_numbers2_again`. And if we choose a different value for the seed—say, 4235—we obtain a different sequence of random numbers.

```
set.seed(4235)
random_numbers <- sample(0:9, 10, replace=TRUE)
random_numbers
```

```
## [1] 8 3 1 4 6 8 8 4 1 7
```

```
random_numbers <- sample(0:9, 10, replace=TRUE)
random_numbers
```

```
## [1] 3 7 8 2 8 8 6 3 3 8
```

In other words, even though the sequences of numbers that R is generating *look* random, they are totally determined when we set a seed value!

So what does this mean for data analysis? Well, `sample` is certainly not the only function that uses randomness in R. Many of the functions that we use in `tidymodels`, `tidyverse`, and beyond use randomness—many of them without even telling you about it. So at the beginning of every data analysis you do, right after loading packages, you should call the `set.seed` function and pass it an integer that you pick. Also note that when R starts up, it creates its own seed to use. So if you do not explicitly call the `set.seed` function in your code, your results will likely not be reproducible. And finally, be careful to set the seed *only once* at the beginning of a data analysis. Each time you set the seed, you are inserting your own human input, thereby influencing the analysis. If you use `set.seed` many times throughout your analysis, the randomness that R uses will not look as random as it should.

In summary: if you want your analysis to be reproducible, i.e., produce *the same result* each time you run it, make sure to use `set.seed` exactly once at the beginning of the analysis. Different argument values in `set.seed` lead to different patterns of randomness, but as long as you pick the same argument value your result will be the same. In the remainder of the textbook, we will set the seed once at the beginning of each chapter.

6.5 Evaluating accuracy with `tidymodels`

Back to evaluating classifiers now! In R, we can use the `tidymodels` package not only to perform K-nearest neighbors classification, but also to assess how well our classification worked. Let's work through an example of how to use tools from `tidymodels` to evaluate a classifier using the breast cancer data set from the previous chapter. We begin the analysis by loading the packages we require, reading in the breast cancer data, and then making a quick scatter plot visualization of tumor cell concavity versus smoothness colored by diagnosis in Figure 6.3. You will also notice that we set the random seed here at the beginning of the analysis using the `set.seed` function, as described in Section 6.4.

```
# load packages
library(tidyverse)
library(tidymodels)

# set the seed
set.seed(1)

# load data
cancer <- read_csv("data/unscaled_wdbc.csv") |>
  # convert the character Class variable to the factor datatype
  mutate(Class = as_factor(Class))

# create scatter plot of tumor cell concavity versus smoothness,
# labeling the points be diagnosis class
perim_concav <- cancer |>
```

```
  ggplot(aes(x = Smoothness, y = Concavity, color = Class)) +
  geom_point(alpha = 0.5) +
  labs(color = "Diagnosis") +
  scale_color_manual(labels = c("Malignant", "Benign"),
                     values = c("orange2", "steelblue2")) +
  theme(text = element_text(size = 12))

perim_concav
```

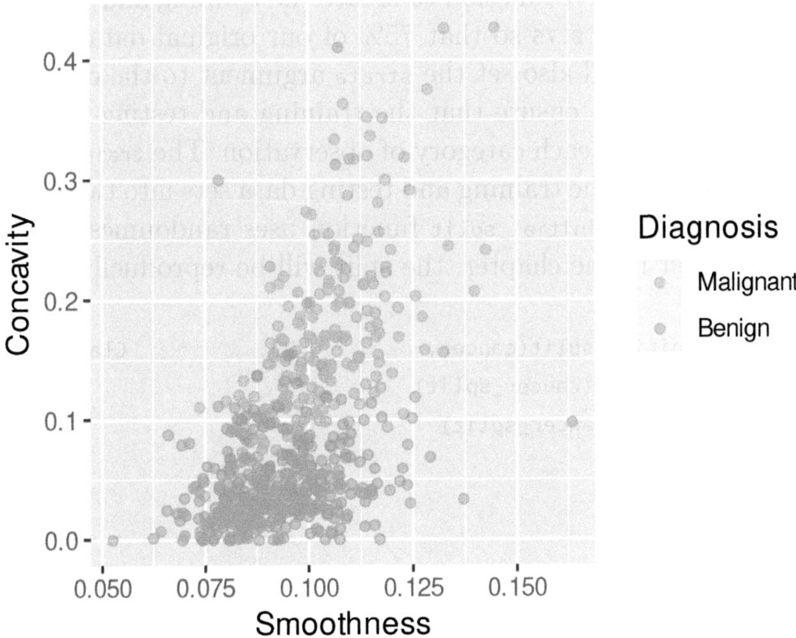

FIGURE 6.3: Scatter plot of tumor cell concavity versus smoothness colored by diagnosis label.

6.5.1 Create the train / test split

Once we have decided on a predictive question to answer and done some preliminary exploration, the very next thing to do is to split the data into the training and test sets. Typically, the training set is between 50% and 95% of the data, while the test set is the remaining 5% to 50%; the intuition is that you want to trade off between training an accurate model (by using a larger training data set) and getting an accurate evaluation of its performance (by using a larger test data set). Here, we will use 75% of the data for training, and 25% for testing.

The `initial_split` function from `tidymodels` handles the procedure of splitting the data for us. It also applies two very important steps when splitting to ensure that the accuracy estimates from the test data are reasonable. First, it **shuffles** the data before splitting, which ensures that any ordering present in the data does not influence the data that ends up in the training and testing sets. Second, it **stratifies** the data by the class label, to ensure that roughly the same proportion of each class ends up in both the training and testing sets. For example, in our data set, roughly 63% of the observations are from the benign class (`B`), and 37% are from the malignant class (`M`), so `initial_split` ensures that roughly 63% of the training data are benign, 37% of the training data are malignant, and the same proportions exist in the testing data.

Let's use the `initial_split` function to create the training and testing sets. We will specify that `prop = 0.75` so that 75% of our original data set ends up in the training set. We will also set the `strata` argument to the categorical label variable (here, `Class`) to ensure that the training and testing subsets contain the right proportions of each category of observation. The `training` and `testing` functions then extract the training and testing data sets into two separate data frames. Note that the `initial_split` function uses randomness, but since we set the seed earlier in the chapter, the split will be reproducible.

```
cancer_split <- initial_split(cancer, prop = 0.75, strata = Class)
cancer_train <- training(cancer_split)
cancer_test <- testing(cancer_split)
```

```
glimpse(cancer_train)
```

```
## Rows: 426
## Columns: 12
## $ ID                <dbl> 8510426, 8510653, 8510824, 857373, 857810, 858477, 8~
## $ Class             <fct> B, B, B, B, B, B, B, B, B, B, B, B, B, B, B, B, B, B~
## $ Radius            <dbl> 13.540, 13.080, 9.504, 13.640, 13.050, 8.618, 10.170~
## $ Texture           <dbl> 14.36, 15.71, 12.44, 16.34, 19.31, 11.79, 14.88, 20.~
## $ Perimeter         <dbl> 87.46, 85.63, 60.34, 87.21, 82.61, 54.34, 64.55, 54.~
## $ Area              <dbl> 566.3, 520.0, 273.9, 571.8, 527.2, 224.5, 311.9, 221~
## $ Smoothness        <dbl> 0.09779, 0.10750, 0.10240, 0.07685, 0.08060, 0.09752~
## $ Compactness       <dbl> 0.08129, 0.12700, 0.06492, 0.06059, 0.03789, 0.05272~
## $ Concavity         <dbl> 0.066640, 0.045680, 0.029560, 0.018570, 0.000692, 0.~
## $ Concave_Points    <dbl> 0.047810, 0.031100, 0.020760, 0.017230, 0.004167, 0.~
## $ Symmetry          <dbl> 0.1885, 0.1967, 0.1815, 0.1353, 0.1819, 0.1683, 0.27~
## $ Fractal_Dimension <dbl> 0.05766, 0.06811, 0.06905, 0.05953, 0.05501, 0.07187~
```

```
glimpse(cancer_test)
```

```
## Rows: 143
## Columns: 12
## $ ID                <dbl> 84501001, 846381, 84799002, 849014, 852763, 853401, ~
## $ Class              <fct> M, M, M, M, M, M, M, B, M, M, M, B, B, B, B, B, B, M~
## $ Radius            <dbl> 12.460, 15.850, 14.540, 19.810, 14.580, 18.630, 16.7~
## $ Texture           <dbl> 24.04, 23.95, 27.54, 22.15, 21.53, 25.11, 21.59, 18.~
## $ Perimeter         <dbl> 83.97, 103.70, 96.73, 130.00, 97.41, 124.80, 110.10,~
## $ Area              <dbl> 475.9, 782.7, 658.8, 1260.0, 644.8, 1088.0, 869.5, 5~
## $ Smoothness        <dbl> 0.11860, 0.08401, 0.11390, 0.09831, 0.10540, 0.10640~
## $ Compactness       <dbl> 0.23960, 0.10020, 0.15950, 0.10270, 0.18680, 0.18870~
## $ Concavity         <dbl> 0.22730, 0.09938, 0.16390, 0.14790, 0.14250, 0.23190~
## $ Concave_Points     <dbl> 0.085430, 0.053640, 0.073640, 0.094980, 0.087830, 0.~
## $ Symmetry          <dbl> 0.2030, 0.1847, 0.2303, 0.1582, 0.2252, 0.2183, 0.18~
## $ Fractal_Dimension <dbl> 0.08243, 0.05338, 0.07077, 0.05395, 0.06924, 0.06197~
```

We can see from `glimpse` in the code above that the training set contains 426 observations, while the test set contains 143 observations. This corresponds to a train / test split of 75% / 25%, as desired. Recall from Chapter 5 that we use the `glimpse` function to view data with a large number of columns, as it prints the data such that the columns go down the page (instead of across).

We can use `group_by` and `summarize` to find the percentage of malignant and benign classes in `cancer_train` and we see about 63% of the training data are benign and 37% are malignant, indicating that our class proportions were roughly preserved when we split the data.

```
cancer_proportions <- cancer_train |>
                      group_by(Class) |>
                      summarize(n = n()) |>
                      mutate(percent = 100*n/nrow(cancer_train))

cancer_proportions
```

```
## # A tibble: 2 x 3
##   Class      n percent
##   <fct> <int>   <dbl>
## 1 M       159    37.3
## 2 B       267    62.7
```

6.5.2 Preprocess the data

As we mentioned in the last chapter, K-nearest neighbors is sensitive to the scale of the predictors, so we should perform some preprocessing to standardize them. An additional consideration we need to take when doing this is that we should create the standardization preprocessor using **only the training data**. This ensures that our test data does not influence any aspect of our model training. Once we have created the standardization preprocessor, we can then apply it separately to both the training and test data sets.

Fortunately, the `recipe` framework from `tidymodels` helps us handle this properly. Below we construct and prepare the recipe using only the training data (due to `data = cancer_train` in the first line).

```
cancer_recipe <- recipe(Class ~ Smoothness + Concavity, data = cancer_train) |>
  step_scale(all_predictors()) |>
  step_center(all_predictors())
```

6.5.3 Train the classifier

Now that we have split our original data set into training and test sets, we can create our K-nearest neighbors classifier with only the training set using the technique we learned in the previous chapter. For now, we will just choose the number K of neighbors to be 3, and use concavity and smoothness as the predictors. As before we need to create a model specification, combine the model specification and recipe into a workflow, and then finally use `fit` with the training data `cancer_train` to build the classifier.

```
knn_spec <- nearest_neighbor(weight_func = "rectangular", neighbors = 3) |>
  set_engine("kknn") |>
  set_mode("classification")

knn_fit <- workflow() |>
  add_recipe(cancer_recipe) |>
  add_model(knn_spec) |>
  fit(data = cancer_train)

knn_fit
```

```
## == Workflow [trained] ==========================================================
## Preprocessor: Recipe
## Model: nearest_neighbor()
##
## -- Preprocessor -------------------------------------------------------
-----
## 2 Recipe Steps
```

```
##
## * step_scale()
## * step_center()
##
## -- Model -------------------------------------------------------------------
-----
##
## Call:
## kknn::train.kknn(formula = ..y ~ ., data = data, ks = min_rows(3,     data, 5),
## kernel = ~"rectangular")
##
## Type of response variable: nominal
## Minimal misclassification: 0.1150235
## Best kernel: rectangular
## Best k: 3
```

6.5.4 Predict the labels in the test set

Now that we have a K-nearest neighbors classifier object, we can use it to predict the class labels for our test set. We use the `bind_cols` to add the column of predictions to the original test data, creating the `cancer_test_predictions` data frame. The `Class` variable contains the true diagnoses, while the `.pred_class` contains the predicted diagnoses from the classifier.

```
cancer_test_predictions <- predict(knn_fit, cancer_test) |>
  bind_cols(cancer_test)

cancer_test_predictions
```

```
## # A tibble: 143 x 13
##    .pred_class       ID Class Radius Texture Perimeter  Area Smoothness
##    <fct>          <dbl> <fct>  <dbl>   <dbl>     <dbl> <dbl>      <dbl>
##  1 M           84501001 M       12.5    24.0      84.0  476.     0.119
##  2 B             846381 M       15.8    24.0     104.   783.     0.0840
##  3 M           84799002 M       14.5    27.5      96.7  659.     0.114
##  4 M             849014 M       19.8    22.2     130   1260      0.0983
##  5 M             852763 M       14.6    21.5      97.4  645.     0.105
##  6 M             853401 M       18.6    25.1     125.  1088      0.106
##  7 B             854253 M       16.7    21.6     110.   870.     0.0961
##  8 B             854941 B       13.0    18.4      82.6  524.     0.0898
##  9 M             855138 M       13.5    20.8      88.4  559.     0.102
## 10 B             855167 M       13.4    21.6      86.2  563      0.0816
```

```
## # ... with 133 more rows, and 5 more variables: Compactness <dbl>,
## #   Concavity <dbl>, Concave_Points <dbl>, Symmetry <dbl>,
## #   Fractal_Dimension <dbl>
```

6.5.5 Compute the accuracy

Finally, we can assess our classifier's accuracy. To do this we use the `metrics` function from `tidymodels` to get the statistics about the quality of our model, specifying the `truth` and `estimate` arguments:

```
cancer_test_predictions |>
  metrics(truth = Class, estimate = .pred_class) |>
  filter(.metric == "accuracy")
```

```
## # A tibble: 1 x 3
##   .metric  .estimator .estimate
##   <chr>    <chr>          <dbl>
## 1 accuracy binary         0.860
```

In the metrics data frame, we filtered the `.metric` column since we are interested in the `accuracy` row. Other entries involve more advanced metrics that are beyond the scope of this book. Looking at the value of the `.estimate` variable shows that the estimated accuracy of the classifier on the test data was 86%.

We can also look at the *confusion matrix* for the classifier, which shows the table of predicted labels and correct labels, using the `conf_mat` function:

```
confusion <- cancer_test_predictions |>
             conf_mat(truth = Class, estimate = .pred_class)
confusion
```

```
##           Truth
## Prediction  M  B
##          M 39  6
##          B 14 84
```

The confusion matrix shows 39 observations were correctly predicted as malignant, and 84 were correctly predicted as benign. Therefore the classifier labeled $39 + 84 = 123$ observations correctly. It also shows that the classifier made some mistakes; in particular, it classified 14 observations as benign when they were truly malignant, and 6 observations as malignant when they were truly benign.

6.5.6 Critically analyze performance

We now know that the classifier was 86% accurate on the test data set. That sounds pretty good! Wait, *is* it good? Or do we need something higher?

In general, what a *good* value for accuracy is depends on the application. For instance, suppose you are predicting whether a tumor is benign or malignant for a type of tumor that is benign 99% of the time. It is very easy to obtain a 99% accuracy just by guessing benign for every observation. In this case, 99% accuracy is probably not good enough. And beyond just accuracy, sometimes the *kind* of mistake the classifier makes is important as well. In the previous example, it might be very bad for the classifier to predict "benign" when the true class is "malignant", as this might result in a patient not receiving appropriate medical attention. On the other hand, it might be less bad for the classifier to guess "malignant" when the true class is "benign", as the patient will then likely see a doctor who can provide an expert diagnosis. This is why it is important not only to look at accuracy, but also the confusion matrix.

However, there is always an easy baseline that you can compare to for any classification problem: the *majority classifier*. The majority classifier *always* guesses the majority class label from the training data, regardless of the predictor variables' values. It helps to give you a sense of scale when considering accuracies. If the majority classifier obtains a 90% accuracy on a problem, then you might hope for your K-nearest neighbors classifier to do better than that. If your classifier provides a significant improvement upon the majority classifier, this means that at least your method is extracting some useful information from your predictor variables. Be careful though: improving on the majority classifier does not *necessarily* mean the classifier is working well enough for your application.

As an example, in the breast cancer data, recall the proportions of benign and malignant observations in the training data are as follows:

```
cancer_proportions
```

```
## # A tibble: 2 x 3
##   Class      n percent
##   <fct> <int>   <dbl>
## 1 M       159    37.3
## 2 B       267    62.7
```

Since the benign class represents the majority of the training data, the majority classifier would *always* predict that a new observation is benign. The estimated accuracy of the majority classifier is usually fairly close to the majority class proportion in the training data. In this case, we would suspect that the majority classifier will have an accuracy of around 63%. The K-nearest neighbors classifier we built does quite a bit better than this, with an accuracy of 86%. This means that from the perspective of accuracy, the K-nearest neighbors classifier improved quite a bit on the basic majority classifier. Hooray! But we still need to be cautious; in this application, it is likely very important not to misdiagnose any malignant tumors to avoid missing patients who actually need medical care. The confusion matrix above shows that the classifier does, indeed, misdiagnose a significant number of malignant tumors as benign (14 out of 53 malignant tumors, or 26%!). Therefore, even though the accuracy improved upon the majority classifier, our critical analysis suggests that this classifier may not have appropriate performance for the application.

6.6 Tuning the classifier

The vast majority of predictive models in statistics and machine learning have *parameters*. A *parameter* is a number you have to pick in advance that determines some aspect of how the model behaves. For example, in the K-nearest neighbors classification algorithm, K is a parameter that we have to pick that determines how many neighbors participate in the class vote. By picking different values of K, we create different classifiers that make different predictions.

So then, how do we pick the *best* value of K, i.e., *tune* the model? And is it possible to make this selection in a principled way? Ideally, we want somehow to maximize the performance of our classifier on data *it hasn't seen yet*. But we cannot use our test data set in the process of building our model. So we will play the same trick we did before when evaluating our classifier: we'll split our *training data itself* into two subsets, use one to train the model, and then use the other to evaluate it. In this section, we will cover the details of this procedure, as well as how to use it to help you pick a good parameter value for your classifier.

And remember: don't touch the test set during the tuning process. Tuning is a part of model training!

6.6.1 Cross-validation

The first step in choosing the parameter K is to be able to evaluate the classifier using only the training data. If this is possible, then we can compare the classifier's performance for different values of K—and pick the best—using only the training data. As suggested at the beginning of this section, we will accomplish this by splitting the training data, training on one subset, and evaluating on the other. The subset of training data used for evaluation is often called the **validation set**.

There is, however, one key difference from the train/test split that we performed earlier. In particular, we were forced to make only a *single split* of the data. This is because at the end of the day, we have to produce a single classifier. If we had multiple different splits of the data into training and testing data, we would produce multiple different classifiers. But while we are tuning the classifier, we are free to create multiple classifiers based on multiple splits of the training data, evaluate them, and then choose a parameter value based on ***all*** of the different results. If we just split our overall training data *once*, our best parameter choice will depend strongly on whatever data was lucky enough to end up in the validation set. Perhaps using multiple different train/validation splits, we'll get a better estimate of accuracy, which will lead to a better choice of the number of neighbors K for the overall set of training data.

Let's investigate this idea in R! In particular, we will generate five different train/validation splits of our overall training data, train five different K-nearest neighbors models, and evaluate their accuracy. We will start with just a single split.

```
# create the 25/75 split of the training data into training and validation
cancer_split <- initial_split(cancer_train, prop = 0.75, strata = Class)
cancer_subtrain <- training(cancer_split)
cancer_validation <- testing(cancer_split)

# recreate the standardization recipe from before
# (since it must be based on the training data)
cancer_recipe <- recipe(Class ~ Smoothness + Concavity,
                        data = cancer_subtrain) |>
  step_scale(all_predictors()) |>
  step_center(all_predictors())

# fit the knn model (we can reuse the old knn_spec model from before)
knn_fit <- workflow() |>
  add_recipe(cancer_recipe) |>
  add_model(knn_spec) |>
  fit(data = cancer_subtrain)

# get predictions on the validation data
```

```
validation_predicted <- predict(knn_fit, cancer_validation) |>
  bind_cols(cancer_validation)

# compute the accuracy
acc <- validation_predicted |>
  metrics(truth = Class, estimate = .pred_class) |>
  filter(.metric == "accuracy") |>
  select(.estimate) |>
  pull()

acc
```

```
## [1] 0.8878505
```

The accuracy estimate using this split is 88.8%. Now we repeat the above code 4 more times, which generates 4 more splits. Therefore we get five different shuffles of the data, and therefore five different values for accuracy: 88.8%, 86.9%, 83.2%, 88.8%, 87.9%. None of these values are necessarily "more correct" than any other; they're just five estimates of the true, underlying accuracy of our classifier built using our overall training data. We can combine the estimates by taking their average (here 87%) to try to get a single assessment of our classifier's accuracy; this has the effect of reducing the influence of any one (un)lucky validation set on the estimate.

In practice, we don't use random splits, but rather use a more structured splitting procedure so that each observation in the data set is used in a validation set only a single time. The name for this strategy is **cross-validation**. In **cross-validation**, we split our **overall training data** into C evenly sized chunks. Then, iteratively use 1 chunk as the **validation set** and combine the remaining $C - 1$ chunks as the **training set**. This procedure is shown in Figure 6.4. Here, $C = 5$ different chunks of the data set are used, resulting in 5 different choices for the **validation set**; we call this *5-fold* cross-validation.

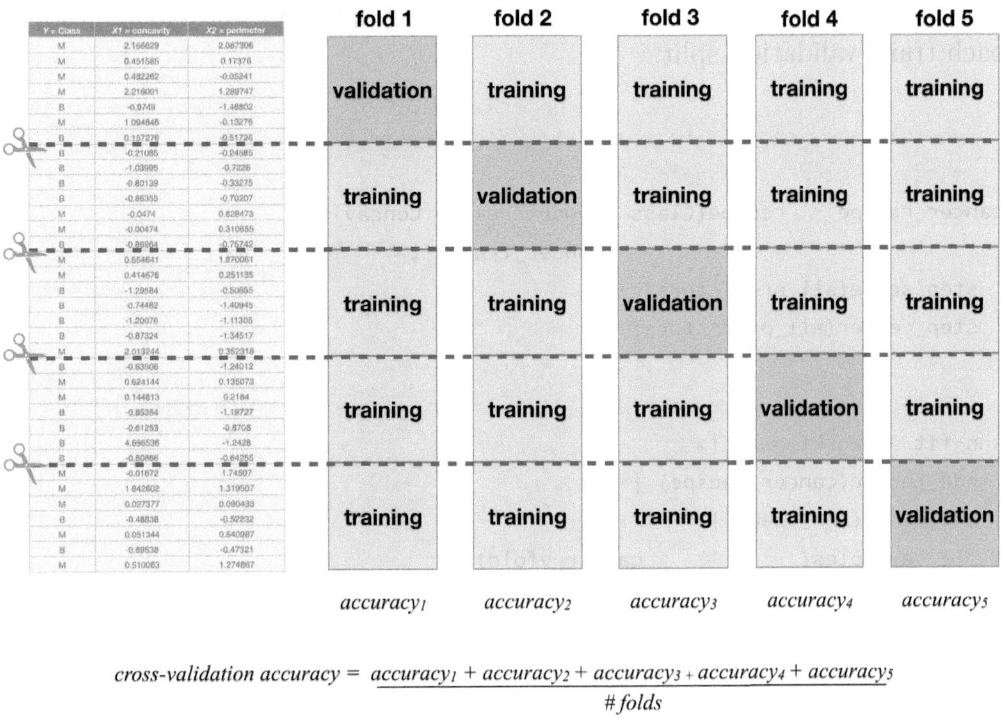

fold 1 fold 2 fold 3 fold 4 fold 5

$$\text{cross-validation accuracy} = \frac{accuracy_1 + accuracy_2 + accuracy_3 + accuracy_4 + accuracy_5}{\#folds}$$

FIGURE 6.4: 5-fold cross-validation.

To perform 5-fold cross-validation in R with `tidymodels`, we use another function: `vfold_cv`. This function splits our training data into `v` folds automatically. We set the `strata` argument to the categorical label variable (here, `Class`) to ensure that the training and validation subsets contain the right proportions of each category of observation.

```
cancer_vfold <- vfold_cv(cancer_train, v = 5, strata = Class)
cancer_vfold
```

```
## #  5-fold cross-validation using stratification
## # A tibble: 5 x 2
##   splits            id
##   <list>            <chr>
## 1 <split [340/86]> Fold1
## 2 <split [340/86]> Fold2
## 3 <split [341/85]> Fold3
## 4 <split [341/85]> Fold4
## 5 <split [342/84]> Fold5
```

Then, when we create our data analysis workflow, we use the `fit_resamples` function instead of the `fit` function for training. This runs cross-validation on each train/validation split.

```
# recreate the standardization recipe from before
# (since it must be based on the training data)
cancer_recipe <- recipe(Class ~ Smoothness + Concavity,
                        data = cancer_train) |>
  step_scale(all_predictors()) |>
  step_center(all_predictors())

# fit the knn model (we can reuse the old knn_spec model from before)
knn_fit <- workflow() |>
  add_recipe(cancer_recipe) |>
  add_model(knn_spec) |>
  fit_resamples(resamples = cancer_vfold)

knn_fit
```

```
## # Resampling results
## # 5-fold cross-validation using stratification
## # A tibble: 5 x 4
##   splits           id    .metrics         .notes
##   <list>           <chr> <list>           <list>
## 1 <split [340/86]> Fold1 <tibble [2 x 4]> <tibble [0 x 1]>
## 2 <split [340/86]> Fold2 <tibble [2 x 4]> <tibble [0 x 1]>
## 3 <split [341/85]> Fold3 <tibble [2 x 4]> <tibble [0 x 1]>
## 4 <split [341/85]> Fold4 <tibble [2 x 4]> <tibble [0 x 1]>
## 5 <split [342/84]> Fold5 <tibble [2 x 4]> <tibble [0 x 1]>
```

The `collect_metrics` function is used to aggregate the *mean* and *standard error* of the classifier's validation accuracy across the folds. You will find results related to the accuracy in the row with `accuracy` listed under the `.metric` column. You should consider the mean (`mean`) to be the estimated accuracy, while the standard error (`std_err`) is a measure of how uncertain we are in the mean value. A detailed treatment of this is beyond the scope of this chapter; but roughly, if your estimated mean is 0.87 and standard error is 0.02, you can expect the *true* average accuracy of the classifier to be somewhere roughly between 85% and 89% (although it may fall outside this range). You may ignore the other columns in the metrics data frame, as they do not provide any additional insight. You can also ignore the entire second row with `roc_auc` in the `.metric` column, as it is beyond the scope of this book.

```
knn_fit |>
  collect_metrics()
```

```
## # A tibble: 2 x 6
##   .metric  .estimator  mean     n std_err .config
##   <chr>    <chr>      <dbl> <int>   <dbl> <chr>
## 1 accuracy binary     0.871     5  0.0212 Preprocessor1_Model1
## 2 roc_auc  binary     0.905     5  0.0188 Preprocessor1_Model1
```

We can choose any number of folds, and typically the more we use the better our accuracy estimate will be (lower standard error). However, we are limited by computational power: the more folds we choose, the more computation it takes, and hence the more time it takes to run the analysis. So when you do cross-validation, you need to consider the size of the data, the speed of the algorithm (e.g., K-nearest neighbors), and the speed of your computer. In practice, this is a trial-and-error process, but typically C is chosen to be either 5 or 10. Here we will try 10-fold cross-validation to see if we get a lower standard error:

```
cancer_vfold <- vfold_cv(cancer_train, v = 10, strata = Class)

vfold_metrics <- workflow() |>
                  add_recipe(cancer_recipe) |>
                  add_model(knn_spec) |>
                  fit_resamples(resamples = cancer_vfold) |>
                  collect_metrics()

vfold_metrics
```

```
## # A tibble: 2 x 6
##   .metric  .estimator  mean     n std_err .config
##   <chr>    <chr>      <dbl> <int>   <dbl> <chr>
## 1 accuracy binary     0.887    10  0.0207 Preprocessor1_Model1
## 2 roc_auc  binary     0.917    10  0.0134 Preprocessor1_Model1
```

In this case, using 10-fold instead of 5-fold cross validation did reduce the standard error, although by only an insignificant amount. In fact, due to the randomness in how the data are split, sometimes you might even end up with a *higher* standard error when increasing the number of folds! We can make the reduction in standard error more dramatic by increasing the number of folds by a large amount. In the following code we show the result when $C = 50$;

picking such a large number of folds often takes a long time to run in practice, so we usually stick to 5 or 10.

```
cancer_vfold_50 <- vfold_cv(cancer_train, v = 50, strata = Class)

vfold_metrics_50 <- workflow() |>
                  add_recipe(cancer_recipe) |>
                  add_model(knn_spec) |>
                  fit_resamples(resamples = cancer_vfold_50) |>
                  collect_metrics()
vfold_metrics_50
```

```
## # A tibble: 2 x 6
##   .metric  .estimator  mean     n std_err .config
##   <chr>    <chr>      <dbl> <int>   <dbl> <chr>
## 1 accuracy binary     0.881    50  0.0158 Preprocessor1_Model1
## 2 roc_auc  binary     0.911    50  0.0154 Preprocessor1_Model1
```

6.6.2 Parameter value selection

Using 5- and 10-fold cross-validation, we have estimated that the prediction accuracy of our classifier is somewhere around 89%. Whether that is good or not depends entirely on the downstream application of the data analysis. In the present situation, we are trying to predict a tumor diagnosis, with expensive, damaging chemo/radiation therapy or patient death as potential consequences of misprediction. Hence, we might like to do better than 89% for this application.

In order to improve our classifier, we have one choice of parameter: the number of neighbors, K. Since cross-validation helps us evaluate the accuracy of our classifier, we can use cross-validation to calculate an accuracy for each value of K in a reasonable range, and then pick the value of K that gives us the best accuracy. The `tidymodels` package collection provides a very simple syntax for tuning models: each parameter in the model to be tuned should be specified as `tune()` in the model specification rather than given a particular value.

```
knn_spec <- nearest_neighbor(weight_func = "rectangular",
                             neighbors = tune()) |>
  set_engine("kknn") |>
  set_mode("classification")
```

Then instead of using `fit` or `fit_resamples`, we will use the `tune_grid` function to fit the model for each value in a range of parameter values. In particular, we first create a data frame with a `neighbors` variable that contains the sequence of values of K to try; below we create the `k_vals` data frame with the `neighbors` variable containing values from 1 to 100 (stepping by 5) using the `seq` function. Then we pass that data frame to the `grid` argument of `tune_grid`.

```
k_vals <- tibble(neighbors = seq(from = 1, to = 100, by = 5))

knn_results <- workflow() |>
  add_recipe(cancer_recipe) |>
  add_model(knn_spec) |>
  tune_grid(resamples = cancer_vfold, grid = k_vals) |>
  collect_metrics()

accuracies <- knn_results |>
  filter(.metric == "accuracy")

accuracies
```

```
## # A tibble: 20 x 7
##    neighbors .metric  .estimator  mean     n std_err .config
##        <dbl> <chr>    <chr>      <dbl> <int>   <dbl> <chr>
##  1         1 accuracy binary     0.850    10  0.0209 Preprocessor1_Model01
##  2         6 accuracy binary     0.868    10  0.0216 Preprocessor1_Model02
##  3        11 accuracy binary     0.873    10  0.0201 Preprocessor1_Model03
##  4        16 accuracy binary     0.880    10  0.0175 Preprocessor1_Model04
##  5        21 accuracy binary     0.882    10  0.0179 Preprocessor1_Model05
##  6        26 accuracy binary     0.887    10  0.0168 Preprocessor1_Model06
##  7        31 accuracy binary     0.887    10  0.0180 Preprocessor1_Model07
##  8        36 accuracy binary     0.884    10  0.0166 Preprocessor1_Model08
##  9        41 accuracy binary     0.892    10  0.0152 Preprocessor1_Model09
## 10        46 accuracy binary     0.889    10  0.0156 Preprocessor1_Model10
## 11        51 accuracy binary     0.889    10  0.0155 Preprocessor1_Model11
## 12        56 accuracy binary     0.889    10  0.0155 Preprocessor1_Model12
## 13        61 accuracy binary     0.882    10  0.0174 Preprocessor1_Model13
## 14        66 accuracy binary     0.887    10  0.0170 Preprocessor1_Model14
## 15        71 accuracy binary     0.882    10  0.0167 Preprocessor1_Model15
## 16        76 accuracy binary     0.882    10  0.0167 Preprocessor1_Model16
## 17        81 accuracy binary     0.877    10  0.0161 Preprocessor1_Model17
## 18        86 accuracy binary     0.875    10  0.0188 Preprocessor1_Model18
## 19        91 accuracy binary     0.882    10  0.0158 Preprocessor1_Model19
## 20        96 accuracy binary     0.871    10  0.0165 Preprocessor1_Model20
```

We can decide which number of neighbors is best by plotting the accuracy versus K, as shown in Figure 6.5.

```
accuracy_vs_k <- ggplot(accuracies, aes(x = neighbors, y = mean)) +
  geom_point() +
  geom_line() +
  labs(x = "Neighbors", y = "Accuracy Estimate") +
  theme(text = element_text(size = 12))

accuracy_vs_k
```

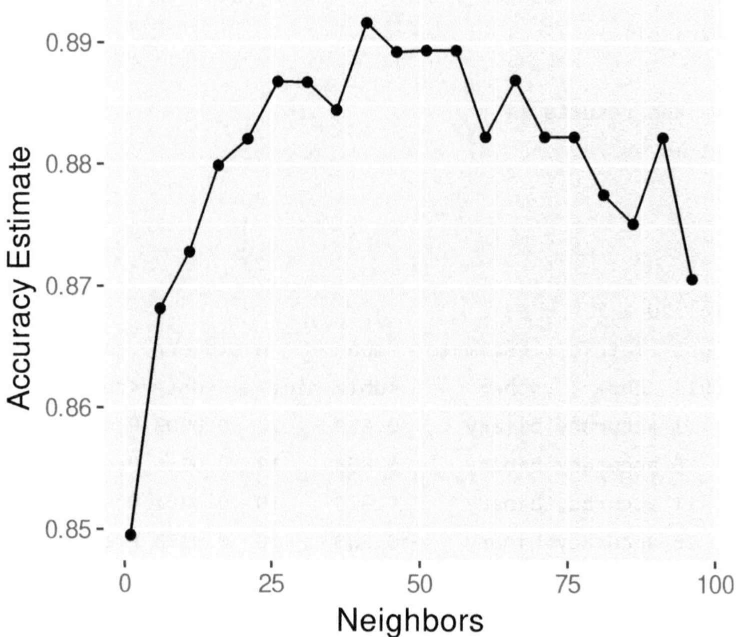

FIGURE 6.5: Plot of estimated accuracy versus the number of neighbors.

Setting the number of neighbors to $K = 41$ provides the highest accuracy (89.16%). But there is no exact or perfect answer here; any selection from $K = 30$ and 60 would be reasonably justified, as all of these differ in classifier accuracy by a small amount. Remember: the values you see on this plot are *estimates* of the true accuracy of our classifier. Although the $K = 41$ value is higher than the others on this plot, that doesn't mean the classifier is actually more accurate with this parameter value! Generally, when selecting K (and other parameters for other predictive models), we are looking for a value where:

- we get roughly optimal accuracy, so that our model will likely be accurate;
- changing the value to a nearby one (e.g., adding or subtracting a small

number) doesn't decrease accuracy too much, so that our choice is reliable in the presence of uncertainty;
- the cost of training the model is not prohibitive (e.g., in our situation, if K is too large, predicting becomes expensive!).

We know that $K = 41$ provides the highest estimated accuracy. Further, Figure 6.5 shows that the estimated accuracy changes by only a small amount if we increase or decrease K near $K = 41$. And finally, $K = 41$ does not create a prohibitively expensive computational cost of training. Considering these three points, we would indeed select $K = 41$ for the classifier.

6.6.3 Under/Overfitting

To build a bit more intuition, what happens if we keep increasing the number of neighbors K? In fact, the accuracy actually starts to decrease! Let's specify a much larger range of values of K to try in the `grid` argument of `tune_grid`. Figure 6.6 shows a plot of estimated accuracy as we vary K from 1 to almost the number of observations in the data set.

```
k_lots <- tibble(neighbors = seq(from = 1, to = 385, by = 10))

knn_results <- workflow() |>
  add_recipe(cancer_recipe) |>
  add_model(knn_spec) |>
  tune_grid(resamples = cancer_vfold, grid = k_lots) |>
  collect_metrics()

accuracies <- knn_results |>
  filter(.metric == "accuracy")

accuracy_vs_k_lots <- ggplot(accuracies, aes(x = neighbors, y = mean)) +
  geom_point() +
  geom_line() +
  labs(x = "Neighbors", y = "Accuracy Estimate") +
  theme(text = element_text(size = 12))

accuracy_vs_k_lots
```

Underfitting: What is actually happening to our classifier that causes this? As we increase the number of neighbors, more and more of the training observations (and those that are farther and farther away from the point) get a "say" in what the class of a new observation is. This causes a sort of "averaging effect" to take place, making the boundary between where our classifier would

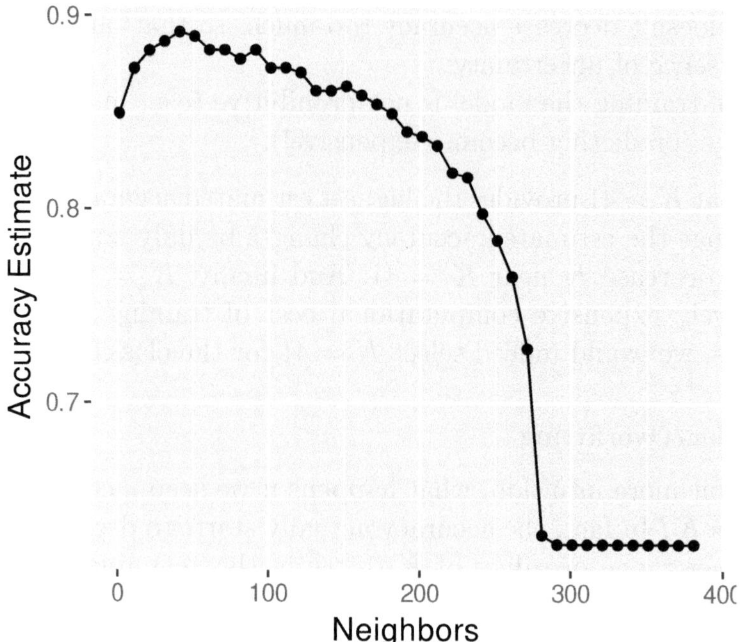

FIGURE 6.6: Plot of accuracy estimate versus number of neighbors for many K values.

predict a tumor to be malignant versus benign to smooth out and become *simpler.* If you take this to the extreme, setting K to the total training data set size, then the classifier will always predict the same label regardless of what the new observation looks like. In general, if the model *isn't influenced enough* by the training data, it is said to **underfit** the data.

Overfitting: In contrast, when we decrease the number of neighbors, each individual data point has a stronger and stronger vote regarding nearby points. Since the data themselves are noisy, this causes a more "jagged" boundary corresponding to a *less simple* model. If you take this case to the extreme, setting $K = 1$, then the classifier is essentially just matching each new observation to its closest neighbor in the training data set. This is just as problematic as the large K case, because the classifier becomes unreliable on new data: if we had a different training set, the predictions would be completely different. In general, if the model *is influenced too much* by the training data, it is said to **overfit** the data.

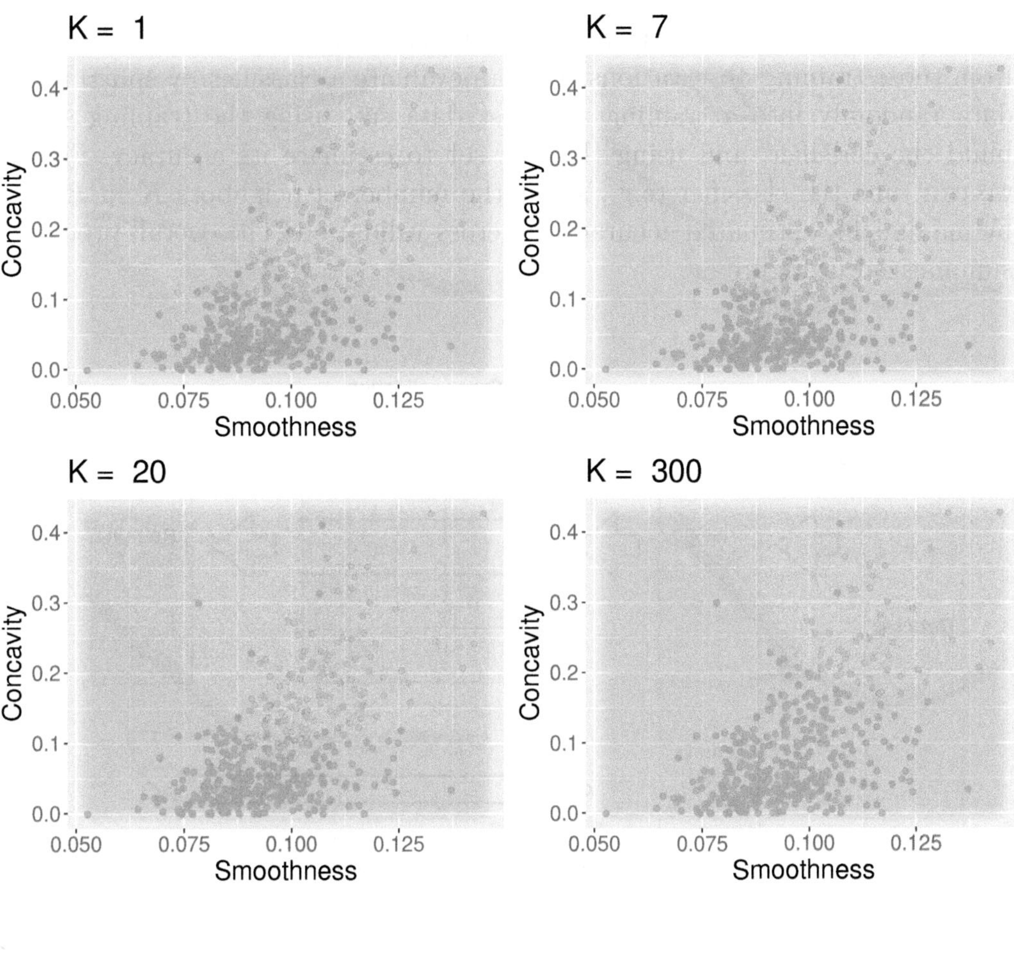

FIGURE 6.7: Effect of K in overfitting and underfitting.

Both overfitting and underfitting are problematic and will lead to a model that does not generalize well to new data. When fitting a model, we need to strike a balance between the two. You can see these two effects in Figure 6.7, which shows how the classifier changes as we set the number of neighbors K to 1, 7, 20, and 300.

6.7 Summary

Classification algorithms use one or more quantitative variables to predict the value of another categorical variable. In particular, the K-nearest neighbors algorithm does this by first finding the K points in the training data nearest to the new observation, and then returning the majority class vote from those training observations. We can evaluate a classifier by splitting the data randomly into a training and test data set, using the training set to build the classifier, and using the test set to estimate its accuracy. Finally, we can tune the classifier (e.g., select the number of neighbors K in K-NN) by maximizing estimated accuracy via cross-validation. The overall process is summarized in Figure 6.8.

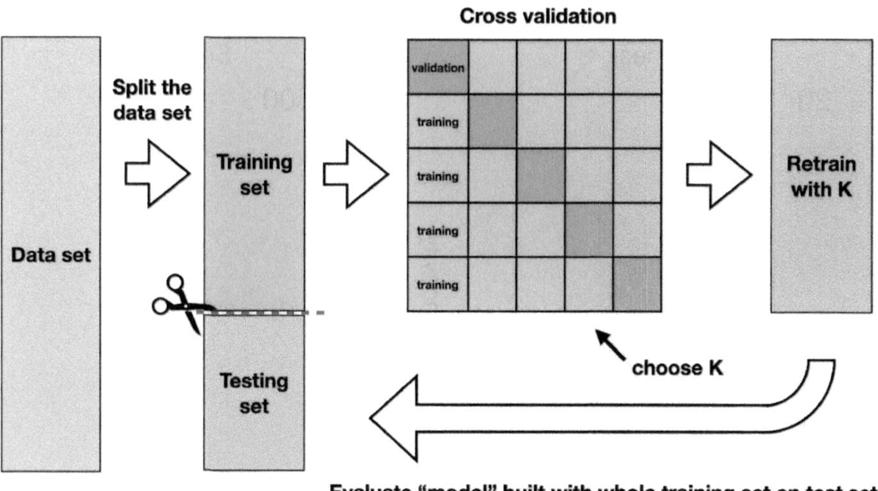

FIGURE 6.8: Overview of KNN classification.

The overall workflow for performing K-nearest neighbors classification using `tidymodels` is as follows:

1. Use the `initial_split` function to split the data into a training and test set. Set the `strata` argument to the class label variable. Put the test set aside for now.
2. Use the `vfold_cv` function to split up the training data for cross-validation.
3. Create a `recipe` that specifies the class label and predictors, as well as preprocessing steps for all variables. Pass the training data as the `data` argument of the recipe.
4. Create a `nearest_neighbors` model specification, with `neighbors = tune()`.
5. Add the recipe and model specification to a `workflow()`, and use the

`tune_grid` function on the train/validation splits to estimate the classifier accuracy for a range of K values.

6. Pick a value of K that yields a high accuracy estimate that doesn't change much if you change K to a nearby value.
7. Make a new model specification for the best parameter value (i.e., K), and retrain the classifier using the `fit` function.
8. Evaluate the estimated accuracy of the classifier on the test set using the `predict` function.

In these last two chapters, we focused on the K-nearest neighbor algorithm, but there are many other methods we could have used to predict a categorical label. All algorithms have their strengths and weaknesses, and we summarize these for the K-NN here.

Strengths: K-nearest neighbors classification

1. is a simple, intuitive algorithm,
2. requires few assumptions about what the data must look like, and
3. works for binary (two-class) and multi-class (more than 2 classes) classification problems.

Weaknesses: K-nearest neighbors classification

1. becomes very slow as the training data gets larger,
2. may not perform well with a large number of predictors, and
3. may not perform well when classes are imbalanced.

6.8 Predictor variable selection

Note: This section is not required reading for the remainder of the textbook. It is included for those readers interested in learning how irrelevant variables can influence the performance of a classifier, and how to pick a subset of useful variables to include as predictors.

Another potentially important part of tuning your classifier is to choose which variables from your data will be treated as predictor variables. Technically, you can choose anything from using a single predictor variable to using every variable in your data; the K-nearest neighbors algorithm accepts any number of predictors. However, it is **not** the case that using more predictors always yields better predictions! In fact, sometimes including irrelevant predictors can actually negatively affect classifier performance.

6.8.1 The effect of irrelevant predictors

Let's take a look at an example where K-nearest neighbors performs worse when given more predictors to work with. In this example, we modified the breast cancer data to have only the `Smoothness`, `Concavity`, and `Perimeter` variables from the original data. Then, we added irrelevant variables that we created ourselves using a random number generator. The irrelevant variables each take a value of 0 or 1 with equal probability for each observation, regardless of what the value `Class` variable takes. In other words, the irrelevant variables have no meaningful relationship with the `Class` variable.

```
cancer_irrelevant |>
      select(Class, Smoothness, Concavity, Perimeter, Irrelevant1, Irrelevant2)
```

```
## # A tibble: 569 x 6
##    Class Smoothness Concavity Perimeter Irrelevant1 Irrelevant2
##    <fct>      <dbl>     <dbl>     <dbl>       <dbl>       <dbl>
##  1 M         0.118     0.300      123.            1           0
##  2 M         0.0847    0.0869     133.            0           0
##  3 M         0.110     0.197      130             0           0
##  4 M         0.142     0.241       77.6           0           1
##  5 M         0.100     0.198      135.            0           0
##  6 M         0.128     0.158       82.6           1           0
##  7 M         0.0946    0.113      120.            0           1
##  8 M         0.119     0.0937      90.2           1           0
##  9 M         0.127     0.186       87.5           0           0
## 10 M         0.119     0.227       84.0           1           1
## # ... with 559 more rows
```

Next, we build a sequence of K-NN classifiers that include `Smoothness`, `Concavity`, and `Perimeter` as predictor variables, but also increasingly many irrelevant variables. In particular, we create 6 data sets with 0, 5, 10, 15, 20, and 40 irrelevant predictors. Then we build a model, tuned via 5-fold cross-validation, for each data set. Figure 6.9 shows the estimated cross-validation accuracy versus the number of irrelevant predictors. As we add more irrelevant predictor variables, the estimated accuracy of our classifier decreases. This is because the irrelevant variables add a random amount to the distance between each pair of observations; the more irrelevant variables there are, the more (random) influence they have, and the more they corrupt the set of nearest neighbors that vote on the class of the new observation to predict.

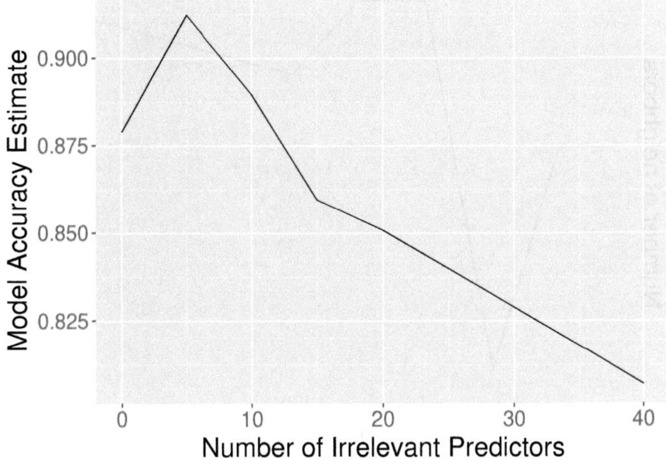

FIGURE 6.9: Effect of inclusion of irrelevant predictors.

Although the accuracy decreases as expected, one surprising thing about Figure 6.9 is that it shows that the method still outperforms the baseline majority classifier (with about 63% accuracy) even with 40 irrelevant variables. How could that be? Figure 6.10 provides the answer: the tuning procedure for the K-nearest neighbors classifier combats the extra randomness from the irrelevant variables by increasing the number of neighbors. Of course, because of all the extra noise in the data from the irrelevant variables, the number of neighbors does not increase smoothly; but the general trend is increasing. Figure 6.11 corroborates this evidence; if we fix the number of neighbors to $K = 3$, the accuracy falls off more quickly.

6.8.2 Finding a good subset of predictors

So then, if it is not ideal to use all of our variables as predictors without consideration, how do we choose which variables we *should* use? A simple method is to rely on your scientific understanding of the data to tell you which variables are not likely to be useful predictors. For example, in the cancer data that we have been studying, the `ID` variable is just a unique identifier for the observation. As it is not related to any measured property of the cells, the `ID` variable should therefore not be used as a predictor. That is, of course, a very clear-cut case. But the decision for the remaining variables is less obvious, as all seem like reasonable candidates. It is not clear which subset of them will create the best classifier. One could use visualizations and other exploratory analyses to try to help understand which variables are potentially relevant, but this process is both time-consuming and error-prone when there are many

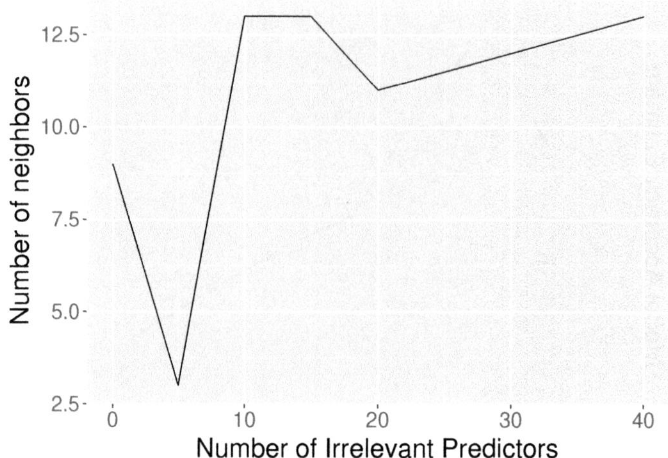

FIGURE 6.10: Tuned number of neighbors for varying number of irrelevant predictors.

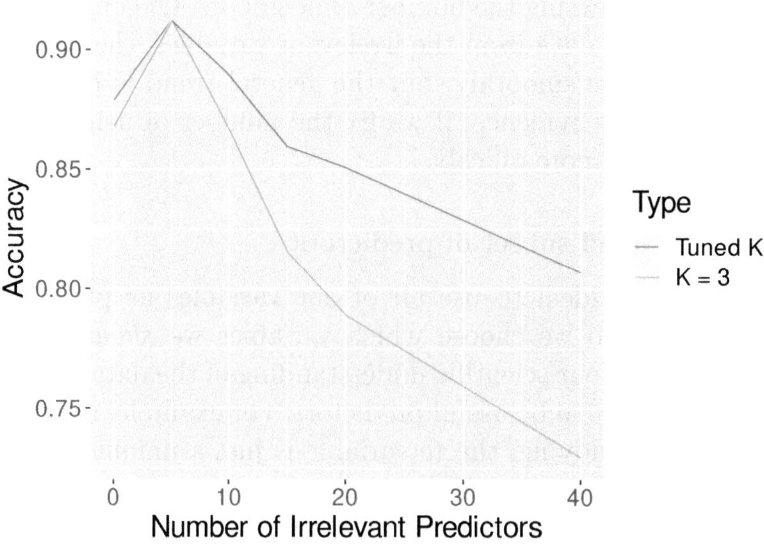

FIGURE 6.11: Accuracy versus number of irrelevant predictors for tuned and untuned number of neighbors.

variables to consider. Therefore we need a more systematic and programmatic way of choosing variables. This is a very difficult problem to solve in general, and there are a number of methods that have been developed that apply in particular cases of interest. Here we will discuss two basic selection methods as an introduction to the topic. See the additional resources at the end of this chapter to find out where you can learn more about variable selection, including more advanced methods.

The first idea you might think of for a systematic way to select predictors is to try all possible subsets of predictors and then pick the set that results in the "best" classifier. This procedure is indeed a well-known variable selection method referred to as *best subset selection* [Beale et al., 1967, Hocking and Leslie, 1967]. In particular, you

1. create a separate model for every possible subset of predictors,
2. tune each one using cross-validation, and
3. pick the subset of predictors that gives you the highest cross-validation accuracy.

Best subset selection is applicable to any classification method (K-NN or otherwise). However, it becomes very slow when you have even a moderate number of predictors to choose from (say, around 10). This is because the number of possible predictor subsets grows very quickly with the number of predictors, and you have to train the model (itself a slow process!) for each one. For example, if we have 2 predictors—let's call them A and B—then we have 3 variable sets to try: A alone, B alone, and finally A and B together. If we have 3 predictors—A, B, and C—then we have 7 to try: A, B, C, AB, BC, AC, and ABC. In general, the number of models we have to train for m predictors is $2^m - 1$; in other words, when we get to 10 predictors we have over *one thousand* models to train, and at 20 predictors we have over *one million* models to train! So although it is a simple method, best subset selection is usually too computationally expensive to use in practice.

Another idea is to iteratively build up a model by adding one predictor variable at a time. This method—known as *forward selection* [Eforymson, 1966, Draper and Smith, 1966]—is also widely applicable and fairly straightforward. It involves the following steps:

1. Start with a model having no predictors.
2. Run the following 3 steps until you run out of predictors:
    1. For each unused predictor, add it to the model to form a *candidate model*.
    2. Tune all of the candidate models.

   3. Update the model to be the candidate model with the highest cross-validation accuracy.
3. Select the model that provides the best trade-off between accuracy and simplicity.

Say you have m total predictors to work with. In the first iteration, you have to make m candidate models, each with 1 predictor. Then in the second iteration, you have to make $m-1$ candidate models, each with 2 predictors (the one you chose before and a new one). This pattern continues for as many iterations as you want. If you run the method all the way until you run out of predictors to choose, you will end up training $\frac{1}{2}m(m+1)$ separate models. This is a *big* improvement from the $2^m - 1$ models that best subset selection requires you to train! For example, while best subset selection requires training over 1000 candidate models with $m = 10$ predictors, forward selection requires training only 55 candidate models. Therefore we will continue the rest of this section using forward selection.

Note: One word of caution before we move on. Every additional model that you train increases the likelihood that you will get unlucky and stumble on a model that has a high cross-validation accuracy estimate, but a low true accuracy on the test data and other future observations. Since forward selection involves training a lot of models, you run a fairly high risk of this happening. To keep this risk low, only use forward selection when you have a large amount of data and a relatively small total number of predictors. More advanced methods do not suffer from this problem as much; see the additional resources at the end of this chapter for where to learn more about advanced predictor selection methods.

6.8.3 Forward selection in R

We now turn to implementing forward selection in R. Unfortunately there is no built-in way to do this using the `tidymodels` framework, so we will have to code it ourselves. First we will use the `select` function to extract the "total" set of predictors that we are willing to work with. Here we will load the modified version of the cancer data with irrelevant predictors, and select `Smoothness`, `Concavity`, `Perimeter`, `Irrelevant1`, `Irrelevant2`, and `Irrelevant3` as potential predictors, and the `Class` variable as the label. We will also extract the column names for the full set of predictor variables.

```
cancer_subset <- cancer_irrelevant |>
  select(Class,
         Smoothness,
         Concavity,
         Perimeter,
         Irrelevant1,
         Irrelevant2,
         Irrelevant3)

names <- colnames(cancer_subset |> select(-Class))

cancer_subset
```

```
## # A tibble: 569 x 7
##    Class Smoothness Concavity Perimeter Irrelevant1 Irrelevant2 Irrelevant3
##    <fct>      <dbl>     <dbl>     <dbl>       <dbl>       <dbl>       <dbl>
##  1 M         0.118     0.300      123.            1           0           1
##  2 M         0.0847    0.0869     133.            0           0           0
##  3 M         0.110     0.197      130             0           0           0
##  4 M         0.142     0.241       77.6           0           1           0
##  5 M         0.100     0.198      135.            0           0           0
##  6 M         0.128     0.158       82.6           1           0           1
##  7 M         0.0946    0.113      120.            0           1           1
##  8 M         0.119     0.0937      90.2           1           0           0
##  9 M         0.127     0.186       87.5           0           0           1
## 10 M         0.119     0.227       84.0           1           1           0
## # ... with 559 more rows
```

The key idea of the forward selection code is to use the `paste` function (which concatenates strings separated by spaces) to create a model formula for each subset of predictors for which we want to build a model. The `collapse` argument tells `paste` what to put between the items in the list; to make a formula, we need to put a `+` symbol between each variable. As an example, let's make a model formula for all the predictors, which should output something like `Class ~ Smoothness + Concavity + Perimeter + Irrelevant1 + Irrelevant2 + Irrelevant3`:

```
example_formula <- paste("Class", "~", paste(names, collapse="+"))
example_formula
```

```
## [1] "Class ~ Smoothness+Concavity+Perimeter+Irrelevant1+Irrelevant2+Irrelevant3"
```

Finally, we need to write some code that performs the task of sequentially finding the best predictor to add to the model. If you recall the end of the wrangling chapter, we mentioned that sometimes one needs more flexible forms of iteration than what we have used earlier, and in these cases one typically resorts to a *for loop*; see the chapter on iteration[2] in *R for Data Science* [Wickham and Grolemund, 2016]. Here we will use two for loops: one over increasing predictor set sizes (where you see `for (i in 1:length(names))` below), and another to check which predictor to add in each round (where you see `for (j in 1:length(names))` below). For each set of predictors to try, we construct a model formula, pass it into a `recipe`, build a `workflow` that tunes a K-NN classifier using 5-fold cross-validation, and finally records the estimated accuracy.

```
# create an empty tibble to store the results
accuracies <- tibble(size = integer(),
                     model_string = character(),
                     accuracy = numeric())

# create a model specification
knn_spec <- nearest_neighbor(weight_func = "rectangular",
                             neighbors = tune()) |>
     set_engine("kknn") |>
     set_mode("classification")

# create a 5-fold cross-validation object
cancer_vfold <- vfold_cv(cancer_subset, v = 5, strata = Class)

# store the total number of predictors
n_total <- length(names)

# stores selected predictors
selected <- c()

# for every size from 1 to the total number of predictors
for (i in 1:n_total) {
    # for every predictor still not added yet
    accs <- list()
    models <- list()
    for (j in 1:length(names)) {
        # create a model string for this combination of predictors
        preds_new <- c(selected, names[[j]])
        model_string <- paste("Class", "~", paste(preds_new, collapse="+"))
```

[2] https://r4ds.had.co.nz/iteration.html

```
    # create a recipe from the model string
    cancer_recipe <- recipe(as.formula(model_string),
                            data = cancer_subset) |>
                     step_scale(all_predictors()) |>
                     step_center(all_predictors())

    # tune the KNN classifier with these predictors,
    # and collect the accuracy for the best K
    acc <- workflow() |>
      add_recipe(cancer_recipe) |>
      add_model(knn_spec) |>
      tune_grid(resamples = cancer_vfold, grid = 10) |>
      collect_metrics() |>
      filter(.metric == "accuracy") |>
      summarize(mx = max(mean))
    acc <- acc$mx |> unlist()

    # add this result to the dataframe
    accs[[j]] <- acc
    models[[j]] <- model_string
  }
  jstar <- which.max(unlist(accs))
  accuracies <- accuracies |>
    add_row(size = i,
            model_string = models[[jstar]],
            accuracy = accs[[jstar]])
  selected <- c(selected, names[[jstar]])
  names <- names[-jstar]
}
accuracies
```

```
## # A tibble: 6 x 3
##    size model_string                                               accuracy
##   <int> <chr>                                                         <dbl>
## 1     1 Class ~ Perimeter                                             0.896
## 2     2 Class ~ Perimeter+Concavity                                   0.916
## 3     3 Class ~ Perimeter+Concavity+Smoothness                        0.931
## 4     4 Class ~ Perimeter+Concavity+Smoothness+Irrelevant1            0.928
## 5     5 Class ~ Perimeter+Concavity+Smoothness+Irrelevant1+Irrelevant3 0.924
## 6     6 Class ~ Perimeter+Concavity+Smoothness+Irrelevant1+Irrelevant3~ 0.902
```

Interesting! The forward selection procedure first added the three meaningful variables `Perimeter`, `Concavity`, and `Smoothness`, followed by the irrelevant variables. Figure 6.12 visualizes the accuracy versus the number of predictors in the model. You can see that as meaningful predictors are added, the estimated accuracy increases substantially; and as you add irrelevant variables, the accuracy either exhibits small fluctuations or decreases as the model attempts to tune the number of neighbors to account for the extra noise. In order to pick the right model from the sequence, you have to balance high accuracy and model simplicity (i.e., having fewer predictors and a lower chance of overfitting). The way to find that balance is to look for the *elbow* in Figure 6.12, i.e., the place on the plot where the accuracy stops increasing dramatically and levels off or begins to decrease. The elbow in Figure 6.12 appears to occur at the model with 3 predictors; after that point the accuracy levels off. So here the right trade-off of accuracy and number of predictors occurs with 3 variables: `Class ~ Perimeter + Concavity + Smoothness`. In other words, we have successfully removed irrelevant predictors from the model! It is always worth remembering, however, that what cross-validation gives you is an *estimate* of the true accuracy; you have to use your judgement when looking at this plot to decide where the elbow occurs, and whether adding a variable provides a meaningful increase in accuracy.

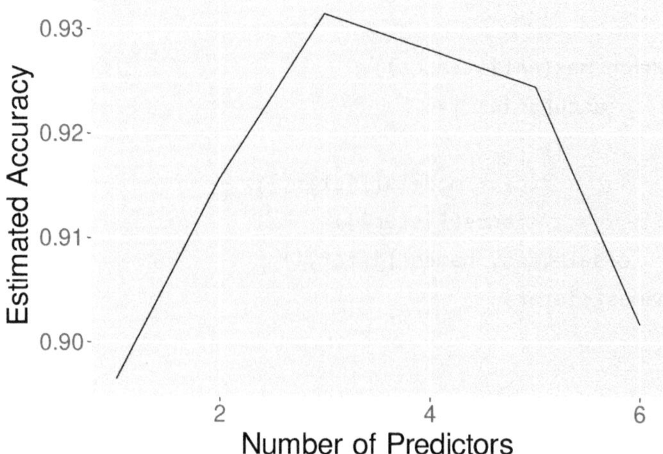

FIGURE 6.12: Estimated accuracy versus the number of predictors for the sequence of models built using forward selection.

Note: Since the choice of which variables to include as predictors is part of tuning your classifier, you *cannot use your test data* for this process!

6.9 Exercises

Practice exercises for the material covered in this chapter can be found in the accompanying worksheets repository[3] in the "Classification II: evaluation and tuning" row. You can launch an interactive version of the worksheet in your browser by clicking the "launch binder" button. You can also preview a non-interactive version of the worksheet by clicking "view worksheet." If you instead decide to download the worksheet and run it on your own machine, make sure to follow the instructions for computer setup found in Chapter 13. This will ensure that the automated feedback and guidance that the worksheets provide will function as intended.

6.10 Additional resources

- The `tidymodels` website[4] is an excellent reference for more details on, and advanced usage of, the functions and packages in the past two chapters. Aside from that, it also has a nice beginner's tutorial[5] and an extensive list of more advanced examples[6] that you can use to continue learning beyond the scope of this book. It's worth noting that the `tidymodels` package does a lot more than just classification, and so the examples on the website similarly go beyond classification as well. In the next two chapters, you'll learn about another kind of predictive modeling setting, so it might be worth visiting the website only after reading through those chapters.
- *An Introduction to Statistical Learning* [James et al., 2013] provides a great next stop in the process of learning about classification. Chapter 4 discusses additional basic techniques for classification that we do not cover, such as logistic regression, linear discriminant analysis, and naive Bayes. Chapter 5 goes into much more detail about cross-validation. Chapters 8 and 9 cover decision trees and support vector machines, two very popular but more advanced classification methods. Finally, Chapter 6 covers a number of methods for selecting predictor variables. Note that while this book is still a very accessible introductory text, it requires a bit more mathematical background than we require.

[3]https://github.com/UBC-DSCI/data-science-a-first-intro-worksheets#readme

[4]https://tidymodels.org/packages

[5]https://www.tidymodels.org/start/

[6]https://www.tidymodels.org/learn/

7

Regression I: K-nearest neighbors

7.1 Overview

This chapter continues our foray into answering predictive questions. Here we will focus on predicting *numerical* variables and will use *regression* to perform this task. This is unlike the past two chapters, which focused on predicting categorical variables via classification. However, regression does have many similarities to classification: for example, just as in the case of classification, we will split our data into training, validation, and test sets, we will use `tidymodels` workflows, we will use a K-nearest neighbors (KNN) approach to make predictions, and we will use cross-validation to choose K. Because of how similar these procedures are, make sure to read Chapters 5 and 6 before reading this one—we will move a little bit faster here with the concepts that have already been covered. This chapter will primarily focus on the case where there is a single predictor, but the end of the chapter shows how to perform regression with more than one predictor variable, i.e., *multivariable regression*. It is important to note that regression can also be used to answer inferential and causal questions, however that is beyond the scope of this book.

7.2 Chapter learning objectives

By the end of the chapter, readers will be able to do the following:

- Recognize situations where a simple regression analysis would be appropriate for making predictions.
- Explain the K-nearest neighbor (KNN) regression algorithm and describe how it differs from KNN classification.
- Interpret the output of a KNN regression.
- In a dataset with two or more variables, perform K-nearest neighbor regression in R using a `tidymodels` workflow.
- Execute cross-validation in R to choose the number of neighbors.

- Evaluate KNN regression prediction accuracy in R using a test data set and the root mean squared prediction error (RMSPE).
- In the context of KNN regression, compare and contrast goodness of fit and prediction properties (namely RMSE vs RMSPE).
- Describe the advantages and disadvantages of K-nearest neighbors regression.

7.3 The regression problem

Regression, like classification, is a predictive problem setting where we want to use past information to predict future observations. But in the case of regression, the goal is to predict *numerical* values instead of *categorical* values. The variable that you want to predict is often called the *response variable*. For example, we could try to use the number of hours a person spends on exercise each week to predict their race time in the annual Boston marathon. As another example, we could try to use the size of a house to predict its sale price. Both of these response variables—race time and sale price—are numerical, and so predicting them given past data is considered a regression problem.

Just like in the classification setting, there are many possible methods that we can use to predict numerical response variables. In this chapter we will focus on the **K-nearest neighbors** algorithm [Fix and Hodges, 1951, Cover and Hart, 1967], and in the next chapter we will study **linear regression**. In your future studies, you might encounter regression trees, splines, and general local regression methods; see the additional resources section at the end of the next chapter for where to begin learning more about these other methods.

Many of the concepts from classification map over to the setting of regression. For example, a regression model predicts a new observation's response variable based on the response variables for similar observations in the data set of past observations. When building a regression model, we first split the data into training and test sets, in order to ensure that we assess the performance of our method on observations not seen during training. And finally, we can use cross-validation to evaluate different choices of model parameters (e.g., K in a K-nearest neighbors model). The major difference is that we are now predicting numerical variables instead of categorical variables.

Note: You can usually tell whether a variable is numerical or categorical—and therefore whether you need to perform regression or classification—by taking two response variables X and Y from your data, and asking the question, "is response variable X *more* than response variable Y?" If the variable is categorical, the question will make no sense. (Is blue more than red? Is benign more than malignant?) If the variable is numerical, it will make sense. (Is 1.5 hours more than 2.25 hours? Is $500,000 more than $400,000?) Be careful when applying this heuristic, though: sometimes categorical variables will be encoded as numbers in your data (e.g., "1" represents "benign", and "0" represents "malignant"). In these cases you have to ask the question about the *meaning* of the labels ("benign" and "malignant"), not their values ("1" and "0").

7.4 Exploring a data set

In this chapter and the next, we will study a data set of 932 real estate transactions in Sacramento, California[1] originally reported in the *Sacramento Bee* newspaper. We first need to formulate a precise question that we want to answer. In this example, our question is again predictive: Can we use the size of a house in the Sacramento, CA area to predict its sale price? A rigorous, quantitative answer to this question might help a realtor advise a client as to whether the price of a particular listing is fair, or perhaps how to set the price of a new listing. We begin the analysis by loading and examining the data, and setting the seed value.

```
library(tidyverse)
library(tidymodels)
library(gridExtra)

set.seed(5)

sacramento <- read_csv("data/sacramento.csv")
sacramento
```

```
## # A tibble: 932 x 9
##    city            zip    beds baths  sqft type       price latitude longitude
##    <chr>           <chr> <dbl> <dbl> <dbl> <chr>      <dbl>    <dbl>     <dbl>
```

[1] https://support.spatialkey.com/spatialkey-sample-csv-data/

```
## 1 SACRAMENTO     z95838     2     1   836 Residential 59222    38.6    -121.
## 2 SACRAMENTO     z95823     3     1  1167 Residential 68212    38.5    -121.
## 3 SACRAMENTO     z95815     2     1   796 Residential 68880    38.6    -121.
## 4 SACRAMENTO     z95815     2     1   852 Residential 69307    38.6    -121.
## 5 SACRAMENTO     z95824     2     1   797 Residential 81900    38.5    -121.
## 6 SACRAMENTO     z95841     3     1  1122 Condo       89921    38.7    -121.
## 7 SACRAMENTO     z95842     3     2  1104 Residential 90895    38.7    -121.
## 8 SACRAMENTO     z95820     3     1  1177 Residential 91002    38.5    -121.
## 9 RANCHO_CORDOVA z95670     2     2   941 Condo       94905    38.6    -121.
## 10 RIO_LINDA     z95673     3     2  1146 Residential 98937    38.7    -121.
## # ... with 922 more rows
```

The scientific question guides our initial exploration: the columns in the data that we are interested in are `sqft` (house size, in livable square feet) and `price` (house sale price, in US dollars (USD)). The first step is to visualize the data as a scatter plot where we place the predictor variable (house size) on the x-axis, and we place the target/response variable that we want to predict (sale price) on the y-axis.

Note: Given that the y-axis unit is dollars in Figure 7.1, we format the axis labels to put dollar signs in front of the house prices, as well as commas to increase the readability of the larger numbers. We can do this in R by passing the `dollar_format` function (from the `scales` package) to the `labels` argument of the `scale_y_continuous` function.

```
eda <- ggplot(sacramento, aes(x = sqft, y = price)) +
  geom_point(alpha = 0.4) +
  xlab("House size (square feet)") +
  ylab("Price (USD)") +
  scale_y_continuous(labels = dollar_format()) +
  theme(text = element_text(size = 12))

eda
```

The plot is shown in Figure 7.1. We can see that in Sacramento, CA, as the size of a house increases, so does its sale price. Thus, we can reason that we may be able to use the size of a not-yet-sold house (for which we don't know the sale price) to predict its final sale price. Note that we do not suggest here that a larger house size *causes* a higher sale price; just that house price tends

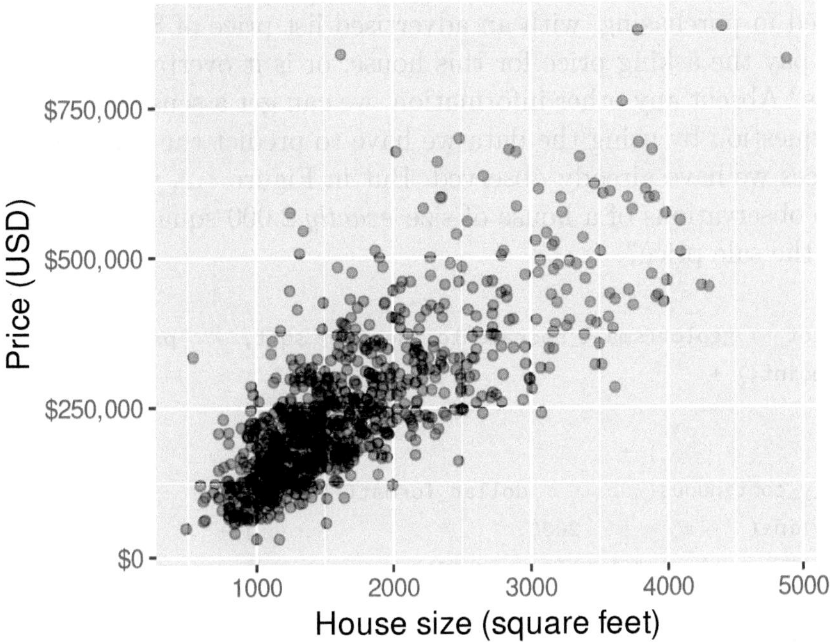

FIGURE 7.1: Scatter plot of price (USD) versus house size (square feet).

to increase with house size, and that we may be able to use the latter to predict the former.

7.5 K-nearest neighbors regression

Much like in the case of classification, we can use a K-nearest neighbors-based approach in regression to make predictions. Let's take a small sample of the data in Figure 7.1 and walk through how K-nearest neighbors (KNN) works in a regression context before we dive in to creating our model and assessing how well it predicts house sale price. This subsample is taken to allow us to illustrate the mechanics of KNN regression with a few data points; later in this chapter we will use all the data.

To take a small random sample of size 30, we'll use the function `slice_sample`, and input the data frame to sample from and the number of rows to randomly select.

```
small_sacramento <- slice_sample(sacramento, n = 30)
```

Next let's say we come across a 2,000 square-foot house in Sacramento we are interested in purchasing, with an advertised list price of $350,000. Should we offer to pay the asking price for this house, or is it overpriced and we should offer less? Absent any other information, we can get a sense for a good answer to this question by using the data we have to predict the sale price given the sale prices we have already observed. But in Figure 7.2, you can see that we have no observations of a house of size *exactly* 2,000 square feet. How can we predict the sale price?

```
small_plot <- ggplot(small_sacramento, aes(x = sqft, y = price)) +
  geom_point() +
  xlab("House size (square feet)") +
  ylab("Price (USD)") +
  scale_y_continuous(labels = dollar_format()) +
  geom_vline(xintercept = 2000, linetype = "dotted") +
  theme(text = element_text(size = 12))

small_plot
```

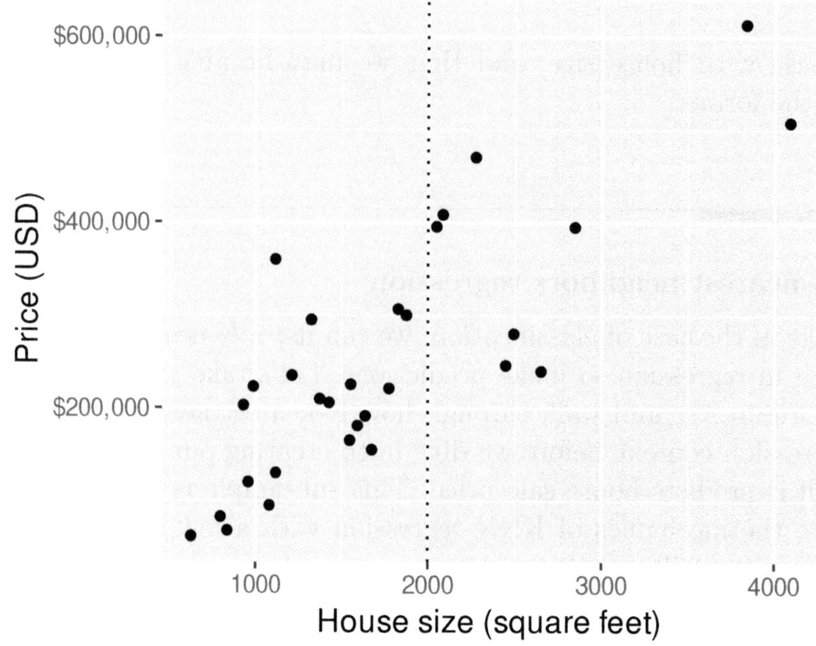

FIGURE 7.2: Scatter plot of price (USD) versus house size (square feet) with vertical line indicating 2,000 square feet on x-axis.

We will employ the same intuition from the classification chapter, and use the neighboring points to the new point of interest to suggest/predict what its sale price might be. For the example shown in Figure 7.2, we find and label the 5 nearest neighbors to our observation of a house that is 2,000 square feet.

```
nearest_neighbors <- small_sacramento |>
  mutate(diff = abs(2000 - sqft)) |>
  arrange(diff) |>
  slice(1:5) #subset the first 5 rows

nearest_neighbors
```

```
## # A tibble: 5 x 10
##   city       zip     beds baths  sqft type          price latitude longitude  diff
##   <chr>      <chr>  <dbl> <dbl> <dbl> <chr>         <dbl>    <dbl>     <dbl> <dbl>
## 1 ROSEVILLE  z95661     3   2    2049 Residenti~ 395500     38.7     -121.    49
## 2 ANTELOPE   z95843     4   3    2085 Residenti~ 408431     38.7     -121.    85
## 3 SACRAMENTO z95823     4   2    1876 Residenti~ 299940     38.5     -121.   124
## 4 ROSEVILLE  z95747     3   2.5  1829 Residenti~ 306500     38.8     -121.   171
## 5 SACRAMENTO z95825     4   2    1776 Multi_Fam~ 221250     38.6     -121.   224
```

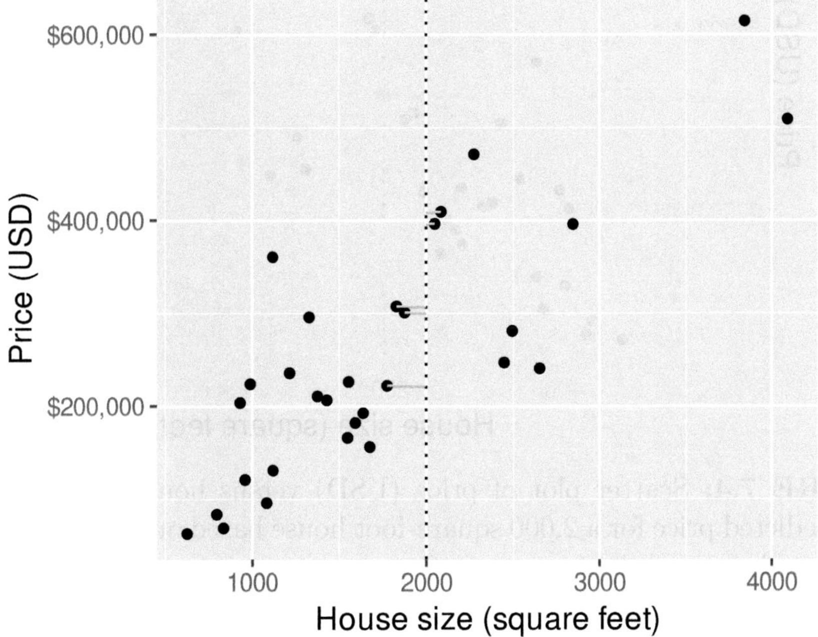

FIGURE 7.3: Scatter plot of price (USD) versus house size (square feet) with lines to 5 nearest neighbors.

Figure 7.3 illustrates the difference between the house sizes of the 5 nearest neighbors (in terms of house size) to our new 2,000 square-foot house of interest. Now that we have obtained these nearest neighbors, we can use their values to predict the sale price for the new home. Specifically, we can take the mean (or average) of these 5 values as our predicted value, as illustrated by the red point in Figure 7.4.

```
prediction <- nearest_neighbors |>
  summarise(predicted = mean(price))

prediction
```

```
## # A tibble: 1 x 1
##   predicted
##       <dbl>
## 1   326324.
```

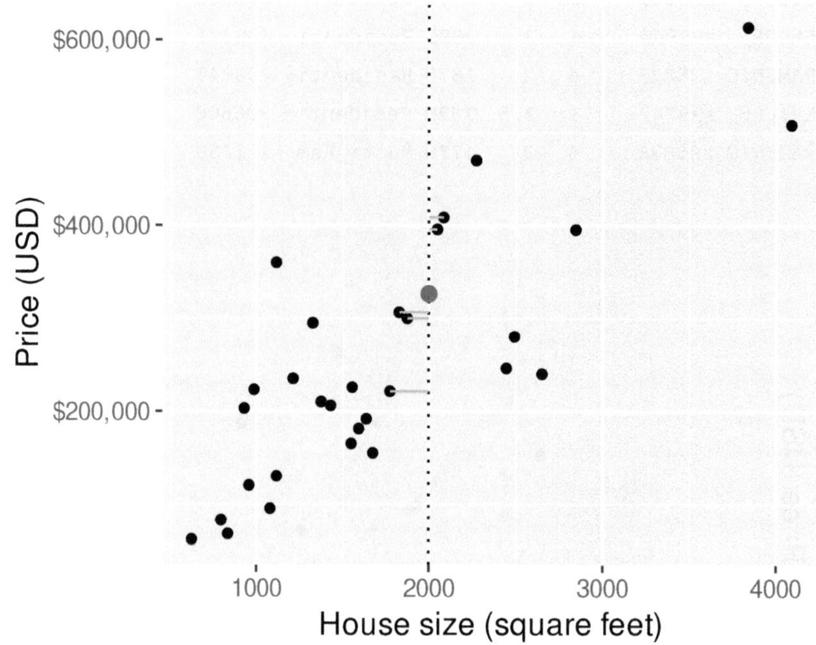

FIGURE 7.4: Scatter plot of price (USD) versus house size (square feet) with predicted price for a 2,000 square-foot house based on 5 nearest neighbors represented as a red dot.

Our predicted price is \$326,324 (shown as a red point in Figure 7.4), which is much less than \$350,000; perhaps we might want to offer less than the list price at which the house is advertised. But this is only the very beginning of the story. We still have all the same unanswered questions here with KNN regression that we had with KNN classification: which K do we choose, and is our model any good at making predictions? In the next few sections, we will address these questions in the context of KNN regression.

One strength of the KNN regression algorithm that we would like to draw attention to at this point is its ability to work well with non-linear relationships (i.e., if the relationship is not a straight line). This stems from the use of nearest neighbors to predict values. The algorithm really has very few assumptions about what the data must look like for it to work.

7.6 Training, evaluating, and tuning the model

As usual, we must start by putting some test data away in a lock box that we will come back to only after we choose our final model. Let's take care of that now. Note that for the remainder of the chapter we'll be working with the entire Sacramento data set, as opposed to the smaller sample of 30 points that we used earlier in the chapter (Figure 7.2).

```
sacramento_split <- initial_split(sacramento, prop = 0.75, strata = price)
sacramento_train <- training(sacramento_split)
sacramento_test <- testing(sacramento_split)
```

Next, we'll use cross-validation to choose K. In KNN classification, we used accuracy to see how well our predictions matched the true labels. We cannot use the same metric in the regression setting, since our predictions will almost never *exactly* match the true response variable values. Therefore in the context of KNN regression we will use root mean square prediction error (RMSPE) instead. The mathematical formula for calculating RMSPE is:

$$\text{RMSPE} = \sqrt{\frac{1}{n}\sum_{i=1}^{n}(y_i - \hat{y}_i)^2}$$

where:

- n is the number of observations,
- y_i is the observed value for the ith observation, and
- $\hat{y}_i$ is the forecasted/predicted value for the ith observation.

In other words, we compute the *squared* difference between the predicted and true response value for each observation in our test (or validation) set, compute the average, and then finally take the square root. The reason we use the *squared* difference (and not just the difference) is that the differences can be positive or negative, i.e., we can overshoot or undershoot the true response value. Figure 7.5 illustrates both positive and negative differences between predicted and true response values. So if we want to measure error—a notion of distance between our predicted and true response values—we want to make sure that we are only adding up positive values, with larger positive values representing larger mistakes. If the predictions are very close to the true values, then RMSPE will be small. If, on the other-hand, the predictions are very different from the true values, then RMSPE will be quite large. When we use cross-validation, we will choose the K that gives us the smallest RMSPE.

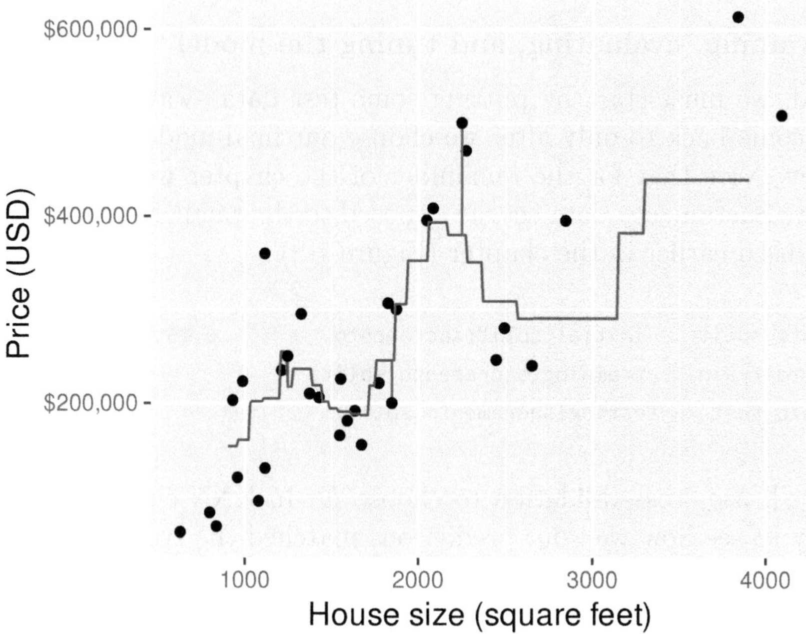

FIGURE 7.5: Scatter plot of price (USD) versus house size (square feet) with example predictions (blue line) and the error in those predictions compared with true response values for three selected observations (vertical red lines).

Note: When using many code packages (`tidymodels` included), the evaluation output we will get to assess the prediction quality of our KNN regression models is labeled "RMSE", or "root mean squared error". Why is this so, and why not RMSPE? In statistics, we try to be very precise with our language to indicate whether we are calculating the prediction error on the training data (*in-sample* prediction) versus on the testing data (*out-of-sample* prediction). When predicting and evaluating prediction quality on the training data, we say RMSE. By contrast, when predicting and evaluating prediction quality on the testing or validation data, we say RMSPE. The equation for calculating RMSE and RMSPE is exactly the same; all that changes is whether the ys are training or testing data. But many people just use RMSE for both, and rely on context to denote which data the root mean squared error is being calculated on.

Now that we know how we can assess how well our model predicts a numerical value, let's use R to perform cross-validation and to choose the optimal K. First, we will create a recipe for preprocessing our data. Note that we include standardization in our preprocessing to build good habits, but since we only have one predictor, it is technically not necessary; there is no risk of comparing two predictors of different scales. Next we create a model specification for K-nearest neighbors regression. Note that we use `set_mode("regression")` now in the model specification to denote a regression problem, as opposed to the classification problems from the previous chapters. The use of `set_mode("regression")` essentially tells `tidymodels` that we need to use different metrics (RMSPE, not accuracy) for tuning and evaluation. Then we create a 5-fold cross-validation object, and put the recipe and model specification together in a workflow.

```
sacr_recipe <- recipe(price ~ sqft, data = sacramento_train) |>
  step_scale(all_predictors()) |>
  step_center(all_predictors())

sacr_spec <- nearest_neighbor(weight_func = "rectangular",
                              neighbors = tune()) |>
  set_engine("kknn") |>
  set_mode("regression")

sacr_vfold <- vfold_cv(sacramento_train, v = 5, strata = price)

sacr_wkflw <- workflow() |>
  add_recipe(sacr_recipe) |>
  add_model(sacr_spec)

sacr_wkflw
```

```
## == Workflow ================================================================
## Preprocessor: Recipe
## Model: nearest_neighbor()
##
## -- Preprocessor ------------------------------------------------------------
-----
## 2 Recipe Steps
##
## * step_scale()
```

```
## * step_center()
##
## -- Model ---------------------------------------------------------------
-----
## K-Nearest Neighbor Model Specification (regression)
##
## Main Arguments:
##   neighbors = tune()
##   weight_func = rectangular
##
## Computational engine: kknn
```

Next we run cross-validation for a grid of numbers of neighbors ranging from 1 to 200. The following code tunes the model and returns the RMSPE for each number of neighbors. In the output of the `sacr_results` results data frame, we see that the `neighbors` variable contains the value of K, the mean (`mean`) contains the value of the RMSPE estimated via cross-validation, and the standard error (`std_err`) contains a value corresponding to a measure of how uncertain we are in the mean value. A detailed treatment of this is beyond the scope of this chapter; but roughly, if your estimated mean is 100,000 and standard error is 1,000, you can expect the *true* RMSPE to be somewhere roughly between 99,000 and 101,000 (although it may fall outside this range). You may ignore the other columns in the metrics data frame, as they do not provide any additional insight.

```
gridvals <- tibble(neighbors = seq(from = 1, to = 200, by = 3))

sacr_results <- sacr_wkflw |>
  tune_grid(resamples = sacr_vfold, grid = gridvals) |>
  collect_metrics() |>
  filter(.metric == "rmse")

# show the results
sacr_results
```

```
## # A tibble: 67 x 7
##    neighbors .metric .estimator    mean     n std_err .config
##        <dbl> <chr>   <chr>        <dbl> <int>   <dbl> <chr>
##  1         1 rmse    standard   113086.     5    996. Preprocessor1_Model01
##  2         4 rmse    standard    93911.     5   3078. Preprocessor1_Model02
##  3         7 rmse    standard    87395.     5   2882. Preprocessor1_Model03
##  4        10 rmse    standard    86203.     5   3014. Preprocessor1_Model04
```

```
## 5          13 rmse    standard    85533.      5   2568. Preprocessor1_Model05
## 6          16 rmse    standard    85707.      5   2279. Preprocessor1_Model06
## 7          19 rmse    standard    85950.      5   2130. Preprocessor1_Model07
## 8          22 rmse    standard    85789.      5   2132. Preprocessor1_Model08
## 9          25 rmse    standard    85373.      5   2284. Preprocessor1_Model09
## 10         28 rmse    standard    85308.      5   2153. Preprocessor1_Model10
## # ... with 57 more rows
```

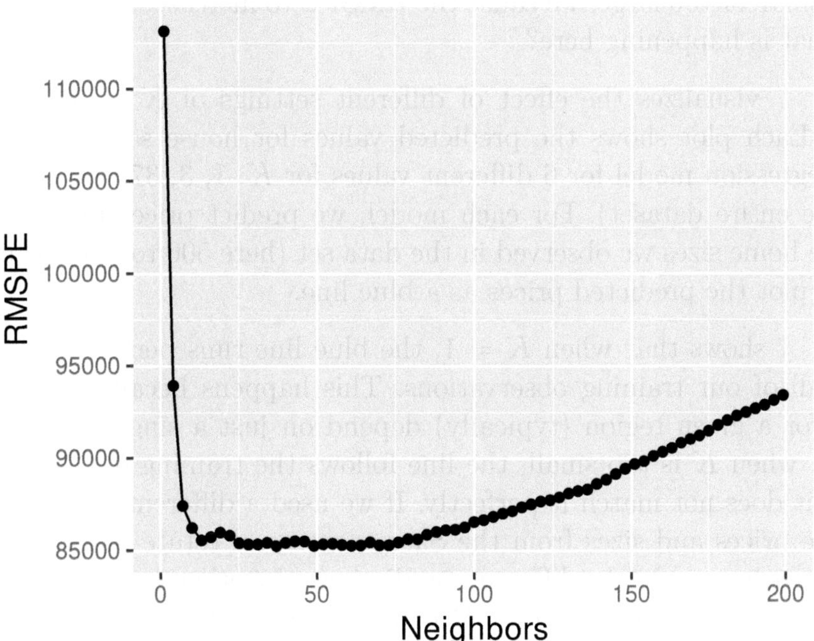

FIGURE 7.6: Effect of the number of neighbors on the RMSPE.

Figure 7.6 visualizes how the RMSPE varies with the number of neighbors K. We take the *minimum* RMSPE to find the best setting for the number of neighbors:

```
# show only the row of minimum RMSPE
sacr_min <- sacr_results |>
  filter(mean == min(mean))

sacr_min
```

```
## # A tibble: 1 x 7
##   neighbors .metric .estimator   mean     n std_err .config
##       <dbl> <chr>   <chr>       <dbl> <int>   <dbl> <chr>
## 1        37 rmse    standard   85227.     5   2177. Preprocessor1_Model13
```

The smallest RMSPE occurs when $K = 37$.

7.7 Underfitting and overfitting

Similar to the setting of classification, by setting the number of neighbors to be too small or too large, we cause the RMSPE to increase, as shown in Figure 7.6. What is happening here?

Figure 7.7 visualizes the effect of different settings of K on the regression model. Each plot shows the predicted values for house sale price from our KNN regression model for 6 different values for K: 1, 3, 37, 41, 250, and 932 (i.e., the entire dataset). For each model, we predict prices for the range of possible home sizes we observed in the data set (here 500 to 5,000 square feet) and we plot the predicted prices as a blue line.

Figure 7.7 shows that when $K = 1$, the blue line runs perfectly through (almost) all of our training observations. This happens because our predicted values for a given region (typically) depend on just a single observation. In general, when K is too small, the line follows the training data quite closely, even if it does not match it perfectly. If we used a different training data set of house prices and sizes from the Sacramento real estate market, we would end up with completely different predictions. In other words, the model is *influenced too much* by the data. Because the model follows the training data so closely, it will not make accurate predictions on new observations which, generally, will not have the same fluctuations as the original training data. Recall from the classification chapters that this behavior—where the model is influenced too much by the noisy data—is called *overfitting*; we use this same term in the context of regression.

What about the plots in Figure 7.7 where K is quite large, say, $K = 250$ or 932? In this case the blue line becomes extremely smooth, and actually becomes flat once K is equal to the number of datapoints in the entire data set. This happens because our predicted values for a given x value (here, home size), depend on many neighboring observations; in the case where K is equal to the size of the dataset, the prediction is just the mean of the house prices in the dataset (completely ignoring the house size). In contrast to the $K = 1$ example, the smooth, inflexible blue line does not follow the training observations very closely. In other words, the model is *not influenced enough* by the training data. Recall from the classification chapters that this behavior is called *underfitting*; we again use this same term in the context of regression.

Ideally, what we want is neither of the two situations discussed above. Instead,

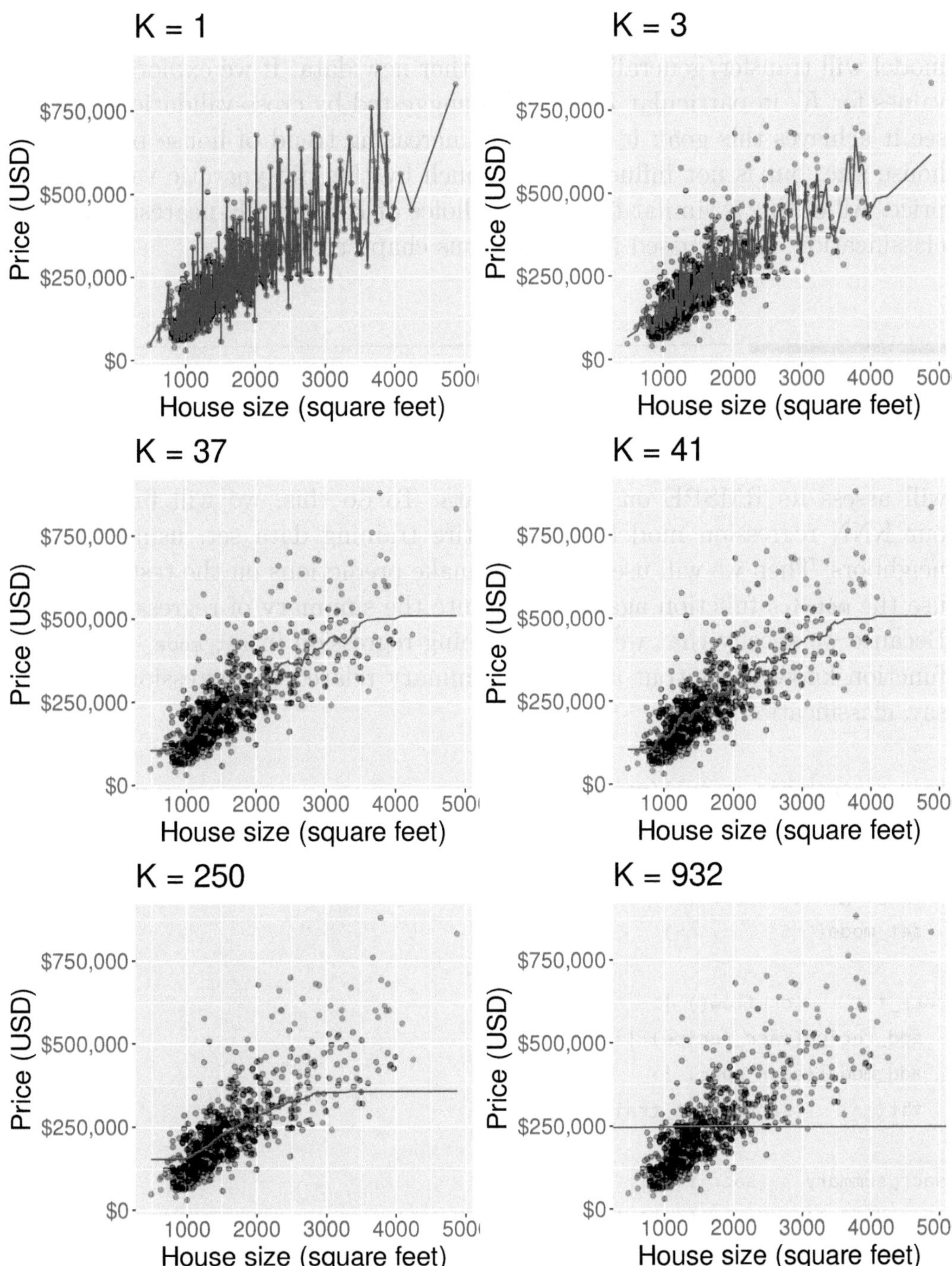

FIGURE 7.7: Predicted values for house price (represented as a blue line) from KNN regression models for six different values for K.

we would like a model that (1) follows the overall "trend" in the training data, so the model actually uses the training data to learn something useful, and (2) does not follow the noisy fluctuations, so that we can be confident that our model will transfer/generalize well to other new data. If we explore the other values for K, in particular $K = 37$ (as suggested by cross-validation), we can see it achieves this goal: it follows the increasing trend of house price versus house size, but is not influenced too much by the idiosyncratic variations in price. All of this is similar to how the choice of K affects K-nearest neighbors classification, as discussed in the previous chapter.

7.8 Evaluating on the test set

To assess how well our model might do at predicting on unseen data, we will assess its RMSPE on the test data. To do this, we will first re-train our KNN regression model on the entire training data set, using $K = 37$ neighbors. Then we will use `predict` to make predictions on the test data, and use the `metrics` function again to compute the summary of regression quality. Because we specify that we are performing regression in `set_mode`, the `metrics` function knows to output a quality summary related to regression, and not, say, classification.

```
kmin <- sacr_min |> pull(neighbors)

sacr_spec <- nearest_neighbor(weight_func = "rectangular", neighbors = kmin) |>
  set_engine("kknn") |>
  set_mode("regression")

sacr_fit <- workflow() |>
  add_recipe(sacr_recipe) |>
  add_model(sacr_spec) |>
  fit(data = sacramento_train)

sacr_summary <- sacr_fit |>
  predict(sacramento_test) |>
  bind_cols(sacramento_test) |>
  metrics(truth = price, estimate = .pred) |>
  filter(.metric == 'rmse')

sacr_summary
```

```
## # A tibble: 1 x 3
##   .metric .estimator .estimate
##   <chr>   <chr>          <dbl>
## 1 rmse    standard      89279.
```

Our final model's test error as assessed by RMSPE is $89,279. Note that RMSPE is measured in the same units as the response variable. In other words, on new observations, we expect the error in our prediction to be *roughly* $89,279. From one perspective, this is good news: this is about the same as the cross-validation RMSPE estimate of our tuned model (which was $85,227), so we can say that the model appears to generalize well to new data that it has never seen before. However, much like in the case of KNN classification, whether this value for RMSPE is *good*—i.e., whether an error of around $89,279 is acceptable—depends entirely on the application. In this application, this error is not prohibitively large, but it is not negligible either; $89,279 might represent a substantial fraction of a home buyer's budget, and could make or break whether or not they could afford put an offer on a house.

Finally, Figure 7.8 shows the predictions that our final model makes across the range of house sizes we might encounter in the Sacramento area—from 500 to 5000 square feet. You have already seen a few plots like this in this chapter, but here we also provide the code that generated it as a learning challenge.

```
sacr_preds <- tibble(sqft = seq(from = 500, to = 5000, by = 10))

sacr_preds <- sacr_fit |>
  predict(sacr_preds) |>
  bind_cols(sacr_preds)

plot_final <- ggplot(sacramento_train, aes(x = sqft, y = price)) +
  geom_point(alpha = 0.4) +
  geom_line(data = sacr_preds,
            mapping = aes(x = sqft, y = .pred),
            color = "blue") +
  xlab("House size (square feet)") +
  ylab("Price (USD)") +
  scale_y_continuous(labels = dollar_format()) +
  ggtitle(paste0("K = ", kmin)) +
  theme(text = element_text(size = 12))

plot_final
```

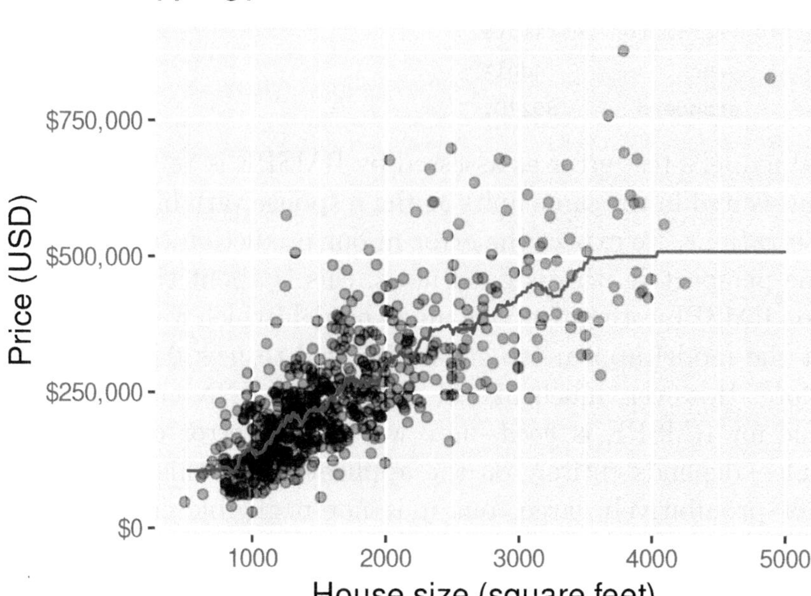

FIGURE 7.8: Predicted values of house price (blue line) for the final KNN regression model.

7.9 Multivariable KNN regression

As in KNN classification, we can use multiple predictors in KNN regression. In this setting, we have the same concerns regarding the scale of the predictors. Once again, predictions are made by identifying the K observations that are nearest to the new point we want to predict; any variables that are on a large scale will have a much larger effect than variables on a small scale. But since the `recipe` we built above scales and centers all predictor variables, this is handled for us.

Note that we also have the same concern regarding the selection of predictors in KNN regression as in KNN classification: having more predictors is **not** always better, and the choice of which predictors to use has a potentially large influence on the quality of predictions. Fortunately, we can use the predictor selection algorithm from the classification chapter in KNN regression as well. As the algorithm is the same, we will not cover it again in this chapter.

We will now demonstrate a multivariable KNN regression analysis of the Sacramento real estate data using `tidymodels`. This time we will use house size (measured in square feet) as well as number of bedrooms as our predictors, and continue to use house sale price as our outcome/target variable that we are trying to predict. It is always a good practice to do exploratory data analysis, such as visualizing the data, before we start modeling the data. Figure 7.9 shows that the number of bedrooms might provide useful information to help predict the sale price of a house.

```
plot_beds <- sacramento |>
  ggplot(aes(x = beds, y = price)) +
  geom_point(alpha = 0.4) +
  labs(x = 'Number of Bedrooms', y = 'Price (USD)') +
  theme(text = element_text(size = 12))
plot_beds
```

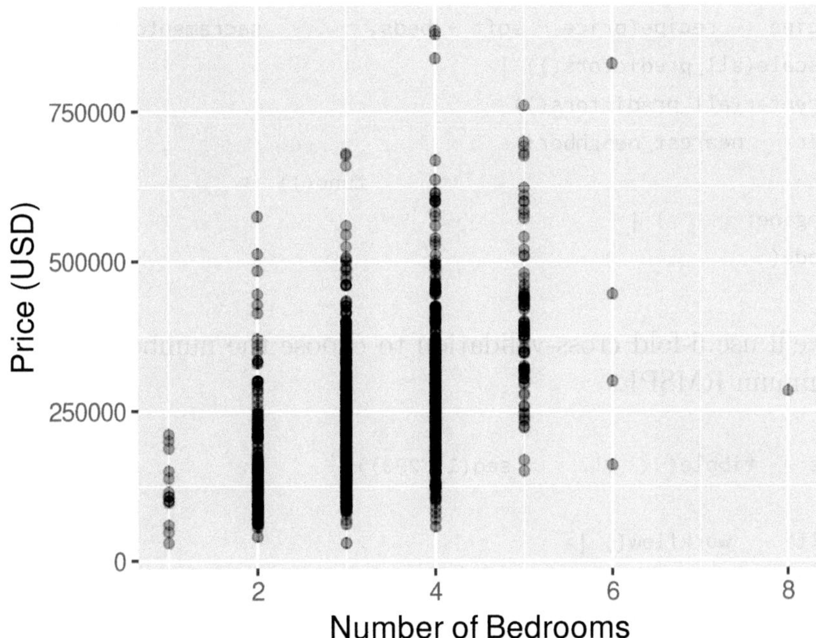

FIGURE 7.9: Scatter plot of the sale price of houses versus the number of bedrooms.

Figure 7.9 shows that as the number of bedrooms increases, the house sale price tends to increase as well, but that the relationship is quite weak. Does adding the number of bedrooms to our model improve our ability to predict price? To answer that question, we will have to create a new KNN regression model using house size and number of bedrooms, and then we can compare it to the model we previously came up with that only used house size. Let's do that now!

First we'll build a new model specification and recipe for the analysis. Note that we use the formula `price ~ sqft + beds` to denote that we have two predictors, and set `neighbors = tune()` to tell `tidymodels` to tune the number of neighbors for us.

```
sacr_recipe <- recipe(price ~ sqft + beds, data = sacramento_train) |>
  step_scale(all_predictors()) |>
  step_center(all_predictors())
sacr_spec <- nearest_neighbor(weight_func = "rectangular",
                              neighbors = tune()) |>
  set_engine("kknn") |>
  set_mode("regression")
```

Next, we'll use 5-fold cross-validation to choose the number of neighbors via the minimum RMSPE:

```
gridvals <- tibble(neighbors = seq(1, 200))

sacr_multi <- workflow() |>
  add_recipe(sacr_recipe) |>
  add_model(sacr_spec) |>
  tune_grid(sacr_vfold, grid = gridvals) |>
  collect_metrics() |>
  filter(.metric == "rmse") |>
  filter(mean == min(mean))

sacr_k <- sacr_multi |>
            pull(neighbors)
sacr_multi
```

```
## # A tibble: 1 x 7
##   neighbors .metric .estimator   mean     n std_err .config
##       <int> <chr>   <chr>       <dbl> <int>   <dbl> <chr>
## 1        12 rmse    standard   82648.     5   2365. Preprocessor1_Model012
```

Here we see that the smallest estimated RMSPE from cross-validation occurs when $K = 12$. If we want to compare this multivariable KNN regression model to the model with only a single predictor *as part of the model tuning process* (e.g., if we are running forward selection as described in the chapter on evaluating and tuning classification models), then we must compare the accuracy estimated using only the training data via cross-validation. Looking back, the estimated cross-validation accuracy for the single-predictor model was 85,227. The estimated cross-validation accuracy for the multivariable model is 82,648. Thus in this case, we did not improve the model by a large amount by adding this additional predictor.

Regardless, let's continue the analysis to see how we can make predictions with a multivariable KNN regression model and evaluate its performance on test data. We first need to re-train the model on the entire training data set with $K = 12$, and then use that model to make predictions on the test data.

```
sacr_spec <- nearest_neighbor(weight_func = "rectangular",
                              neighbors = sacr_k) |>
  set_engine("kknn") |>
  set_mode("regression")

knn_mult_fit <- workflow() |>
  add_recipe(sacr_recipe) |>
  add_model(sacr_spec) |>
  fit(data = sacramento_train)

knn_mult_preds <- knn_mult_fit |>
  predict(sacramento_test) |>
  bind_cols(sacramento_test)

knn_mult_mets <- metrics(knn_mult_preds, truth = price, estimate = .pred) |>
                     filter(.metric == 'rmse')

knn_mult_mets
```

```
## # A tibble: 1 x 3
##   .metric .estimator .estimate
##   <chr>   <chr>          <dbl>
## 1 rmse    standard      90953.
```

This time, when we performed KNN regression on the same data set, but also included number of bedrooms as a predictor, we obtained a RMSPE test error of 90,953. Figure 7.10 visualizes the model's predictions overlaid on top of the data. This time the predictions are a surface in 3D space, instead of a line in 2D space, as we have 2 predictors instead of 1.

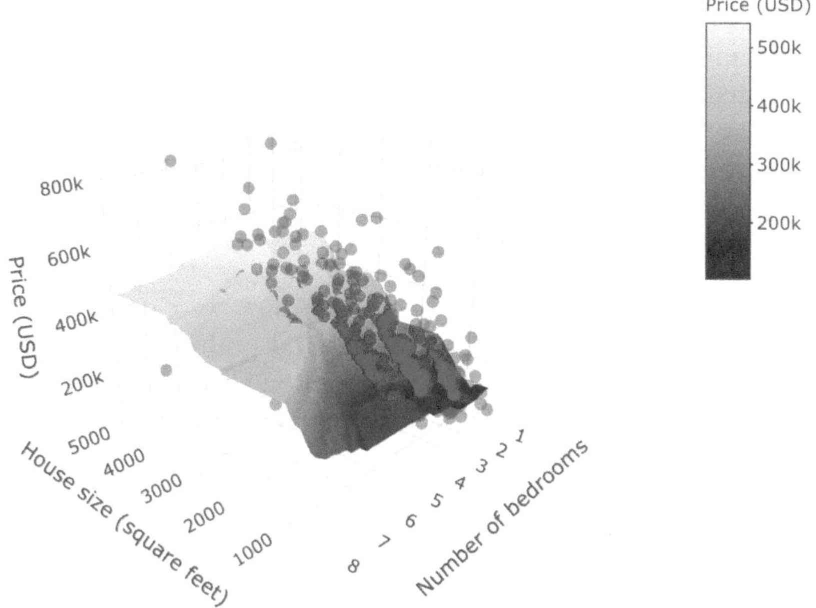

FIGURE 7.10: KNN regression model's predictions represented as a surface in 3D space overlaid on top of the data using three predictors (price, house size, and the number of bedrooms). Note that in general we recommend against using 3D visualizations; here we use a 3D visualization only to illustrate what the surface of predictions looks like for learning purposes.

We can see that the predictions in this case, where we have 2 predictors, form a surface instead of a line. Because the newly added predictor (number of bedrooms) is related to price (as price changes, so does number of bedrooms) and is not totally determined by house size (our other predictor), we get additional and useful information for making our predictions. For example, in this model we would predict that the cost of a house with a size of 2,500 square feet generally increases slightly as the number of bedrooms increases. Without having the additional predictor of number of bedrooms, we would predict the same price for these two houses.

7.10 Strengths and limitations of KNN regression

As with KNN classification (or any prediction algorithm for that matter), KNN regression has both strengths and weaknesses. Some are listed here:

Strengths: K-nearest neighbors regression

1. is a simple, intuitive algorithm,
2. requires few assumptions about what the data must look like, and
3. works well with non-linear relationships (i.e., if the relationship is not a straight line).

Weaknesses: K-nearest neighbors regression

1. becomes very slow as the training data gets larger,
2. may not perform well with a large number of predictors, and
3. may not predict well beyond the range of values input in your training data.

7.11 Exercises

Practice exercises for the material covered in this chapter can be found in the accompanying worksheets repository[2] in the "Regression I: K-nearest neighbors" row. You can launch an interactive version of the worksheet in your browser by clicking the "launch binder" button. You can also preview a non-interactive version of the worksheet by clicking "view worksheet." If you instead decide to download the worksheet and run it on your own machine, make sure to follow the instructions for computer setup found in Chapter 13. This will ensure that the automated feedback and guidance that the worksheets provide will function as intended.

[2] https://github.com/UBC-DSCI/data-science-a-first-intro-worksheets#readme

8

Regression II: linear regression

8.1 Overview

Up to this point, we have solved all of our predictive problems—both classification and regression—using K-nearest neighbors (KNN)-based approaches. In the context of regression, there is another commonly used method known as *linear regression*. This chapter provides an introduction to the basic concept of linear regression, shows how to use `tidymodels` to perform linear regression in R, and characterizes its strengths and weaknesses compared to KNN regression. The focus is, as usual, on the case where there is a single predictor and single response variable of interest; but the chapter concludes with an example using *multivariable linear regression* when there is more than one predictor.

8.2 Chapter learning objectives

By the end of the chapter, readers will be able to do the following:

- Use R and `tidymodels` to fit a linear regression model on training data.
- Evaluate the linear regression model on test data.
- Compare and contrast predictions obtained from K-nearest neighbor regression to those obtained using linear regression from the same data set.
- In R, overlay predictions from linear regression on a scatter plot of data using `geom_smooth`.

8.3 Simple linear regression

At the end of the previous chapter, we noted some limitations of KNN regression. While the method is simple and easy to understand, KNN regression does not predict well beyond the range of the predictors in the training data, and the method gets significantly slower as the training data set grows.

Fortunately, there is an alternative to KNN regression—*linear regression*—that addresses both of these limitations. Linear regression is also very commonly used in practice because it provides an interpretable mathematical equation that describes the relationship between the predictor and response variables. In this first part of the chapter, we will focus on *simple* linear regression, which involves only one predictor variable and one response variable; later on, we will consider *multivariable* linear regression, which involves multiple predictor variables. Like KNN regression, simple linear regression involves predicting a numerical response variable (like race time, house price, or height); but *how* it makes those predictions for a new observation is quite different from KNN regression. Instead of looking at the K nearest neighbors and averaging over their values for a prediction, in simple linear regression, we create a straight line of best fit through the training data and then "look up" the prediction using the line.

Note: Although we did not cover it in earlier chapters, there is another popular method for classification called *logistic regression* (it is used for classification even though the name, somewhat confusingly, has the word "regression" in it). In logistic regression—similar to linear regression—you "fit" the model to the training data and then "look up" the prediction for each new observation. Logistic regression and KNN classification have an advantage/disadvantage comparison similar to that of linear regression and KNN regression. It is useful to have a good understanding of linear regression before learning about logistic regression. After reading this chapter, see the "Additional Resources" section at the end of the classification chapters to learn more about logistic regression.

Let's return to the Sacramento housing data from Chapter 7 to learn how to apply linear regression and compare it to KNN regression. For now, we will consider a smaller version of the housing data to help make our visualizations clear. Recall our predictive question: can we use the size of a house in the Sacramento, CA area to predict its sale price? In particular, recall that we have come across a new 2,000 square-foot house we are interested in purchasing with an advertised list price of $350,000. Should we offer the list price, or is that over/undervalued? To answer this question using simple linear regression, we use the data we have to draw the straight line of best fit through our existing data points. The small subset of data as well as the line of best fit are shown in Figure 8.1.

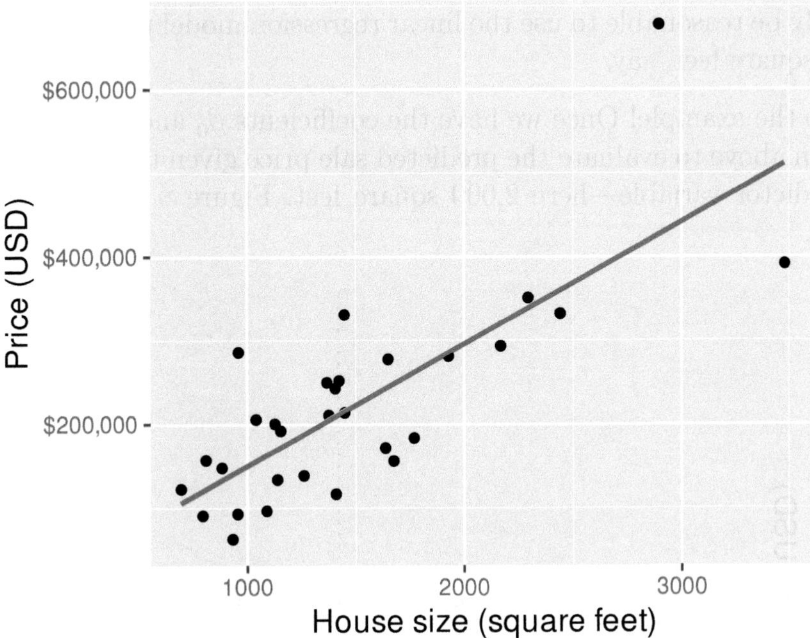

FIGURE 8.1: Scatter plot of sale price versus size with line of best fit for subset of the Sacramento housing data.

The equation for the straight line is:

$$\text{house sale price} = \beta_0 + \beta_1 \cdot (\text{house size}),$$

where

- β_0 is the *vertical intercept* of the line (the price when house size is 0)
- β_1 is the *slope* of the line (how quickly the price increases as you increase house size)

Therefore using the data to find the line of best fit is equivalent to finding coefficients β_0 and β_1 that *parametrize* (correspond to) the line of best fit. Now of course, in this particular problem, the idea of a 0 square-foot house is a bit silly; but you can think of β_0 here as the "base price," and β_1 as the increase in price for each square foot of space. Let's push this thought even further: what would happen in the equation for the line if you tried to evaluate the price of a house with size 6 *million* square feet? Or what about *negative* 2,000 square feet? As it turns out, nothing in the formula breaks; linear regression will happily make predictions for nonsensical predictor values if you ask it to. But even though you *can* make these wild predictions, you shouldn't. You should only make predictions roughly within the range of your original data, and perhaps a bit beyond it only if it makes sense. For example, the data in Figure 8.1 only reaches around 800 square feet on the low end, but it would probably be reasonable to use the linear regression model to make a prediction at 600 square feet, say.

Back to the example! Once we have the coefficients β_0 and β_1, we can use the equation above to evaluate the predicted sale price given the value we have for the predictor variable—here 2,000 square feet. Figure 8.2 demonstrates this process.

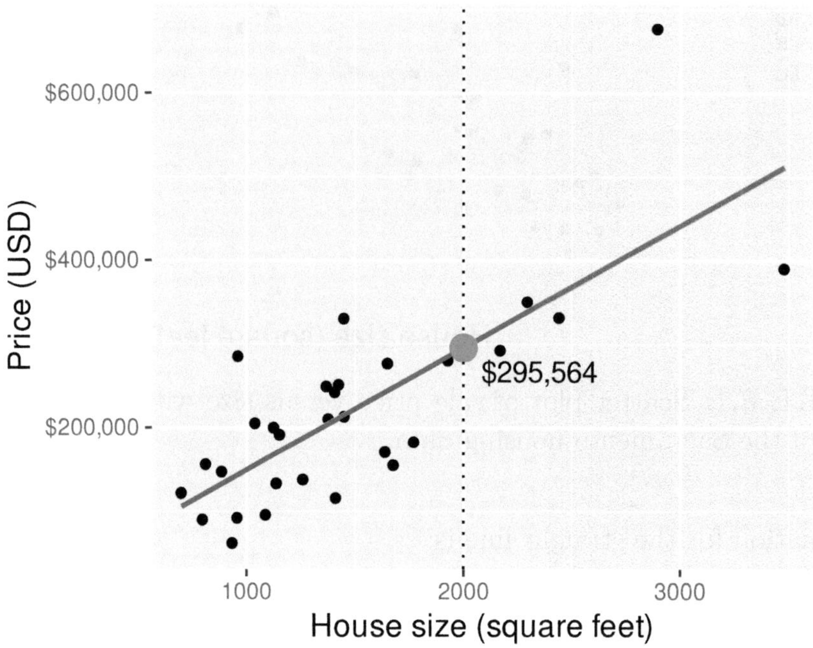

FIGURE 8.2: Scatter plot of sale price versus size with line of best fit and a red dot at the predicted sale price for a 2,000 square-foot home.

By using simple linear regression on this small data set to predict the sale price for a 2,000 square-foot house, we get a predicted value of \$295,564. But wait a minute... how exactly does simple linear regression choose the line of best fit? Many different lines could be drawn through the data points. Some plausible examples are shown in Figure 8.3.

Simple linear regression chooses the straight line of best fit by choosing the line that minimizes the **average squared vertical distance** between itself and each of the observed data points in the training data. Figure 8.4 illustrates these vertical distances as red lines. Finally, to assess the predictive accuracy of a simple linear regression model, we use RMSPE—the same measure of predictive performance we used with KNN regression.

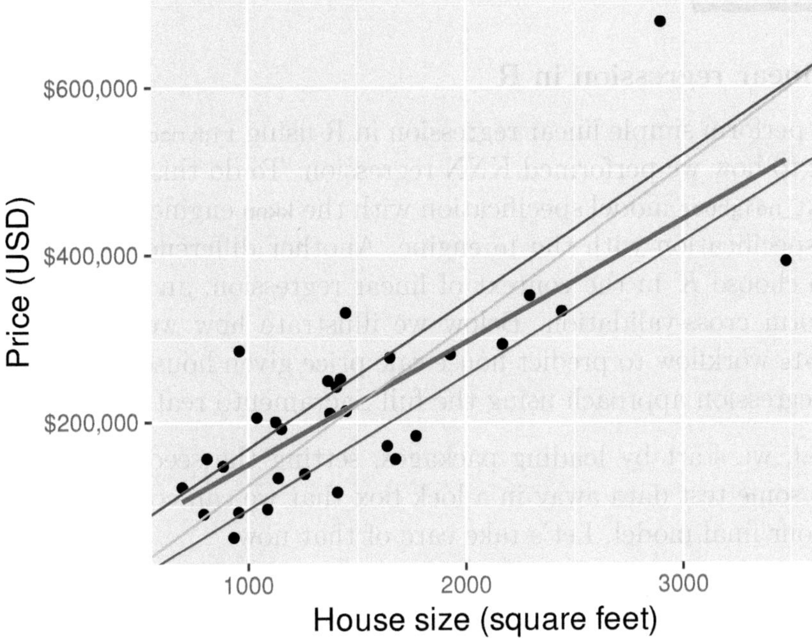

FIGURE 8.3: Scatter plot of sale price versus size with many possible lines that could be drawn through the data points.

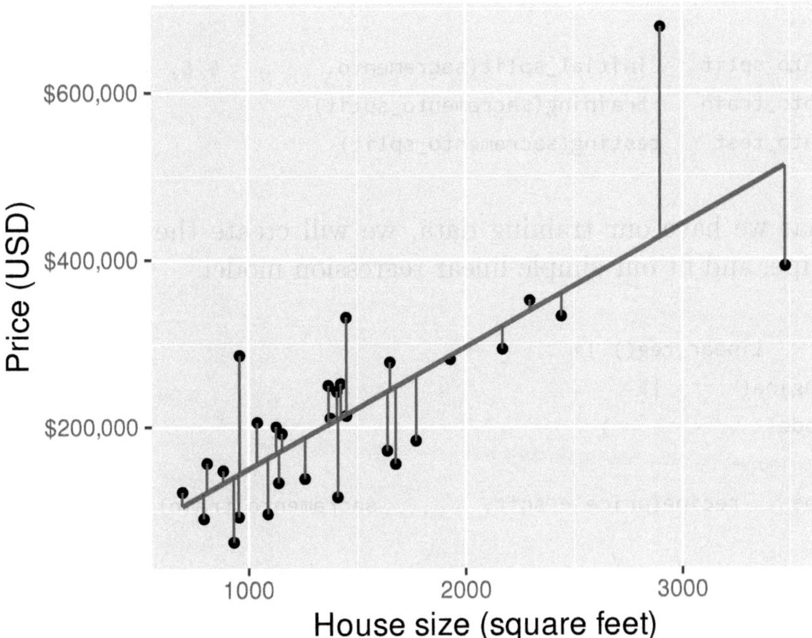

FIGURE 8.4: Scatter plot of sale price versus size with red lines denoting the vertical distances between the predicted values and the observed data points.

8.4 Linear regression in R

We can perform simple linear regression in R using `tidymodels` in a very similar manner to how we performed KNN regression. To do this, instead of creating a `nearest_neighbor` model specification with the `kknn` engine, we use a `linear_reg` model specification with the `lm` engine. Another difference is that we do not need to choose K in the context of linear regression, and so we do not need to perform cross-validation. Below we illustrate how we can use the usual `tidymodels` workflow to predict house sale price given house size using a simple linear regression approach using the full Sacramento real estate data set.

As usual, we start by loading packages, setting the seed, loading data, and putting some test data away in a lock box that we can come back to after we choose our final model. Let's take care of that now.

```
library(tidyverse)
library(tidymodels)

set.seed(1234)

sacramento <- read_csv("data/sacramento.csv")

sacramento_split <- initial_split(sacramento, prop = 0.6, strata = price)
sacramento_train <- training(sacramento_split)
sacramento_test <- testing(sacramento_split)
```

Now that we have our training data, we will create the model specification and recipe, and fit our simple linear regression model:

```
lm_spec <- linear_reg() |>
  set_engine("lm") |>
  set_mode("regression")

lm_recipe <- recipe(price ~ sqft, data = sacramento_train)

lm_fit <- workflow() |>
  add_recipe(lm_recipe) |>
  add_model(lm_spec) |>
  fit(data = sacramento_train)
```

```
lm_fit
```

```
## == Workflow [trained] ==========================================================
## Preprocessor: Recipe
## Model: linear_reg()
##
## -- Preprocessor ------------------------------------------------------------
-----
## 0 Recipe Steps
##
## -- Model --------------------------------------------------------------------
-----
##
## Call:
## stats::lm(formula = ..y ~ ., data = data)
##
## Coefficients:
## (Intercept)         sqft
##     12292.1        139.9
```

Note: An additional difference that you will notice here is that we do not standardize (i.e., scale and center) our predictors. In K-nearest neighbors models, recall that the model fit changes depending on whether we standardize first or not. In linear regression, standardization does not affect the fit (it *does* affect the coefficients in the equation, though!). So you can standardize if you want—it won't hurt anything—but if you leave the predictors in their original form, the best fit coefficients are usually easier to interpret afterward.

Our coefficients are (intercept) $\beta_0 = 12292$ and (slope) $\beta_1 = 140$. This means that the equation of the line of best fit is

$$\text{house sale price} = 12292 + 140 \cdot (\text{house size}).$$

In other words, the model predicts that houses start at $12,292 for 0 square feet, and that every extra square foot increases the cost of the house by $140. Finally, we predict on the test data set to assess how well our model does:

```
lm_test_results <- lm_fit |>
  predict(sacramento_test) |>
  bind_cols(sacramento_test) |>
  metrics(truth = price, estimate = .pred)

lm_test_results
```

```
## # A tibble: 3 x 3
##   .metric .estimator .estimate
##   <chr>   <chr>          <dbl>
## 1 rmse    standard   82342.
## 2 rsq     standard       0.596
## 3 mae     standard   60555.
```

Our final model's test error as assessed by RMSPE is 82,342. Remember that this is in units of the target/response variable, and here that is US Dollars (USD). Does this mean our model is "good" at predicting house sale price based off of the predictor of home size? Again, answering this is tricky and requires knowledge of how you intend to use the prediction.

To visualize the simple linear regression model, we can plot the predicted house sale price across all possible house sizes we might encounter superimposed on a scatter plot of the original housing price data. There is a plotting function in the `tidyverse`, `geom_smooth`, that allows us to add a layer on our plot with the simple linear regression predicted line of best fit. By default `geom_smooth` adds some other information to the plot that we are not interested in at this point; we provide the argument `se = FALSE` to tell `geom_smooth` not to show that information. Figure 8.5 displays the result.

```
lm_plot_final <- ggplot(sacramento_train, aes(x = sqft, y = price)) +
  geom_point(alpha = 0.4) +
  xlab("House size (square feet)") +
  ylab("Price (USD)") +
  scale_y_continuous(labels = dollar_format()) +
  geom_smooth(method = "lm", se = FALSE) +
  theme(text = element_text(size = 12))

lm_plot_final
```

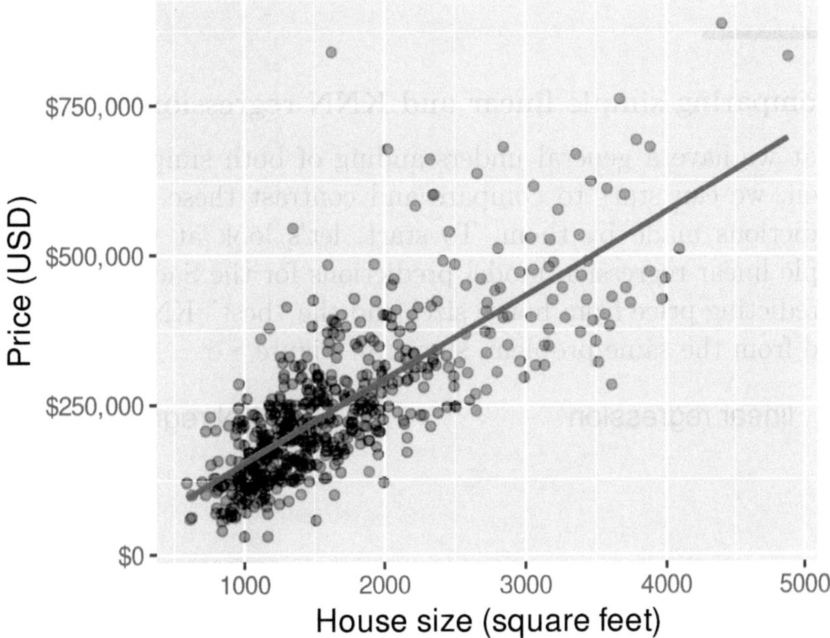

FIGURE 8.5: Scatter plot of sale price versus size with line of best fit for the full Sacramento housing data.

We can extract the coefficients from our model by accessing the fit object that is output by the `fit` function; we first have to extract it from the workflow using the `pull_workflow_fit` function, and then apply the `tidy` function to convert the result into a data frame:

```
coeffs <- lm_fit |>
             pull_workflow_fit() |>
             tidy()

coeffs
```

```
## # A tibble: 2 x 5
##   term        estimate std.error statistic   p.value
##   <chr>          <dbl>     <dbl>     <dbl>     <dbl>
## 1 (Intercept)   12292.   9141.        1.34 1.79e-  1
## 2 sqft            140.      5.00     28.0  4.06e-108
```

8.5 Comparing simple linear and KNN regression

Now that we have a general understanding of both simple linear and KNN regression, we can start to compare and contrast these methods as well as the predictions made by them. To start, let's look at the visualization of the simple linear regression model predictions for the Sacramento real estate data (predicting price from house size) and the "best" KNN regression model obtained from the same problem, shown in Figure 8.6.

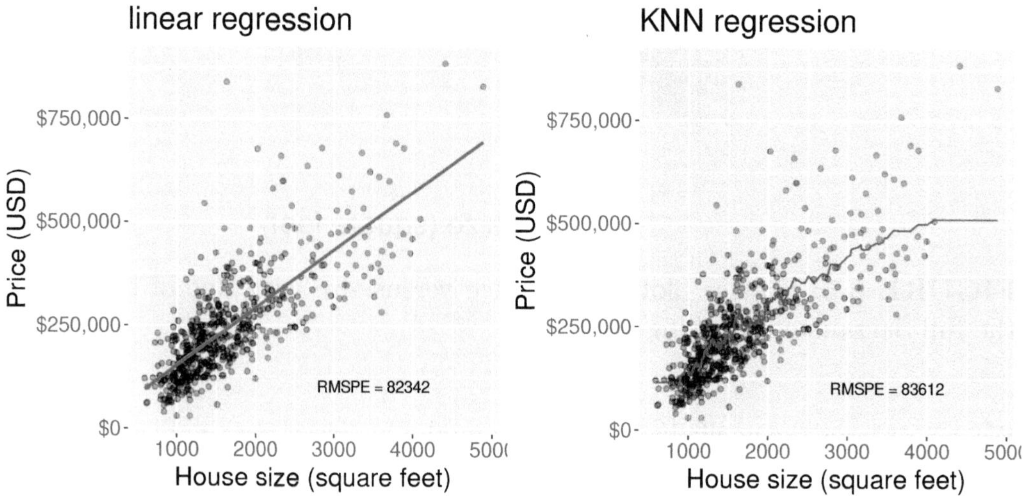

FIGURE 8.6: Comparison of simple linear regression and KNN regression.

What differences do we observe in Figure 8.6? One obvious difference is the shape of the blue lines. In simple linear regression we are restricted to a straight line, whereas in KNN regression our line is much more flexible and can be quite wiggly. But there is a major interpretability advantage in limiting the model to a straight line. A straight line can be defined by two numbers, the vertical intercept and the slope. The intercept tells us what the prediction is when all of the predictors are equal to 0; and the slope tells us what unit increase in the target/response variable we predict given a unit increase in the predictor variable. KNN regression, as simple as it is to implement and understand, has no such interpretability from its wiggly line.

There can, however, also be a disadvantage to using a simple linear regression model in some cases, particularly when the relationship between the target and the predictor is not linear, but instead some other shape (e.g., curved or oscillating). In these cases the prediction model from a simple linear regression will underfit (have high bias), meaning that model/predicted values do not match the actual observed values very well. Such a model would probably have a quite high RMSE when assessing model goodness of fit on the training data and a quite high RMSPE when assessing model prediction quality on a test data set. On such a data set, KNN regression may fare better. Additionally, there are other types of regression you can learn about in future books that may do even better at predicting with such data.

How do these two models compare on the Sacramento house prices data set? In Figure 8.6, we also printed the RMSPE as calculated from predicting on the test data set that was not used to train/fit the models. The RMSPE for the simple linear regression model is slightly lower than the RMSPE for the KNN regression model. Considering that the simple linear regression model is also more interpretable, if we were comparing these in practice we would likely choose to use the simple linear regression model.

Finally, note that the KNN regression model becomes "flat" at the left and right boundaries of the data, while the linear model predicts a constant slope. Predicting outside the range of the observed data is known as *extrapolation*; KNN and linear models behave quite differently when extrapolating. Depending on the application, the flat or constant slope trend may make more sense. For example, if our housing data were slightly different, the linear model may have actually predicted a *negative* price for a small house (if the intercept β_0 was negative), which obviously does not match reality. On the other hand, the trend of increasing house size corresponding to increasing house price probably continues for large houses, so the "flat" extrapolation of KNN likely does not match reality.

8.6 Multivariable linear regression

As in KNN classification and KNN regression, we can move beyond the simple case of only one predictor to the case with multiple predictors, known as *multivariable linear regression*. To do this, we follow a very similar approach to what we did for KNN regression: we just add more predictors to the model formula in the recipe. But recall that we do not need to use cross-validation to choose any parameters, nor do we need to standardize (i.e., center and scale) the data for linear regression. Note once again that we have the same concerns regarding multiple predictors as in the settings of multivariable KNN regression and classification: having more predictors is **not** always better. But because the same predictor selection algorithm from the classification chapter extends to the setting of linear regression, it will not be covered again in this chapter.

We will demonstrate multivariable linear regression using the Sacramento real estate data with both house size (measured in square feet) as well as number of bedrooms as our predictors, and continue to use house sale price as our response variable. We will start by changing the formula in the recipe to include both the `sqft` and `beds` variables as predictors:

```
mlm_recipe <- recipe(price ~ sqft + beds, data = sacramento_train)
```

Now we can build our workflow and fit the model:

```
mlm_fit <- workflow() |>
  add_recipe(mlm_recipe) |>
  add_model(lm_spec) |>
  fit(data = sacramento_train)

mlm_fit
```

```
## == Workflow [trained] ==========================================================
## Preprocessor: Recipe
## Model: linear_reg()
##
## -- Preprocessor ------------------------------------------------------------
-----
## 0 Recipe Steps
##
## -- Model -------------------------------------------------------------------
-----
##
## Call:
## stats::lm(formula = ..y ~ ., data = data)
##
## Coefficients:
## (Intercept)         sqft         beds
##     63474.8        165.7     -28760.9
```

And finally, we make predictions on the test data set to assess the quality of our model:

```
lm_mult_test_results <- mlm_fit |>
  predict(sacramento_test) |>
  bind_cols(sacramento_test) |>
```

```
  metrics(truth = price, estimate = .pred)

lm_mult_test_results
```

```
## # A tibble: 3 x 3
##   .metric .estimator .estimate
##   <chr>   <chr>          <dbl>
## 1 rmse    standard   81418.
## 2 rsq     standard       0.606
## 3 mae     standard   59310.
```

Our model's test error as assessed by RMSPE is 81,418. In the case of two predictors, we can plot the predictions made by our linear regression creates a *plane* of best fit, as shown in Figure 8.7.

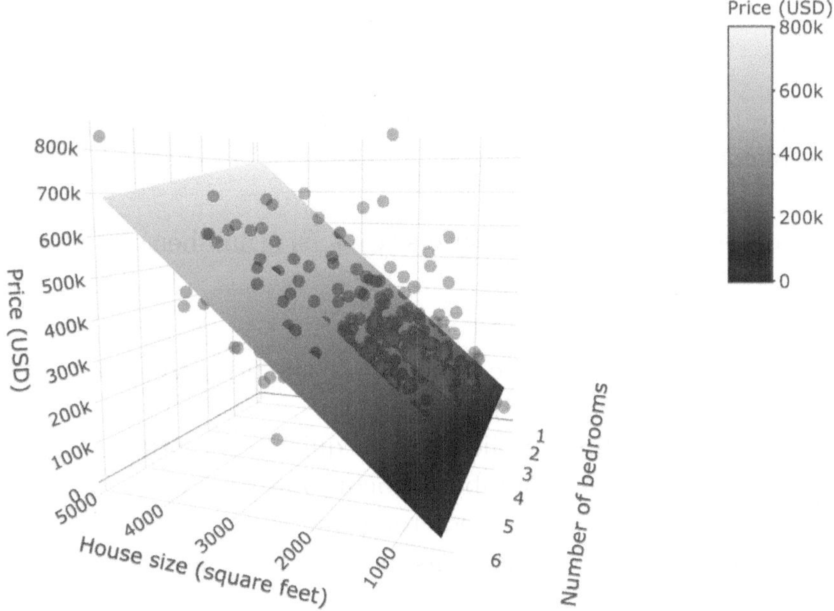

FIGURE 8.7: Linear regression plane of best fit overlaid on top of the data (using price, house size, and number of bedrooms as predictors). Note that in general we recommend against using 3D visualizations; here we use a 3D visualization only to illustrate what the regression plane looks like for learning purposes.

We see that the predictions from linear regression with two predictors form a flat plane. This is the hallmark of linear regression, and differs from the wiggly, flexible surface we get from other methods such as KNN regression. As discussed, this can be advantageous in one aspect, which is that for each predictor, we can get slopes/intercept from linear regression, and thus describe the plane mathematically. We can extract those slope values from our model object as shown below:

```
mcoeffs <- mlm_fit |>
             pull_workflow_fit() |>
             tidy()

mcoeffs
```

```
## # A tibble: 3 x 5
##   term        estimate std.error statistic  p.value
##   <chr>          <dbl>     <dbl>     <dbl>    <dbl>
## 1 (Intercept)   63475.  13426.        4.73 2.88e- 6
## 2 sqft            166.      7.03     23.6  1.07e-85
## 3 beds         -28761.   5628.       -5.11 4.43e- 7
```

And then use those slopes to write a mathematical equation to describe the prediction plane:

$$\text{house sale price} = \beta_0 + \beta_1 \cdot (\text{house size}) + \beta_2 \cdot (\text{number of bedrooms}),$$

where:

- β_0 is the *vertical intercept* of the hyperplane (the price when both house size and number of bedrooms are 0)
- β_1 is the *slope* for the first predictor (how quickly the price changes as you increase house size, holding number of bedrooms constant)
- β_2 is the *slope* for the second predictor (how quickly the price changes as you increase the number of bedrooms, holding house size constant)

Finally, we can fill in the values for β_0, β_1 and β_2 from the model output above to create the equation of the plane of best fit to the data:

$$\text{house sale price} = 63475 + 166 \cdot (\text{house size}) - 28761 \cdot (\text{number of bedrooms})$$

This model is more interpretable than the multivariable KNN regression model; we can write a mathematical equation that explains how each predictor is affecting the predictions. But as always, we should question how well multivariable linear regression is doing compared to the other tools we have, such as simple linear regression and multivariable KNN regression. If this comparison is part of the model tuning process—for example, if we are trying out many different sets of predictors for multivariable linear and KNN regression—we must perform this comparison using cross-validation on only our training data. But if we have already decided on a small number (e.g., 2 or 3) of tuned candidate models and we want to make a final comparison, we can do so by comparing the prediction error of the methods on the test data.

```
lm_mult_test_results
```

```
## # A tibble: 3 x 3
##   .metric .estimator .estimate
##   <chr>   <chr>          <dbl>
## 1 rmse    standard   81418.
## 2 rsq     standard       0.606
## 3 mae     standard   59310.
```

We obtain an RMSPE for the multivariable linear regression model of 81,417.89. This prediction error is less than the prediction error for the multivariable KNN regression model, indicating that we should likely choose linear regression for predictions of house sale price on this data set. Revisiting the simple linear regression model with only a single predictor from earlier in this chapter, we see that the RMSPE for that model was 82,342.28, which is slightly higher than that of our more complex model. Our model with two predictors provided a slightly better fit on test data than our model with just one. As mentioned earlier, this is not always the case: sometimes including more predictors can negatively impact the prediction performance on unseen test data.

8.7 Multicollinearity and outliers

What can go wrong when performing (possibly multivariable) linear regression? This section will introduce two common issues—*outliers* and *collinear predictors*—and illustrate their impact on predictions.

8.7.1 Outliers

Outliers are data points that do not follow the usual pattern of the rest of the data. In the setting of linear regression, these are points that have a vertical distance to the line of best fit that is either much higher or much lower than you might expect based on the rest of the data. The problem with outliers is that they can have *too much influence* on the line of best fit. In general, it is very difficult to judge accurately which data are outliers without advanced techniques that are beyond the scope of this book.

But to illustrate what can happen when you have outliers, Figure 8.8 shows a small subset of the Sacramento housing data again, except we have added a *single* data point (highlighted in red). This house is 5,000 square feet in size, and sold for only $50,000. Unbeknownst to the data analyst, this house was sold by a parent to their child for an absurdly low price. Of course, this is not representative of the real housing market values that the other data points follow; the data point is an *outlier*. In blue we plot the original line of best fit, and in red we plot the new line of best fit including the outlier. You can see how different the red line is from the blue line, which is entirely caused by that one extra outlier data point.

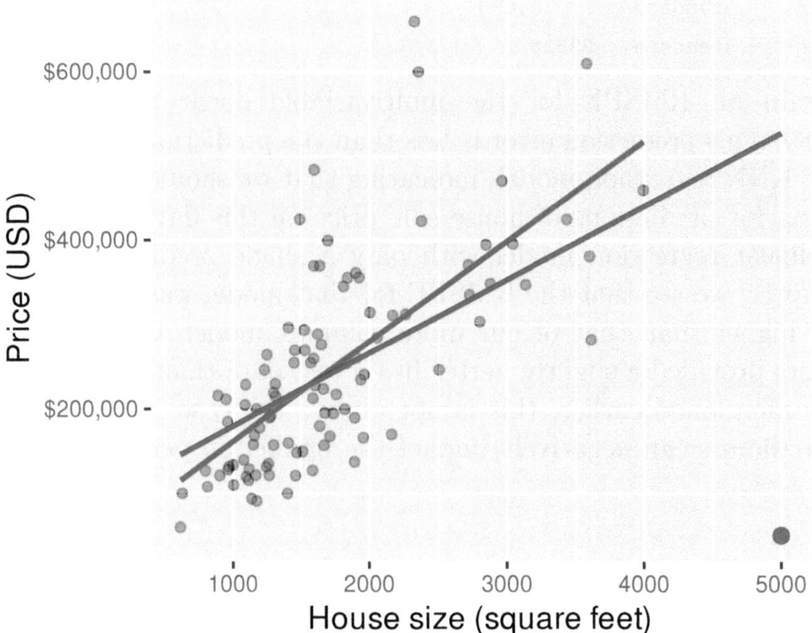

FIGURE 8.8: Scatter plot of a subset of the data, with outlier highlighted in red.

Fortunately, if you have enough data, the inclusion of one or two outliers—as long as their values are not *too* wild—will typically not have a large effect on the line of best fit. Figure 8.9 shows how that same outlier data point from earlier influences the line of best fit when we are working with the entire original Sacramento training data. You can see that with this larger data set, the line changes much less when adding the outlier. Nevertheless, it is still important when working with linear regression to critically think about how much any individual data point is influencing the model.

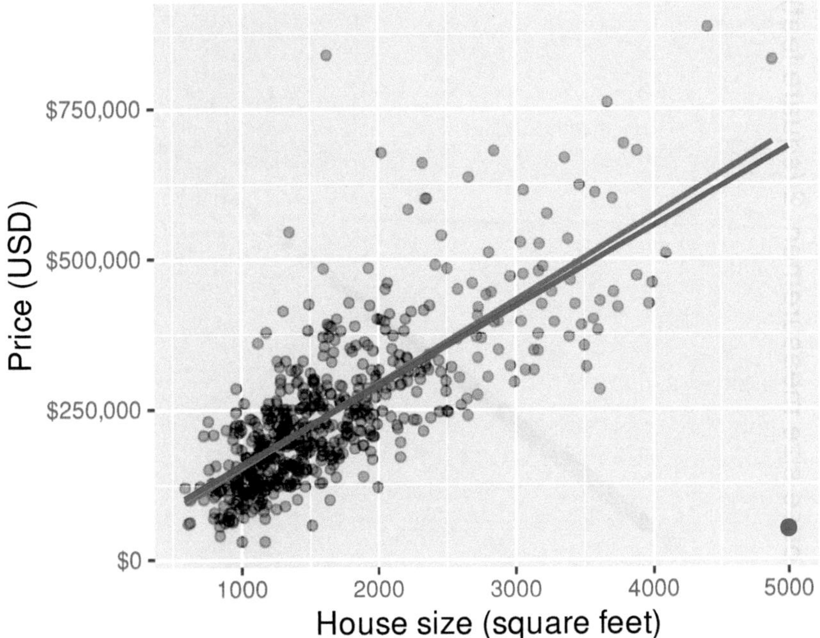

FIGURE 8.9: Scatter plot of the full data, with outlier highlighted in red.

8.7.2 Multicollinearity

The second, and much more subtle, issue can occur when performing multivariable linear regression. In particular, if you include multiple predictors that are strongly linearly related to one another, the coefficients that describe the plane of best fit can be very unreliable—small changes to the data can result in large changes in the coefficients. Consider an extreme example using the Sacramento housing data where the house was measured twice by two people. Since the two people are each slightly inaccurate, the two measurements might not agree exactly, but they are very strongly linearly related to each other, as shown in Figure 8.10.

If we again fit the multivariable linear regression model on this data, then the plane of best fit has regression coefficients that are very sensitive to the exact values in the data. For example, if we change the data ever so slightly—e.g., by running cross-validation, which splits up the data randomly into different chunks—the coefficients vary by large amounts:

Best Fit 1: house sale price $= 3682 + (-43) \cdot$ (house size 1 (ft^2)) $+ (182) \cdot$ (house size 2 (ft^2)).

Best Fit 2: house sale price $= 20596 + (312) \cdot$ (house size 1 (ft^2)) $+ (-172) \cdot$ (house size 2 (ft^2)).

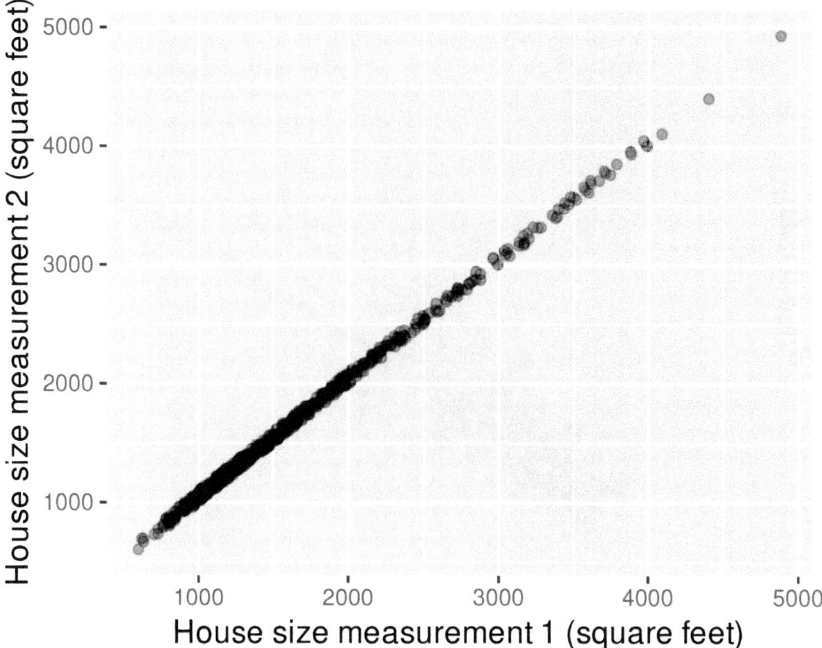

FIGURE 8.10: Scatter plot of house size (in square feet) measured by person 1 versus house size (in square feet) measured by person 2.

Best Fit 3: house sale price $= 6673 + (37) \cdot (\text{house size 1 (ft}^2)) + (104) \cdot (\text{house size 2 (ft}^2))$.

Therefore, when performing multivariable linear regression, it is important to avoid including very linearly related predictors. However, techniques for doing so are beyond the scope of this book; see the list of additional resources at the end of this chapter to find out where you can learn more.

8.8 Designing new predictors

We were quite fortunate in our initial exploration to find a predictor variable (house size) that seems to have a meaningful and nearly linear relationship with our response variable (sale price). But what should we do if we cannot immediately find such a nice variable? Well, sometimes it is just a fact that the variables in the data do not have enough of a relationship with the response variable to provide useful predictions. For example, if the only available predictor was "the current house owner's favorite ice cream flavor", we likely would have little hope of using that variable to predict the house's sale price (barring any future remarkable scientific discoveries about the relationship between the housing market and homeowner ice cream preferences). In cases like these, the only option is to obtain measurements of more useful variables.

There are, however, a wide variety of cases where the predictor variables do have a meaningful relationship with the response variable, but that relationship does not fit the assumptions of the regression method you have chosen. For example, a data frame `df` with two variables—`x` and `y`—with a nonlinear relationship between the two variables will not be fully captured by simple linear regression, as shown in Figure 8.11.

```
df
```

```
## # A tibble: 100 x 2
##          x      y
##      <dbl>  <dbl>
##   1 0.102 0.0720
##   2 0.800 0.532
##   3 0.478 0.148
##   4 0.972 1.01
##   5 0.846 0.677
##   6 0.405 0.157
##   7 0.879 0.768
##   8 0.130 0.0402
##   9 0.852 0.576
##  10 0.180 0.0847
## # ... with 90 more rows
```

Instead of trying to predict the response `y` using a linear regression on `x`, we might have some scientific background about our problem to suggest that `y` should be a cubic function of `x`. So before performing regression, we might *create a new predictor variable* `z` using the `mutate` function:

```
df <- df |>
        mutate(z = x^3)
```

Then we can perform linear regression for `y` using the predictor variable `z`, as shown in Figure 8.12. Here you can see that the transformed predictor `z` helps the linear regression model make more accurate predictions. Note that none of the `y` response values have changed between Figures 8.11 and 8.12; the only change is that the `x` values have been replaced by `z` values.

The process of transforming predictors (and potentially combining multiple predictors in the process) is known as *feature engineering*. In real data analysis

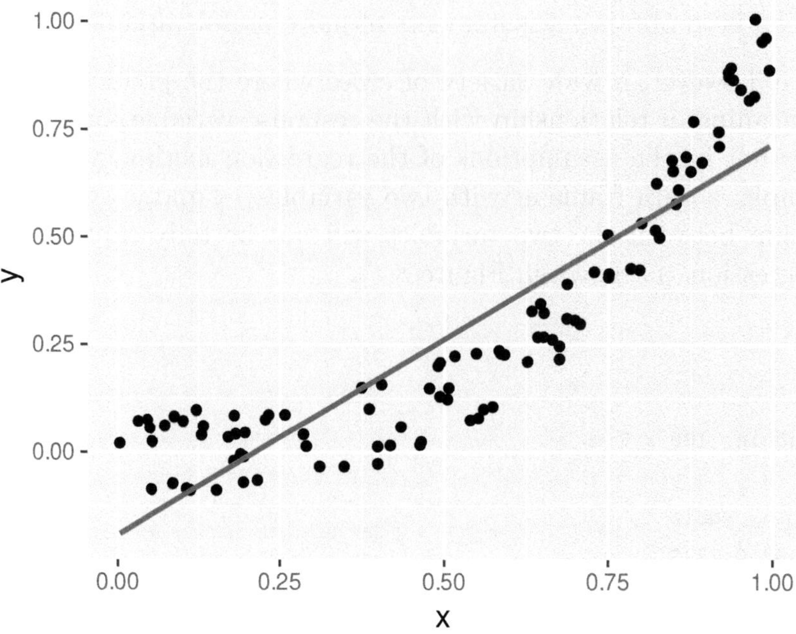

FIGURE 8.11: Example of a data set with a nonlinear relationship between the predictor and the response.

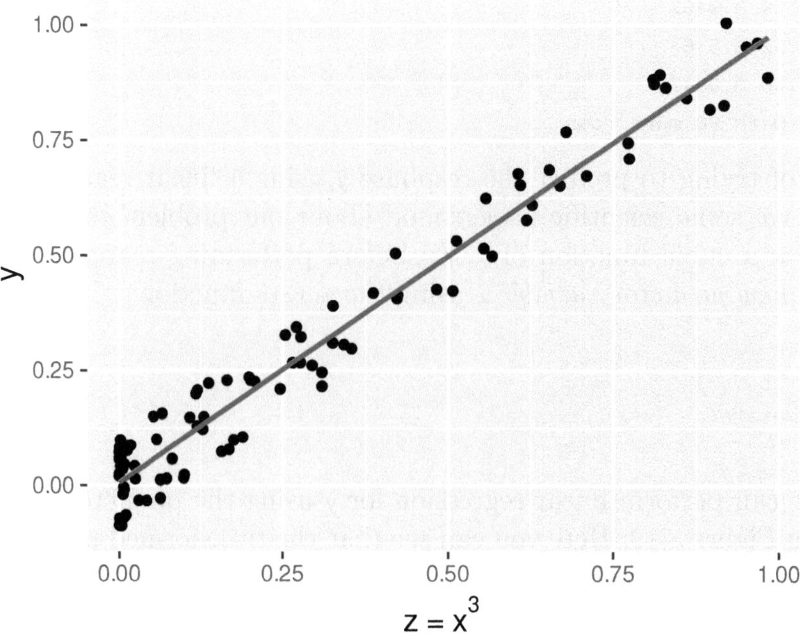

FIGURE 8.12: Relationship between the transformed predictor and the response.

problems, you will need to rely on a deep understanding of the problem—as well as the wrangling tools from previous chapters—to engineer useful new features that improve predictive performance.

Note: Feature engineering is *part of tuning your model*, and as such you must not use your test data to evaluate the quality of the features you produce. You are free to use cross-validation, though!

8.9 The other sides of regression

So far in this textbook we have used regression only in the context of prediction. However, regression can also be seen as a method to understand and quantify the effects of individual variables on a response / outcome of interest. In the housing example from this chapter, beyond just using past data to predict future sale prices, we might also be interested in describing the individual relationships of house size and the number of bedrooms with house price, quantifying how strong each of these relationships are, and assessing how accurately we can estimate their magnitudes. And even beyond that, we may be interested in understanding whether the predictors *cause* changes in the price. These sides of regression are well beyond the scope of this book; but the material you have learned here should give you a foundation of knowledge that will serve you well when moving to more advanced books on the topic.

8.10 Exercises

Practice exercises for the material covered in this chapter can be found in the accompanying worksheets repository[1] in the "Regression II: linear regression" row. You can launch an interactive version of the worksheet in your browser by clicking the "launch binder" button. You can also preview a non-interactive version of the worksheet by clicking "view worksheet." If you instead decide to download the worksheet and run it on your own machine, make sure to follow the instructions for computer setup found in Chapter 13. This will ensure

[1] `https://github.com/UBC-DSCI/data-science-a-first-intro-worksheets#readme`

that the automated feedback and guidance that the worksheets provide will function as intended.

8.11 Additional resources

- The `tidymodels` website[2] is an excellent reference for more details on, and advanced usage of, the functions and packages in the past two chapters. Aside from that, it also has a nice beginner's tutorial[3] and an extensive list of more advanced examples[4] that you can use to continue learning beyond the scope of this book.
- *Modern Dive* [Ismay and Kim, 2020] is another textbook that uses the `tidyverse` / `tidymodels` framework. Chapter 6 complements the material in the current chapter well; it covers some slightly more advanced concepts than we do without getting mathematical. Give this chapter a read before moving on to the next reference. It is also worth noting that this book takes a more "explanatory" / "inferential" approach to regression in general (in Chapters 5, 6, and 10), which provides a nice complement to the predictive tack we take in the present book.
- *An Introduction to Statistical Learning* [James et al., 2013] provides a great next stop in the process of learning about regression. Chapter 3 covers linear regression at a slightly more mathematical level than we do here, but it is not too large a leap and so should provide a good stepping stone. Chapter 6 discusses how to pick a subset of "informative" predictors when you have a data set with many predictors, and you expect only a few of them to be relevant. Chapter 7 covers regression models that are more flexible than linear regression models but still enjoy the computational efficiency of linear regression. In contrast, the KNN methods we covered earlier are indeed more flexible but become very slow when given lots of data.

[2] https://tidymodels.org/packages

[3] https://www.tidymodels.org/start/

[4] https://www.tidymodels.org/learn/

9

Clustering

9.1 Overview

As part of exploratory data analysis, it is often helpful to see if there are meaningful subgroups (or *clusters*) in the data. This grouping can be used for many purposes, such as generating new questions or improving predictive analyses. This chapter provides an introduction to clustering using the K-means algorithm, including techniques to choose the number of clusters.

9.2 Chapter learning objectives

By the end of the chapter, readers will be able to do the following:

- Describe a situation in which clustering is an appropriate technique to use, and what insight it might extract from the data.
- Explain the K-means clustering algorithm.
- Interpret the output of a K-means analysis.
- Differentiate between clustering and classification.
- Identify when it is necessary to scale variables before clustering, and do this using R.
- Perform K-means clustering in R using `kmeans`.
- Use the elbow method to choose the number of clusters for K-means.
- Visualize the output of K-means clustering in R using colored scatter plots.
- Describe the advantages, limitations and assumptions of the K-means clustering algorithm.

9.3 Clustering

Clustering is a data analysis technique involving separating a data set into subgroups of related data. For example, we might use clustering to separate a data set of documents into groups that correspond to topics, a data set of human genetic information into groups that correspond to ancestral subpopulations, or a data set of online customers into groups that correspond to purchasing behaviors. Once the data are separated, we can, for example, use the subgroups to generate new questions about the data and follow up with a predictive modeling exercise. In this course, clustering will be used only for exploratory analysis, i.e., uncovering patterns in the data.

Note that clustering is a fundamentally different kind of task than classification or regression. In particular, both classification and regression are *supervised tasks* where there is a *response variable* (a category label or value), and we have examples of past data with labels/values that help us predict those of future data. By contrast, clustering is an *unsupervised task*, as we are trying to understand and examine the structure of data without any response variable labels or values to help us. This approach has both advantages and disadvantages. Clustering requires no additional annotation or input on the data. For example, while it would be nearly impossible to annotate all the articles on Wikipedia with human-made topic labels, we can cluster the articles without this information to find groupings corresponding to topics automatically. However, given that there is no response variable, it is not as easy to evaluate the "quality" of a clustering. With classification, we can use a test data set to assess prediction performance. In clustering, there is not a single good choice for evaluation. In this book, we will use visualization to ascertain the quality of a clustering, and leave rigorous evaluation for more advanced courses.

As in the case of classification, there are many possible methods that we could use to cluster our observations to look for subgroups. In this book, we will focus on the widely used K-means algorithm [Lloyd, 1982]. In your future studies, you might encounter hierarchical clustering, principal component analysis, multidimensional scaling, and more; see the additional resources section at the end of this chapter for where to begin learning more about these other methods.

Note: There are also so-called *semisupervised* tasks, where only some of the data come with response variable labels/values, but the vast majority don't. The goal is to try to uncover underlying structure in the data that allows one to guess the missing labels. This sort of task is beneficial, for example, when one has an unlabeled data set that is too large to manually label, but one is willing to provide a few informative example labels as a "seed" to guess the labels for all the data.

An illustrative example

Here we will present an illustrative example using a data set from the `palmerpenguins` R package[1] [Horst et al., 2020]. This data set was collected by Dr. Kristen Gorman and the Palmer Station, Antarctica Long Term Ecological Research Site, and includes measurements for adult penguins found near there [Gorman et al., 2014]. We have modified the data set for use in this chapter. Here we will focus on using two variables—penguin bill and flipper length, both in millimeters—to determine whether there are distinct types of penguins in our data. Understanding this might help us with species discovery and classification in a data-driven way.

FIGURE 9.1: Gentoo penguin.

To learn about K-means clustering we will work with `penguin_data` in this chapter. `penguin_data` is a subset of 18 observations of the original data, which has already been standardized (remember from Chapter 5 that scaling is part of the standardization process). We will discuss scaling for K-means in more detail later in this chapter.

[1] https://allisonhorst.github.io/palmerpenguins/

Before we get started, we will load the `tidyverse` metapackage as well as set a random seed. This will ensure we have access to the functions we need and that our analysis will be reproducible. As we will learn in more detail later in the chapter, setting the seed here is important because the K-means clustering algorithm uses random numbers.

```
library(tidyverse)
set.seed(1)
```

Now we can load and preview the data.

```
penguin_data <- read_csv("data/penguins_standardized.csv")
penguin_data
```

```
## # A tibble: 18 x 2
##    flipper_length_standardized bill_length_standardized
##                          <dbl>                    <dbl>
##  1                      -0.190                   -0.641
##  2                      -1.33                    -1.14
##  3                      -0.922                   -1.52
##  4                      -0.922                   -1.11
##  5                      -1.41                    -0.847
##  6                      -0.678                   -0.641
##  7                      -0.271                   -1.24
##  8                      -0.434                   -0.902
##  9                       1.19                     0.720
## 10                       1.36                     0.646
## 11                       1.36                     0.963
## 12                       1.76                     0.440
## 13                       1.11                     1.21
## 14                       0.786                    0.123
## 15                      -0.271                    0.627
## 16                      -0.271                    0.757
## 17                      -0.108                    1.78
## 18                      -0.759                    0.776
```

Next, we can create a scatter plot using this data set to see if we can detect subtypes or groups in our data set.

```
ggplot(data, aes(x = flipper_length_standardized,
                 y = bill_length_standardized)) +
  geom_point() +
  xlab("Flipper Length (standardized)") +
  ylab("Bill Length (standardized)") +
  theme(text = element_text(size = 12))
```

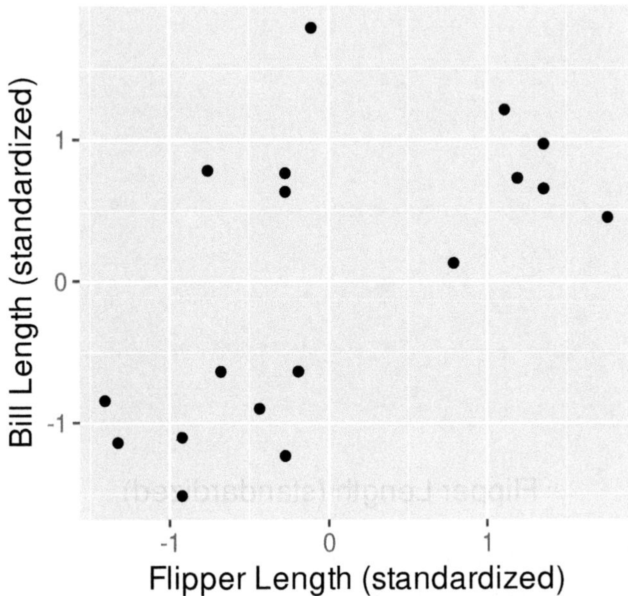

FIGURE 9.2: Scatter plot of standardized bill length versus standardized flipper length.

Based on the visualization in Figure 9.2, we might suspect there are a few subtypes of penguins within our data set. We can see roughly 3 groups of observations in Figure 9.2, including:

1. a small flipper and bill length group,
2. a small flipper length, but large bill length group, and
3. a large flipper and bill length group.

Data visualization is a great tool to give us a rough sense of such patterns when we have a small number of variables. But if we are to group data—and select the number of groups—as part of a reproducible analysis, we need something a bit more automated. Additionally, finding groups via visualization becomes more difficult as we increase the number of variables we consider when clustering. The way to rigorously separate the data into groups is to use a clustering algorithm. In this chapter, we will focus on the *K-means* algorithm, a widely used and often very effective clustering method, combined with the *elbow method* for selecting the number of clusters. This procedure will separate the data into groups; Figure 9.3 shows these groups denoted by colored scatter points.

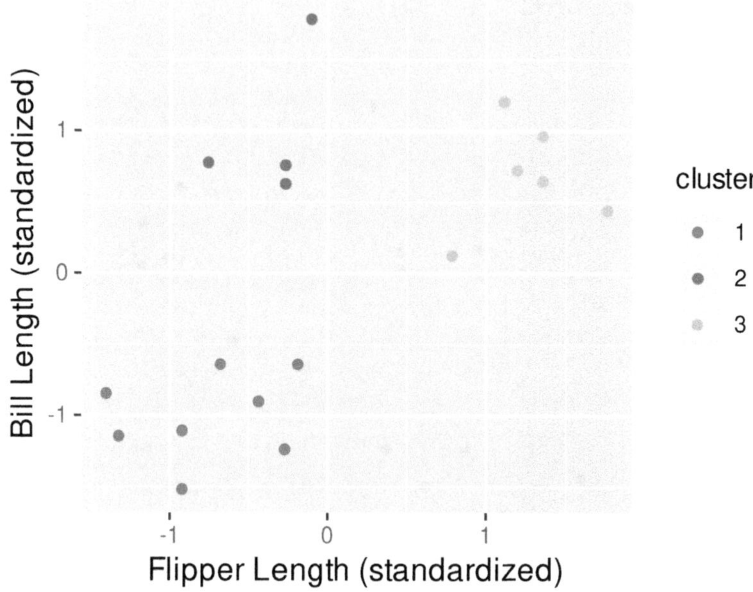

FIGURE 9.3: Scatter plot of standardized bill length versus standardized flipper length with colored groups.

What are the labels for these groups? Unfortunately, we don't have any. K-means, like almost all clustering algorithms, just outputs meaningless "cluster labels" that are typically whole numbers: 1, 2, 3, etc. But in a simple case like this, where we can easily visualize the clusters on a scatter plot, we can give human-made labels to the groups using their positions on the plot:

- small flipper length and small bill length (orange cluster),
- small flipper length and large bill length (blue cluster).
- and large flipper length and large bill length (yellow cluster).

Once we have made these determinations, we can use them to inform our species classifications or ask further questions about our data. For example, we might be interested in understanding the relationship between flipper length and bill length, and that relationship may differ depending on the type of penguin we have.

9.4 K-means

9.4.1 Measuring cluster quality

The K-means algorithm is a procedure that groups data into K clusters. It starts with an initial clustering of the data, and then iteratively improves it by making adjustments to the assignment of data to clusters until it cannot improve any further. But how do we measure the "quality" of a clustering, and what does it mean to improve it? In K-means clustering, we measure the quality of a cluster by its *within-cluster sum-of-squared-distances* (WSSD). Computing this involves two steps. First, we find the cluster centers by computing the mean of each variable over data points in the cluster. For example, suppose we have a cluster containing four observations, and we are using two variables, x and y, to cluster the data. Then we would compute the coordinates, μ_x and μ_y, of the cluster center via

$$\mu_x = \frac{1}{4}(x_1 + x_2 + x_3 + x_4) \quad \mu_y = \frac{1}{4}(y_1 + y_2 + y_3 + y_4).$$

In the first cluster from the example, there are 4 data points. These are shown with their cluster center (flipper_length_standardized = -0.35 and bill_length_standardized = 0.99) highlighted in Figure 9.4.

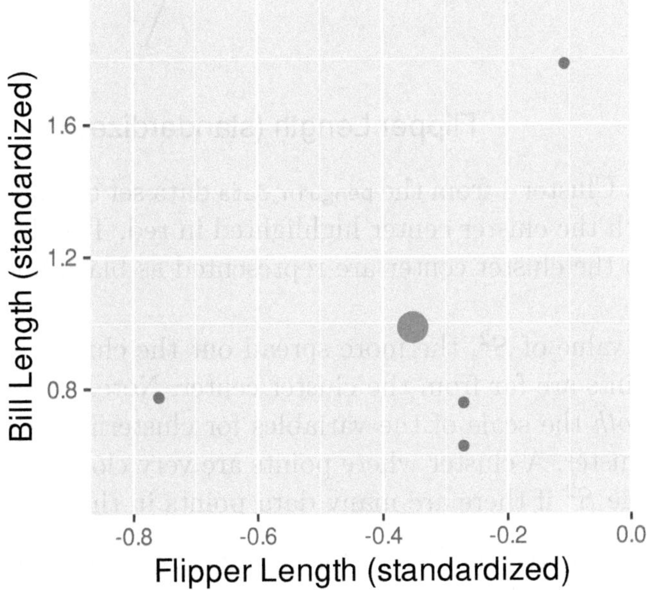

FIGURE 9.4: Cluster 1 from the `penguin_data` data set example. Observations are in blue, with the cluster center highlighted in red.

The second step in computing the WSSD is to add up the squared distance between each point in the cluster and the cluster center. We use the straight-line / Euclidean distance formula that we learned about in Chapter 5. In the 4-observation cluster example above, we would compute the WSSD S^2 via

$$\begin{aligned}S^2 = &\left((x_1 - \mu_x)^2 + (y_1 - \mu_y)^2\right) + \left((x_2 - \mu_x)^2 + (y_2 - \mu_y)^2\right) + \\ &\left((x_3 - \mu_x)^2 + (y_3 - \mu_y)^2\right) + \left((x_4 - \mu_x)^2 + (y_4 - \mu_y)^2\right).\end{aligned}$$

These distances are denoted by lines in Figure 9.5 for the first cluster of the penguin data example.

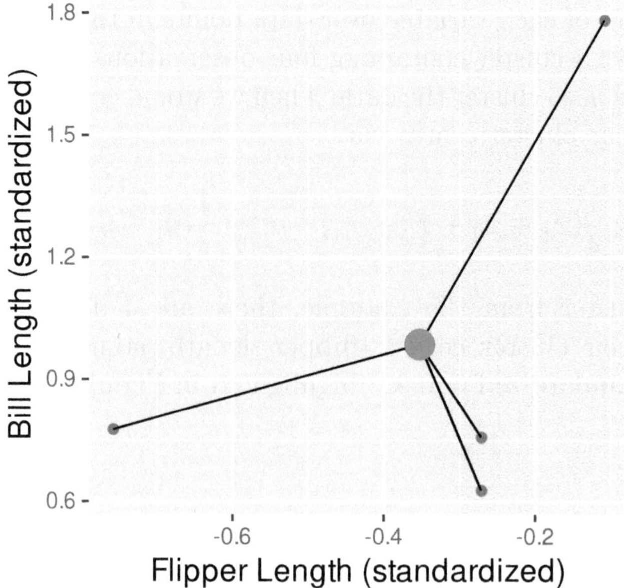

FIGURE 9.5: Cluster 1 from the `penguin_data` data set example. Observations are in blue, with the cluster center highlighted in red. The distances from the observations to the cluster center are represented as black lines.

The larger the value of S^2, the more spread out the cluster is, since large S^2 means that points are far from the cluster center. Note, however, that "large" is relative to *both* the scale of the variables for clustering *and* the number of points in the cluster. A cluster where points are very close to the center might still have a large S^2 if there are many data points in the cluster.

After we have calculated the WSSD for all the clusters, we sum them together to get the *total WSSD*. For our example, this means adding up all the squared distances for the 18 observations. These distances are denoted by black lines in Figure 9.6.

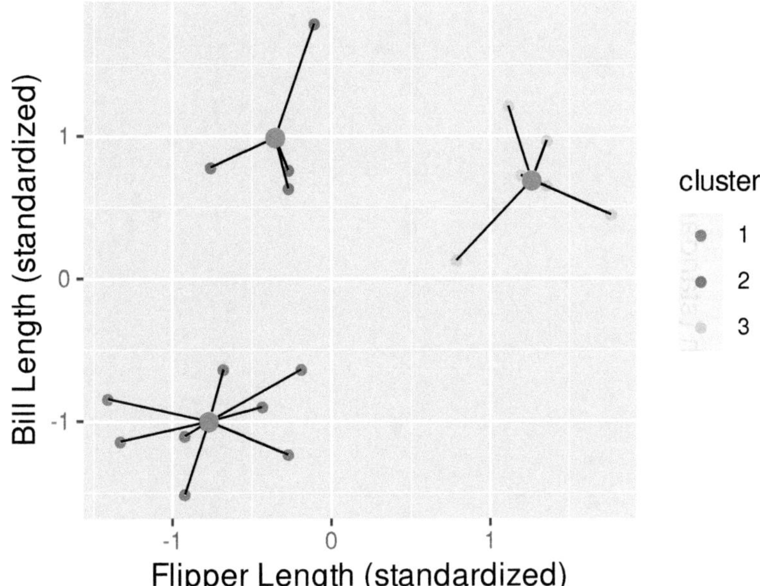

FIGURE 9.6: All clusters from the `penguin_data` data set example. Observations are in orange, blue, and yellow with the cluster center highlighted in red. The distances from the observations to each of the respective cluster centers are represented as black lines.

9.4.2 The clustering algorithm

We begin the K-means algorithm by picking K, and randomly assigning a roughly equal number of observations to each of the K clusters. An example random initialization is shown in Figure 9.7.

Then K-means consists of two major steps that attempt to minimize the sum of WSSDs over all the clusters, i.e., the *total WSSD*:

1. **Center update:** Compute the center of each cluster.
2. **Label update:** Reassign each data point to the cluster with the nearest center.

These two steps are repeated until the cluster assignments no longer change. We show what the first four iterations of K-means would look like in Figure 9.8. There each row corresponds to an iteration, where the left column depicts the center update, and the right column depicts the reassignment of data to clusters.

Note that at this point, we can terminate the algorithm since none of the assignments changed in the fourth iteration; both the centers and labels will remain the same from this point onward.

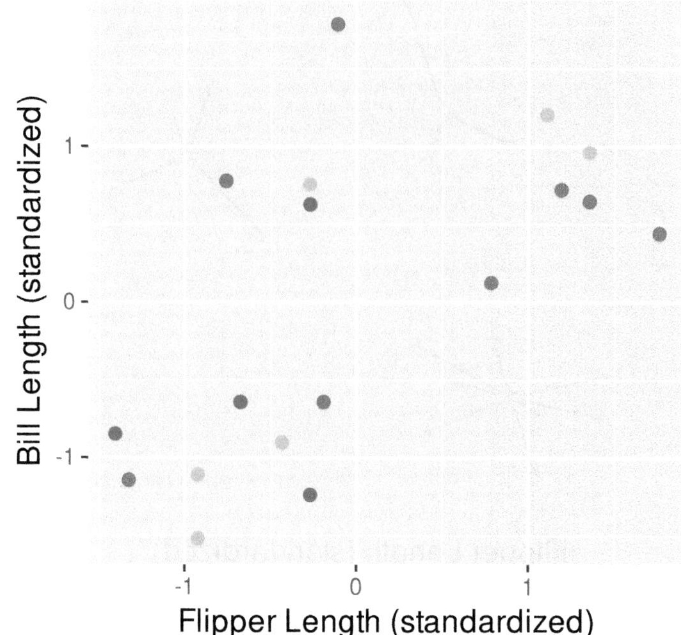

FIGURE 9.7: Random initialization of labels.

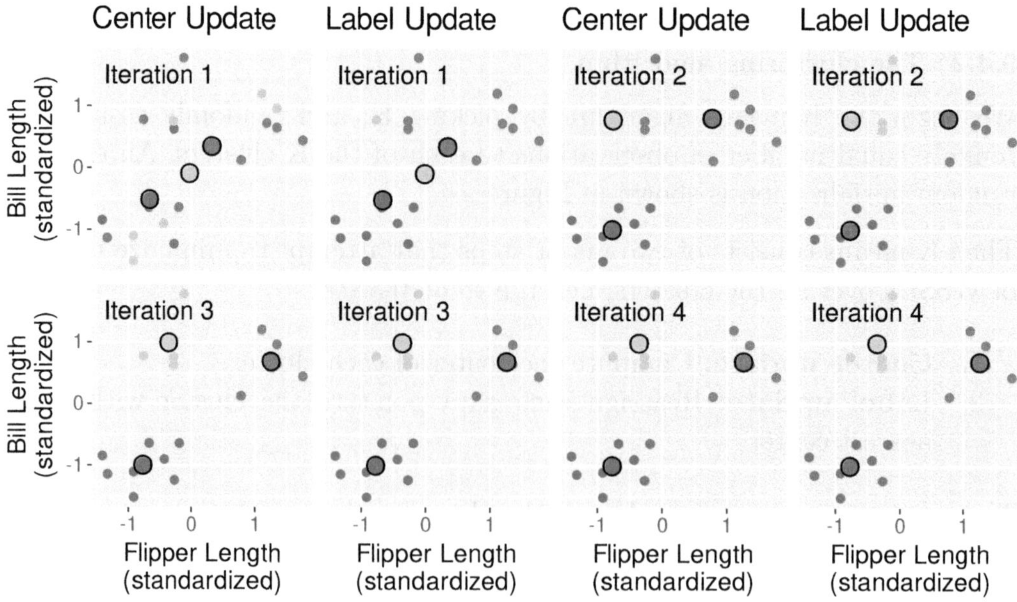

FIGURE 9.8: First four iterations of K-means clustering on the `penguin_data` example data set. Each pair of plots corresponds to an iteration. Within the pair, the first plot depicts the center update, and the second plot depicts the reassignment of data to clusters. Cluster centers are indicated by larger points that are outlined in black.

Note: Is K-means *guaranteed* to stop at some point, or could it iterate forever? As it turns out, thankfully, the answer is that K-means is guaranteed to stop after *some* number of iterations. For the interested reader, the logic for this has three steps: (1) both the label update and the center update decrease total WSSD in each iteration, (2) the total WSSD is always greater than or equal to 0, and (3) there are only a finite number of possible ways to assign the data to clusters. So at some point, the total WSSD must stop decreasing, which means none of the assignments are changing, and the algorithm terminates.

What kind of data is suitable for K-means clustering? In the simplest version of K-means clustering that we have presented here, the straight-line distance is used to measure the distance between observations and cluster centers. This means that only quantitative data should be used with this algorithm. There are variants on the K-means algorithm, as well as other clustering algorithms entirely, that use other distance metrics to allow for non-quantitative data to be clustered. These, however, are beyond the scope of this book.

9.4.3 Random restarts

Unlike the classification and regression models we studied in previous chapters, K-means can get "stuck" in a bad solution. For example, Figure 9.9 illustrates an unlucky random initialization by K-means.

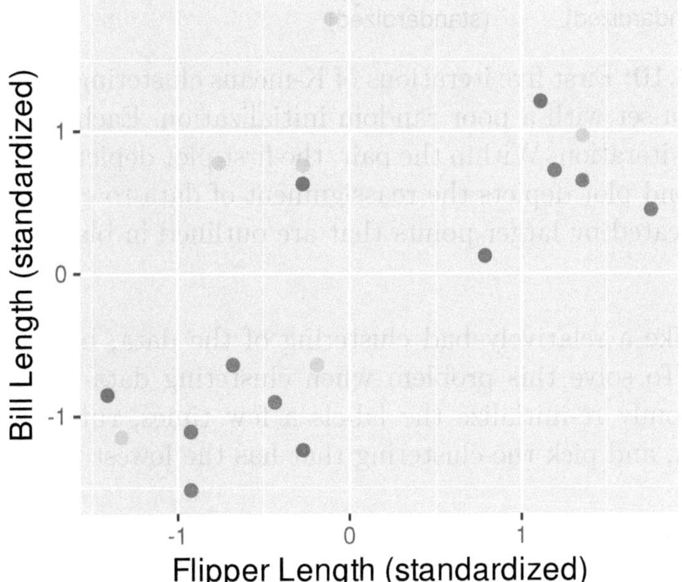

FIGURE 9.9: Random initialization of labels.

Figure 9.10 shows what the iterations of K-means would look like with the unlucky random initialization shown in Figure 9.9.

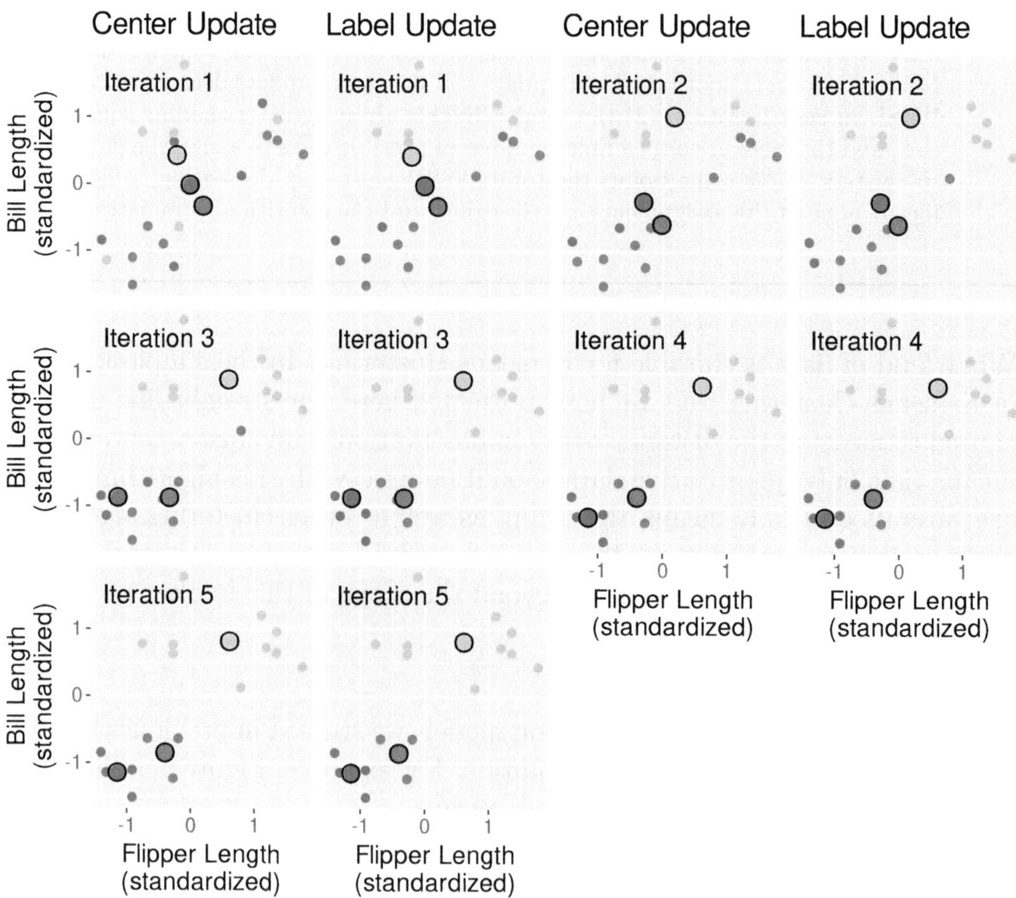

FIGURE 9.10: First five iterations of K-means clustering on the `penguin_data` example data set with a poor random initialization. Each pair of plots corresponds to an iteration. Within the pair, the first plot depicts the center update, and the second plot depicts the reassignment of data to clusters. Cluster centers are indicated by larger points that are outlined in black.

This looks like a relatively bad clustering of the data, but K-means cannot improve it. To solve this problem when clustering data using K-means, we should randomly re-initialize the labels a few times, run K-means for each initialization, and pick the clustering that has the lowest final total WSSD.

9.4.4 Choosing K

In order to cluster data using K-means, we also have to pick the number of clusters, K. But unlike in classification, we have no response variable and cannot perform cross-validation with some measure of model prediction error. Further, if K is chosen too small, then multiple clusters get grouped together; if K is too large, then clusters get subdivided. In both cases, we will potentially miss interesting structure in the data. Figure 9.11 illustrates the impact of K on K-means clustering of our penguin flipper and bill length data by showing the different clusterings for K's ranging from 1 to 9.

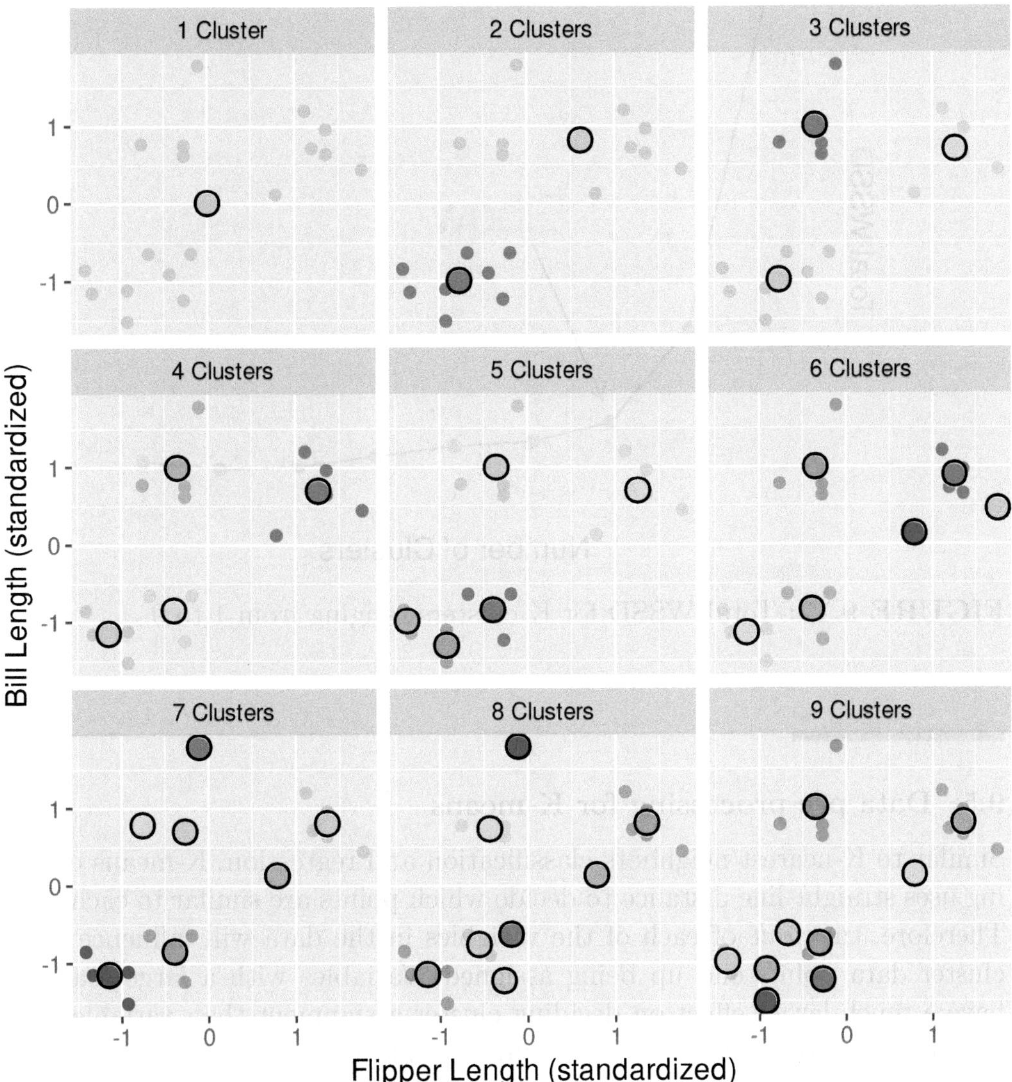

FIGURE 9.11: Clustering of the penguin data for K clusters ranging from 1 to 9. Cluster centers are indicated by larger points that are outlined in black.

If we set K less than 3, then the clustering merges separate groups of data; this causes a large total WSSD, since the cluster center is not close to any of the data in the cluster. On the other hand, if we set K greater than 3, the clustering subdivides subgroups of data; this does indeed still decrease the total WSSD, but by only a *diminishing amount.* If we plot the total WSSD versus the number of clusters, we see that the decrease in total WSSD levels off (or forms an "elbow shape") when we reach roughly the right number of clusters (Figure 9.12).

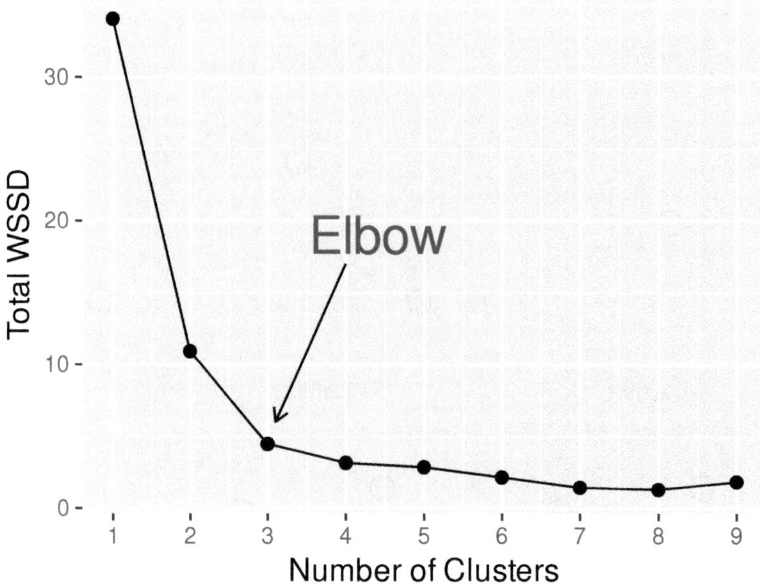

FIGURE 9.12: Total WSSD for K clusters ranging from 1 to 9.

9.5 Data pre-processing for K-means

Similar to K-nearest neighbors classification and regression, K-means clustering uses straight-line distance to decide which points are similar to each other. Therefore, the *scale* of each of the variables in the data will influence which cluster data points end up being assigned. Variables with a large scale will have a much larger effect on deciding cluster assignment than variables with a small scale. To address this problem, we typically standardize our data before clustering, which ensures that each variable has a mean of 0 and standard deviation of 1. The `scale` function in R can be used to do this. We show an example of how to use this function below using an unscaled and unstandardized version of the data set in this chapter.

First, here is what the raw (i.e., not standardized) data looks like:

```
not_standardized_data <- read_csv("data/penguins_not_standardized.csv")
not_standardized_data
```

```
## # A tibble: 18 x 2
##    bill_length_mm flipper_length_mm
##             <dbl>             <dbl>
##  1           39.2               196
##  2           36.5               182
##  3           34.5               187
##  4           36.7               187
##  5           38.1               181
##  6           39.2               190
##  7           36                 195
##  8           37.8               193
##  9           46.5               213
## 10           46.1               215
## 11           47.8               215
## 12           45                 220
## 13           49.1               212
## 14           43.3               208
## 15           46                 195
## 16           46.7               195
## 17           52.2               197
## 18           46.8               189
```

And then we apply the `scale` function to every column in the data frame using `mutate` and `across`.

```
standardized_data <- not_standardized_data |>
  mutate(across(everything(), scale))

standardized_data
```

```
## # A tibble: 18 x 2
##    bill_length_mm[,1] flipper_length_mm[,1]
##                 <dbl>                 <dbl>
##  1             -0.641                -0.190
##  2             -1.14                 -1.33
##  3             -1.52                 -0.922
##  4             -1.11                 -0.922
##  5             -0.847                -1.41
```

```
## 6   -0.641   -0.678
## 7   -1.24    -0.271
## 8   -0.902   -0.434
## 9    0.720    1.19
## 10   0.646    1.36
## 11   0.963    1.36
## 12   0.440    1.76
## 13   1.21     1.11
## 14   0.123    0.786
## 15   0.627   -0.271
## 16   0.757   -0.271
## 17   1.78    -0.108
## 18   0.776   -0.759
```

9.6 K-means in R

To perform K-means clustering in R, we use the `kmeans` function. It takes at least two arguments: the data frame containing the data you wish to cluster, and K, the number of clusters (here we choose K = 3). Note that the K-means algorithm uses a random initialization of assignments; but since we set the random seed earlier, the clustering will be reproducible.

```
penguin_clust <- kmeans(standardized_data, centers = 3)
penguin_clust
```

```
## K-means clustering with 3 clusters of sizes 4, 8, 6
##
## Cluster means:
##   bill_length_mm flipper_length_mm
## 1      0.9858721        -0.3524358
## 2     -1.0050404        -0.7692589
## 3      0.6828058         1.2606357
##
## Clustering vector:
##  [1] 2 2 2 2 2 2 2 2 3 3 3 3 3 3 1 1 1 1
##
## Within cluster sum of squares by cluster:
## [1] 1.098928 2.121932 1.247042
##  (between_SS / total_SS =  86.9 %)
##
```

```
## Available components:
##
## [1] "cluster"      "centers"      "totss"        "withinss"     "tot.withinss"
## [6] "betweenss"    "size"         "iter"         "ifault"
```

As you can see above, the clustering object returned by `kmeans` has a lot of information that can be used to visualize the clusters, pick K, and evaluate the total WSSD. To obtain this information in a tidy format, we will call in help from the `broom` package. Let's start by visualizing the clustering as a colored scatter plot. To do that, we use the `augment` function, which takes in the model and the original data frame, and returns a data frame with the data and the cluster assignments for each point:

```
library(broom)

clustered_data <- augment(penguin_clust, standardized_data)
clustered_data
```

```
## # A tibble: 18 x 3
##    bill_length_mm[,1] flipper_length_mm[,1] .cluster
##                 <dbl>                 <dbl> <fct>
##  1             -0.641                -0.190 2
##  2             -1.14                 -1.33  2
##  3             -1.52                 -0.922 2
##  4             -1.11                 -0.922 2
##  5             -0.847                -1.41  2
##  6             -0.641                -0.678 2
##  7             -1.24                 -0.271 2
##  8             -0.902                -0.434 2
##  9              0.720                 1.19  3
## 10              0.646                 1.36  3
## 11              0.963                 1.36  3
## 12              0.440                 1.76  3
## 13              1.21                  1.11  3
## 14              0.123                 0.786 3
## 15              0.627                -0.271 1
## 16              0.757                -0.271 1
## 17              1.78                 -0.108 1
## 18              0.776                -0.759 1
```

Now that we have this information in a tidy data frame, we can make a visualization of the cluster assignments for each point, as shown in Figure 9.13.

```
cluster_plot <- ggplot(clustered_data,
  aes(x = flipper_length_mm,
      y = bill_length_mm,
      color = .cluster),
  size = 2) +
  geom_point() +
  labs(x = "Flipper Length (standardized)",
       y = "Bill Length (standardized)",
       color = "Cluster") +
  scale_color_manual(values = c("dodgerblue3",
                                "darkorange3",
                                "goldenrod1")) +
  theme(text = element_text(size = 12))

cluster_plot
```

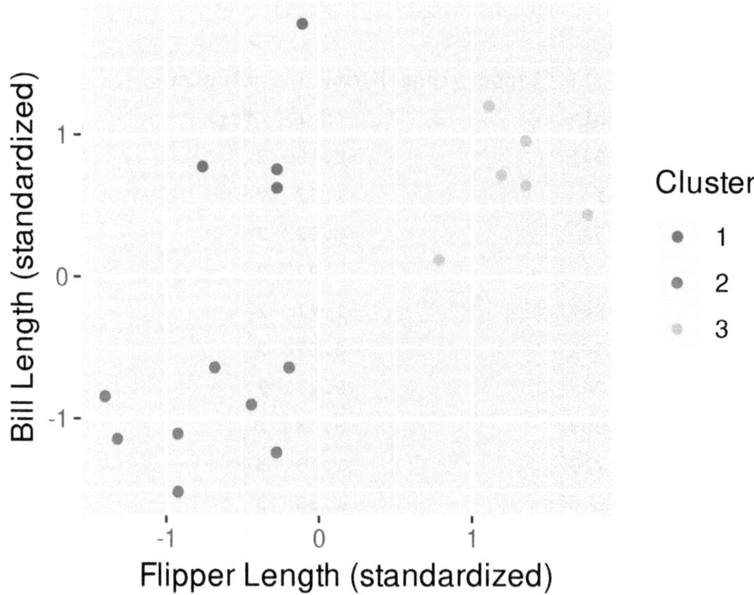

FIGURE 9.13: The data colored by the cluster assignments returned by K-means.

As mentioned above, we also need to select K by finding where the "elbow" occurs in the plot of total WSSD versus the number of clusters. We can obtain the total WSSD (`tot.withinss`) from our clustering using `broom`'s `glance` function. For example:

```
glance(penguin_clust)
```

```
## # A tibble: 1 x 4
##   totss tot.withinss betweenss  iter
##   <dbl>        <dbl>     <dbl> <int>
## 1    34         4.47      29.5     1
```

To calculate the total WSSD for a variety of Ks, we will create a data frame with a column named `k` with rows containing each value of K we want to run K-means with (here, 1 to 9).

```
penguin_clust_ks <- tibble(k = 1:9)
penguin_clust_ks
```

```
## # A tibble: 9 x 1
##       k
##   <int>
## 1     1
## 2     2
## 3     3
## 4     4
## 5     5
## 6     6
## 7     7
## 8     8
## 9     9
```

Then we use `rowwise` + `mutate` to apply the `kmeans` function within each row to each K. However, given that the `kmeans` function returns a model object to us (not a vector), we will need to store the results as a list column. This works because both vectors and lists are legitimate data structures for data frame columns. To make this work, we have to put each model object in a list using the `list` function. We demonstrate how to do this below:

```
penguin_clust_ks <- tibble(k = 1:9) |>
  rowwise() |>
  mutate(penguin_clusts = list(kmeans(standardized_data, k)))
```

If we take a look at our data frame `penguin_clust_ks` now, we see that it has two columns: one with the value for K, and the other holding the clustering model object in a list column.

```
penguin_clust_ks
```

```
## # A tibble: 9 x 2
## # Rowwise:
##       k penguin_clusts
##   <int> <list>
## 1     1 <kmeans>
## 2     2 <kmeans>
## 3     3 <kmeans>
## 4     4 <kmeans>
## 5     5 <kmeans>
## 6     6 <kmeans>
## 7     7 <kmeans>
## 8     8 <kmeans>
## 9     9 <kmeans>
```

If we wanted to get one of the clusterings out of the list column in the data frame, we could use a familiar friend: `pull`. `pull` will return to us a data frame column as a simpler data structure; here, that would be a list. And then to extract the first item of the list, we can use the `pluck` function. We pass it the index for the element we would like to extract (here, `1`).

```
penguin_clust_ks |>
  pull(penguin_clusts) |>
  pluck(1)
```

```
## K-means clustering with 1 clusters of sizes 18
##
## Cluster means:
##   bill_length_mm flipper_length_mm
## 1   6.352943e-16     -8.203315e-16
##
## Clustering vector:
##  [1] 1 1 1 1 1 1 1 1 1 1 1 1 1 1 1 1 1 1
##
## Within cluster sum of squares by cluster:
## [1] 34
##  (between_SS / total_SS =   0.0 %)
##
## Available components:
##
```

```
## [1] "cluster"      "centers"      "totss"        "withinss"     "tot.withinss"
## [6] "betweenss"    "size"         "iter"         "ifault"
```

Next, we use `mutate` again to apply `glance` to each of the K-means clustering objects to get the clustering statistics (including WSSD). The output of `glance` is a data frame, and so we need to create another list column (using `list`) for this to work. This results in a complex data frame with 3 columns, one for K, one for the K-means clustering objects, and one for the clustering statistics:

```
penguin_clust_ks <- tibble(k = 1:9) |>
  rowwise() |>
  mutate(penguin_clusts = list(kmeans(standardized_data, k)),
         glanced = list(glance(penguin_clusts)))

penguin_clust_ks
```

```
## # A tibble: 9 x 3
## # Rowwise:
##       k penguin_clusts glanced
##   <int> <list>         <list>
## 1     1 <kmeans>       <tibble [1 x 4]>
## 2     2 <kmeans>       <tibble [1 x 4]>
## 3     3 <kmeans>       <tibble [1 x 4]>
## 4     4 <kmeans>       <tibble [1 x 4]>
## 5     5 <kmeans>       <tibble [1 x 4]>
## 6     6 <kmeans>       <tibble [1 x 4]>
## 7     7 <kmeans>       <tibble [1 x 4]>
## 8     8 <kmeans>       <tibble [1 x 4]>
## 9     9 <kmeans>       <tibble [1 x 4]>
```

Finally we extract the total WSSD from the column named `glanced`. Given that each item in this list column is a data frame, we will need to use the `unnest` function to unpack the data frames into simpler column data types.

```
clustering_statistics <- penguin_clust_ks |>
  unnest(glanced)

clustering_statistics
```

```
## # A tibble: 9 x 6
##       k penguin_clusts totss tot.withinss betweenss  iter
##   <int> <list>         <dbl>        <dbl>     <dbl> <int>
```

```
## 1      1 <kmeans>       34      34       7.11e-15    1
## 2      2 <kmeans>       34      10.9     2.31e+ 1    1
## 3      3 <kmeans>       34       4.47    2.95e+ 1    1
## 4      4 <kmeans>       34       3.54    3.05e+ 1    1
## 5      5 <kmeans>       34       2.23    3.18e+ 1    2
## 6      6 <kmeans>       34       2.15    3.19e+ 1    3
## 7      7 <kmeans>       34       1.53    3.25e+ 1    2
## 8      8 <kmeans>       34       2.46    3.15e+ 1    1
## 9      9 <kmeans>       34       0.843   3.32e+ 1    2
```

Now that we have `tot.withinss` and `k` as columns in a data frame, we can make a line plot (Figure 9.14) and search for the "elbow" to find which value of K to use.

```
elbow_plot <- ggplot(clustering_statistics, aes(x = k, y = tot.withinss)) +
  geom_point() +
  geom_line() +
  xlab("K") +
  ylab("Total within-cluster sum of squares") +
  scale_x_continuous(breaks = 1:9) +
  theme(text = element_text(size = 12))

elbow_plot
```

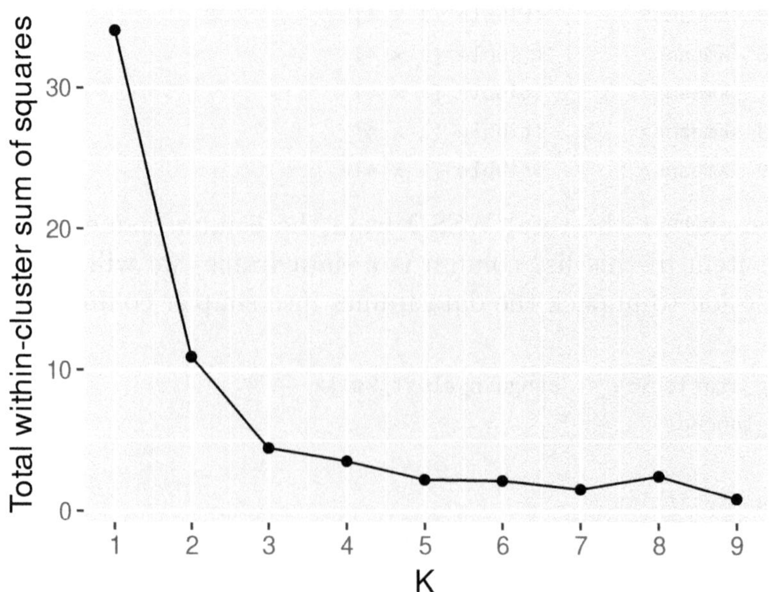

FIGURE 9.14: A plot showing the total WSSD versus the number of clusters.

It looks like 3 clusters is the right choice for this data. But why is there a "bump" in the total WSSD plot here? Shouldn't total WSSD always decrease as we add more clusters? Technically yes, but remember: K-means can get "stuck" in a bad solution. Unfortunately, for K = 8 we had an unlucky initialization and found a bad clustering! We can help prevent finding a bad clustering by trying a few different random initializations via the `nstart` argument (Figure 9.15 shows a setup where we use 10 restarts). When we do this, K-means clustering will be performed the number of times specified by the `nstart` argument, and R will return to us the best clustering from this. The more times we perform K-means clustering, the more likely we are to find a good clustering (if one exists). What value should you choose for `nstart`? The answer is that it depends on many factors: the size and characteristics of your data set, as well as how powerful your computer is. The larger the `nstart` value the better from an analysis perspective, but there is a trade-off that doing many clusterings could take a long time. So this is something that needs to be balanced.

```
penguin_clust_ks <- tibble(k = 1:9) |>
  rowwise() |>
  mutate(penguin_clusts = list(kmeans(standardized_data, nstart = 10, k)),
         glanced = list(glance(penguin_clusts)))

clustering_statistics <- penguin_clust_ks |>
  unnest(glanced)

elbow_plot <- ggplot(clustering_statistics, aes(x = k, y = tot.withinss)) +
  geom_point() +
  geom_line() +
  xlab("K") +
  ylab("Total within-cluster sum of squares") +
  scale_x_continuous(breaks = 1:9) +
  theme(text = element_text(size = 12))

elbow_plot
```

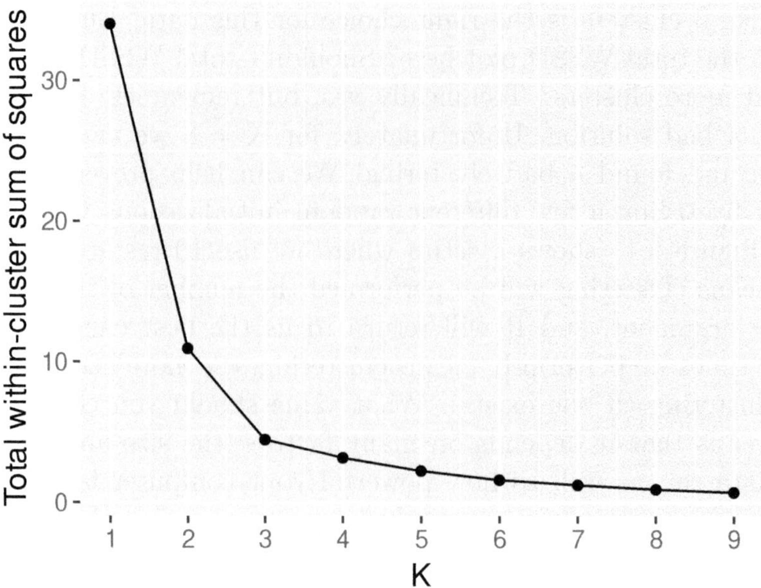

FIGURE 9.15: A plot showing the total WSSD versus the number of clusters when K-means is run with 10 restarts.

9.7 Exercises

Practice exercises for the material covered in this chapter can be found in the accompanying worksheets repository[2] in the "Clustering" row. You can launch an interactive version of the worksheet in your browser by clicking the "launch binder" button. You can also preview a non-interactive version of the worksheet by clicking "view worksheet." If you instead decide to download the worksheet and run it on your own machine, make sure to follow the instructions for computer setup found in Chapter 13. This will ensure that the automated feedback and guidance that the worksheets provide will function as intended.

9.8 Additional resources

- Chapter 10 of *An Introduction to Statistical Learning* [James et al., 2013] provides a great next stop in the process of learning about clustering and unsupervised learning in general. In the realm of clustering specifically, it provides a great companion introduction to K-means, but also covers

[2] https://github.com/UBC-DSCI/data-science-a-first-intro-worksheets#readme

hierarchical clustering for when you expect there to be subgroups, and then subgroups within subgroups, etc., in your data. In the realm of more general unsupervised learning, it covers *principal components analysis (PCA)*, which is a very popular technique for reducing the number of predictors in a dataset.

10

Statistical inference

10.1 Overview

A typical data analysis task in practice is to draw conclusions about some unknown aspect of a population of interest based on observed data sampled from that population; we typically do not get data on the *entire* population. Data analysis questions regarding how summaries, patterns, trends, or relationships in a data set extend to the wider population are called *inferential questions*. This chapter will start with the fundamental ideas of sampling from populations and then introduce two common techniques in statistical inference: *point estimation* and *interval estimation*.

10.2 Chapter learning objectives

By the end of the chapter, readers will be able to do the following:

- Describe real-world examples of questions that can be answered with statistical inference.
- Define common population parameters (e.g., mean, proportion, standard deviation) that are often estimated using sampled data, and estimate these from a sample.
- Define the following statistical sampling terms: population, sample, population parameter, point estimate, and sampling distribution.
- Explain the difference between a population parameter and a sample point estimate.
- Use R to draw random samples from a finite population.
- Use R to create a sampling distribution from a finite population.
- Describe how sample size influences the sampling distribution.
- Define bootstrapping.
- Use R to create a bootstrap distribution to approximate a sampling distribution.
- Contrast the bootstrap and sampling distributions.

10.3 Why do we need sampling?

We often need to understand how quantities we observe in a subset of data relate to the same quantities in the broader population. For example, suppose a retailer is considering selling iPhone accessories, and they want to estimate how big the market might be. Additionally, they want to strategize how they can market their products on North American college and university campuses. This retailer might formulate the following question:

What proportion of all undergraduate students in North America own an iPhone?

In the above question, we are interested in making a conclusion about *all* undergraduate students in North America; this is referred to as the **population**. In general, the population is the complete collection of individuals or cases we are interested in studying. Further, in the above question, we are interested in computing a quantity—the proportion of iPhone owners—based on the entire population. This proportion is referred to as a **population parameter**. In general, a population parameter is a numerical characteristic of the entire population. To compute this number in the example above, we would need to ask every single undergraduate in North America whether they own an iPhone. In practice, directly computing population parameters is often time-consuming and costly, and sometimes impossible.

A more practical approach would be to make measurements for a **sample**, i.e., a subset of individuals collected from the population. We can then compute a **sample estimate**—a numerical characteristic of the sample—that estimates the population parameter. For example, suppose we randomly selected ten undergraduate students across North America (the sample) and computed the proportion of those students who own an iPhone (the sample estimate). In that case, we might suspect that proportion is a reasonable estimate of the proportion of students who own an iPhone in the entire population. Figure 10.1 illustrates this process. In general, the process of using a sample to make a conclusion about the broader population from which it is taken is referred to as **statistical inference**.

Note that proportions are not the *only* kind of population parameter we might be interested in. For example, suppose an undergraduate student studying at the University of British Columbia in Canada is looking for an apartment to rent. They need to create a budget, so they want to know about studio apartment rental prices in Vancouver. This student might formulate the question:

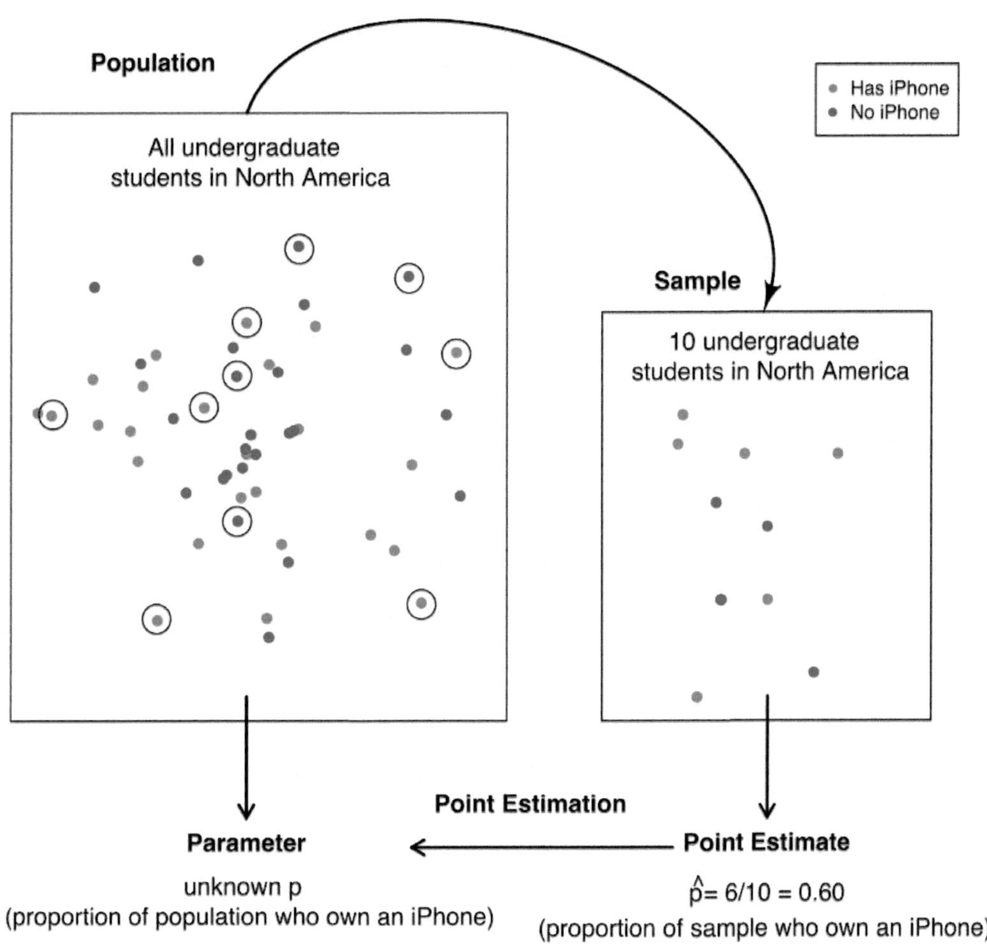

FIGURE 10.1: Population versus sample.

What is the average price per month of studio apartment rentals in Vancouver?

In this case, the population consists of all studio apartment rentals in Vancouver, and the population parameter is the *average price per month.* Here we used the average as a measure of the center to describe the "typical value" of studio apartment rental prices. But even within this one example, we could also be interested in many other population parameters. For instance, we know that not every studio apartment rental in Vancouver will have the same price per month. The student might be interested in how much monthly prices vary and want to find a measure of the rentals' spread (or variability), such as the standard deviation. Or perhaps the student might be interested in the fraction of studio apartment rentals that cost more than $1000 per month. The question we want to answer will help us determine the parameter we want to estimate. If we were somehow able to observe the whole population of studio apartment rental offerings in Vancouver, we could compute each of these numbers exactly; therefore, these are all population parameters. There are many kinds of observations and population parameters that you will run into in practice, but in this chapter, we will focus on two settings:

1. Using categorical observations to estimate the proportion of a category
2. Using quantitative observations to estimate the average (or mean)

10.4 Sampling distributions

10.4.1 Sampling distributions for proportions

We will look at an example using data from Inside Airbnb[1] [Cox, n.d.]. Airbnb is an online marketplace for arranging vacation rentals and places to stay. The data set contains listings for Vancouver, Canada, in September 2020. Our data includes an ID number, neighborhood, type of room, the number of people the rental accommodates, number of bathrooms, bedrooms, beds, and the price per night.

```
library(tidyverse)

set.seed(123)

airbnb <- read_csv("data/listings.csv")
airbnb
```

```
## # A tibble: 4,594 x 8
##       id neighbourhood   room_type  accommodates bathrooms bedrooms  beds price
##    <dbl> <chr>           <chr>             <dbl> <chr>        <dbl> <dbl> <dbl>
## 1      1 Downtown        Entire hom~           5 2 baths          2     2   150
## 2      2 Downtown Easts~ Entire hom~           4 2 baths          2     2   132
## 3      3 West End        Entire hom~           2 1 bath           1     1    85
## 4      4 Kensington-Ced~ Entire hom~           2 1 bath           1     0   146
## 5      5 Kensington-Ced~ Entire hom~           4 1 bath           1     2   110
## 6      6 Hastings-Sunri~ Entire hom~           4 1 bath           2     3   195
```

[1] http://insideairbnb.com/

```
## 7      7 Renfrew-Collin~ Entire hom~          8 3 baths          4     5   130
## 8      8 Mount Pleasant  Entire hom~          2 1 bath           1     1    94
## 9      9 Grandview-Wood~ Private ro~          2 1 privat~        1     1    79
## 10    10 West End        Private ro~          2 1 privat~        1     1    75
## # ... with 4,584 more rows
```

Suppose the city of Vancouver wants information about Airbnb rentals to help plan city bylaws, and they want to know how many Airbnb places are listed as entire homes and apartments (rather than as private or shared rooms). Therefore they may want to estimate the true proportion of all Airbnb listings where the "type of place" is listed as "entire home or apartment." Of course, we usually do not have access to the true population, but here let's imagine (for learning purposes) that our data set represents the population of all Airbnb rental listings in Vancouver, Canada. We can find the proportion of listings where `room_type == "Entire home/apt"`.

```
airbnb |>
  summarize(
    n =  sum(room_type == "Entire home/apt"),
    proportion = sum(room_type == "Entire home/apt") / nrow(airbnb)
  )
```

```
## # A tibble: 1 x 2
##       n proportion
##   <int>      <dbl>
## 1  3434      0.747
```

We can see that the proportion of `Entire home/apt` listings in the data set is 0.747. This value, 0.747, is the population parameter. Remember, this parameter value is usually unknown in real data analysis problems, as it is typically not possible to make measurements for an entire population.

Instead, perhaps we can approximate it with a small subset of data! To investigate this idea, let's try randomly selecting 40 listings (*i.e.,* taking a random sample of size 40 from our population), and computing the proportion for that sample. We will use the `rep_sample_n` function from the `infer` package to take the sample. The arguments of `rep_sample_n` are (1) the data frame to sample from, and (2) the size of the sample to take.

```
library(infer)

sample_1 <- rep_sample_n(tbl = airbnb, size = 40)
```

```
airbnb_sample_1 <- summarize(sample_1,
  n = sum(room_type == "Entire home/apt"),
  prop = sum(room_type == "Entire home/apt") / 40
)

airbnb_sample_1
```

```
## # A tibble: 1 x 3
##   replicate     n  prop
##       <int> <int> <dbl>
## 1         1    28   0.7
```

Here we see that the proportion of entire home/apartment listings in this random sample is 0.7. Wow—that's close to our true population value! But remember, we computed the proportion using a random sample of size 40. This has two consequences. First, this value is only an *estimate*, i.e., our best guess of our population parameter using this sample. Given that we are estimating a single value here, we often refer to it as a **point estimate**. Second, since the sample was random, if we were to take *another* random sample of size 40 and compute the proportion for that sample, we would not get the same answer:

```
sample_2 <- rep_sample_n(airbnb, size = 40)

airbnb_sample_2 <- summarize(sample_2,
  n = sum(room_type == "Entire home/apt"),
  prop = sum(room_type == "Entire home/apt") / 40
)

airbnb_sample_2
```

```
## # A tibble: 1 x 3
##   replicate     n  prop
##       <int> <int> <dbl>
## 1         1    35 0.875
```

Confirmed! We get a different value for our estimate this time. That means that our point estimate might be unreliable. Indeed, estimates vary from sample to sample due to **sampling variability**. But just how much should we expect the estimates of our random samples to vary? Or in other words, how much can we really trust our point estimate based on a single sample?

To understand this, we will simulate many samples (much more than just two) of size 40 from our population of listings and calculate the proportion of entire home/apartment listings in each sample. This simulation will create many sample proportions, which we can visualize using a histogram. The distribution of the estimate for all possible samples of a given size (which we commonly refer to as n) from a population is called a **sampling distribution**. The sampling distribution will help us see how much we would expect our sample proportions from this population to vary for samples of size 40.

We again use the `rep_sample_n` to take samples of size 40 from our population of Airbnb listings. But this time we set the `reps` argument to 20,000 to specify that we want to take 20,000 samples of size 40.

```
samples <- rep_sample_n(airbnb, size = 40, reps = 20000)
samples
```

```
## # A tibble: 800,000 x 9
## # Groups:   replicate [20,000]
##    replicate    id neighbourhood room_type accommodates bathrooms bedrooms  beds
##        <int> <dbl> <chr>         <chr>            <dbl> <chr>        <dbl> <dbl>
##  1         1  4403 Downtown      Entire h~            2 1 bath           1     1
##  2         1   902 Kensington-C~ Private ~            2 1 shared~        1     1
##  3         1  3808 Hastings-Sun~ Entire h~            6 1.5 baths        1     3
##  4         1   561 Kensington-C~ Entire h~            6 1 bath           2     2
##  5         1  3385 Mount Pleasa~ Entire h~            4 1 bath           1     1
##  6         1  4232 Shaughnessy   Entire h~            6 1.5 baths        2     2
##  7         1  1169 Downtown      Entire h~            3 1 bath           1     1
##  8         1   959 Kitsilano     Private ~            1 1.5 shar~        1     1
##  9         1  2171 Downtown      Entire h~            2 1 bath           1     1
## 10         1  1258 Dunbar South~ Entire h~            4 1 bath           2     2
## # ... with 799,990 more rows, and 1 more variable: price <dbl>
```

Notice that the column `replicate` indicates the replicate, or sample, to which each listing belongs. Above, since by default R only prints the first few rows, it looks like all of the listings have `replicate` set to 1. But you can check the last few entries using the `tail()` function to verify that we indeed created 20,000 samples (or replicates).

```
tail(samples)
```

```
## # A tibble: 6 x 9
## # Groups:   replicate [1]
##   replicate    id neighbourhood  room_type accommodates bathrooms bedrooms  beds
```

```
##       <int> <dbl> <chr>           <chr>             <dbl> <chr>       <dbl> <dbl>
## 1     20000  3414 Marpole         Entire h~             4 1 bath          2     2
## 2     20000  1974 Hastings-Sunr~  Private ~             2 1 shared~       1     1
## 3     20000  1846 Riley Park      Entire h~             4 1 bath          2     3
## 4     20000   862 Downtown        Entire h~             5 2 baths         2     2
## 5     20000  3295 Victoria-Fras~  Private ~             2 1 shared~       1     1
## 6     20000   997 Dunbar Southl~  Private ~             1 1.5 shar~       1     1
## # ... with 1 more variable: price <dbl>
```

Now that we have obtained the samples, we need to compute the proportion of entire home/apartment listings in each sample. We first group the data by the `replicate` variable—to group the set of listings in each sample together—and then use `summarize` to compute the proportion in each sample. We print both the first and last few entries of the resulting data frame below to show that we end up with 20,000 point estimates, one for each of the 20,000 samples.

```
sample_estimates <- samples |>
  group_by(replicate) |>
  summarize(sample_proportion = sum(room_type == "Entire home/apt") / 40)

sample_estimates
```

```
## # A tibble: 20,000 x 2
##    replicate sample_proportion
##        <int>             <dbl>
##  1         1             0.85
##  2         2             0.85
##  3         3             0.65
##  4         4             0.7
##  5         5             0.75
##  6         6             0.725
##  7         7             0.775
##  8         8             0.775
##  9         9             0.7
## 10        10             0.675
## # ... with 19,990 more rows
```

```
tail(sample_estimates)
```

```
## # A tibble: 6 x 2
##   replicate sample_proportion
##       <int>             <dbl>
```

```
## 1     19995          0.75
## 2     19996          0.675
## 3     19997          0.625
## 4     19998          0.75
## 5     19999          0.875
## 6     20000          0.65
```

We can now visualize the sampling distribution of sample proportions for samples of size 40 using a histogram in Figure 10.2. Keep in mind: in the real world, we don't have access to the full population. So we can't take many samples and can't actually construct or visualize the sampling distribution. We have created this particular example such that we *do* have access to the full population, which lets us visualize the sampling distribution directly for learning purposes.

```
sampling_distribution <- ggplot(sample_estimates, aes(x = sample_proportion)) +
  geom_histogram(fill = "dodgerblue3", color = "lightgrey", bins = 12) +
  labs(x = "Sample proportions", y = "Count") +
  theme(text = element_text(size = 12))

sampling_distribution
```

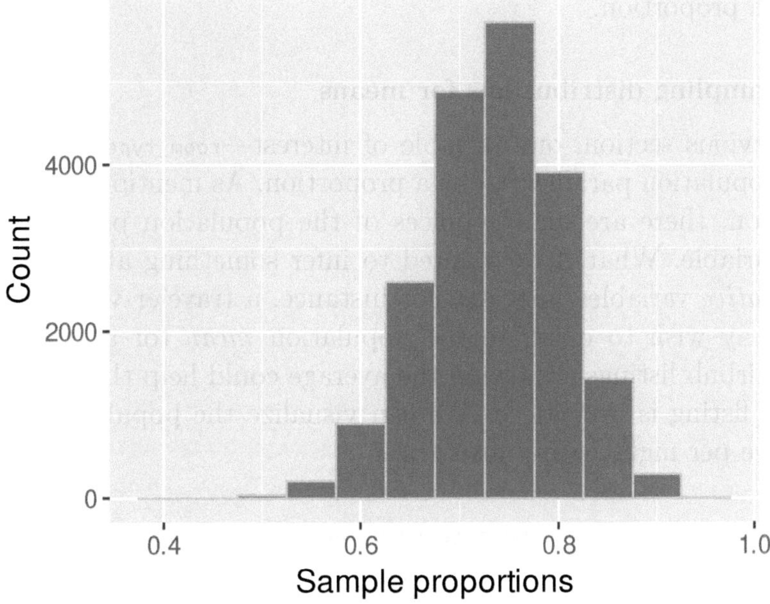

FIGURE 10.2: Sampling distribution of the sample proportion for sample size 40.

The sampling distribution in Figure 10.2 appears to be bell-shaped, is roughly symmetric, and has one peak. It is centered around 0.7 and the sample proportions range from about 0.4 to about 1. In fact, we can calculate the mean of the sample proportions.

```
sample_estimates |>
  summarize(mean = mean(sample_proportion))
```

```
## # A tibble: 1 x 1
##    mean
##   <dbl>
## 1 0.747
```

We notice that the sample proportions are centered around the population proportion value, 0.747! In general, the mean of the sampling distribution should be equal to the population proportion. This is great news because it means that the sample proportion is neither an overestimate nor an underestimate of the population proportion. In other words, if you were to take many samples as we did above, there is no tendency towards over or underestimating the population proportion. In a real data analysis setting where you just have access to your single sample, this implies that you would suspect that your sample point estimate is roughly equally likely to be above or below the true population proportion.

10.4.2 Sampling distributions for means

In the previous section, our variable of interest—`room_type`—was *categorical*, and the population parameter was a proportion. As mentioned in the chapter introduction, there are many choices of the population parameter for each type of variable. What if we wanted to infer something about a population of *quantitative* variables instead? For instance, a traveler visiting Vancouver, Canada may wish to estimate the population *mean* (or average) price per night of Airbnb listings. Knowing the average could help them tell whether a particular listing is overpriced. We can visualize the population distribution of the price per night with a histogram.

```
population_distribution <- ggplot(airbnb, aes(x = price)) +
  geom_histogram(fill = "dodgerblue3", color = "lightgrey") +
  labs(x = "Price per night (Canadian dollars)", y = "Count") +
  theme(text = element_text(size = 12))

population_distribution
```

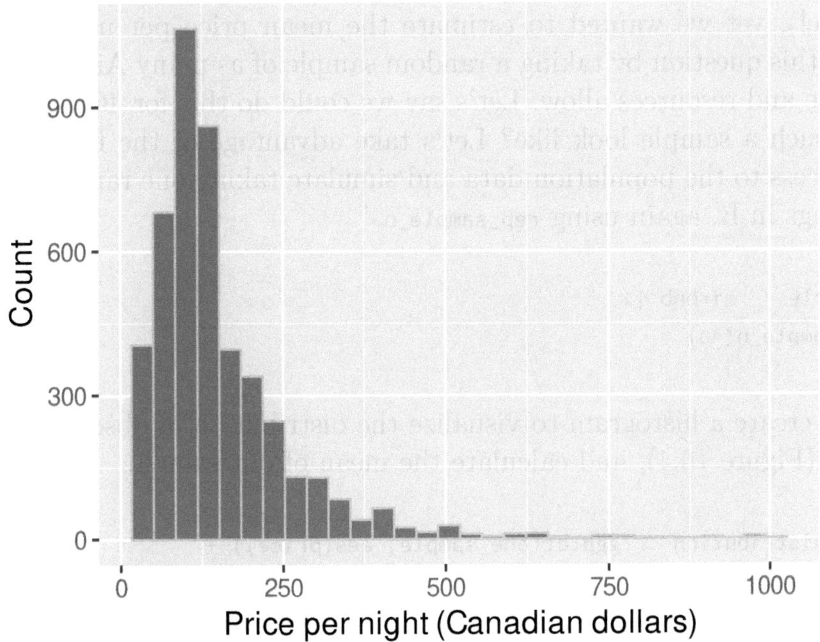

FIGURE 10.3: Population distribution of price per night (Canadian dollars) for all Airbnb listings in Vancouver, Canada.

In Figure 10.3, we see that the population distribution has one peak. It is also skewed (i.e., is not symmetric): most of the listings are less than $250 per night, but a small number of listings cost much more, creating a long tail on the histogram's right side. Along with visualizing the population, we can calculate the population mean, the average price per night for all the Airbnb listings.

```
population_parameters <- airbnb |>
  summarize(pop_mean = mean(price))

population_parameters
```

```
## # A tibble: 1 x 1
##   pop_mean
##      <dbl>
## 1   154.51
```

The price per night of all Airbnb rentals in Vancouver, BC is $154.51, on average. This value is our population parameter since we are calculating it using the population data.

Now suppose we did not have access to the population data (which is usually the case!), yet we wanted to estimate the mean price per night. We could answer this question by taking a random sample of as many Airbnb listings as our time and resources allow. Let's say we could do this for 40 listings. What would such a sample look like? Let's take advantage of the fact that we do have access to the population data and simulate taking one random sample of 40 listings in R, again using `rep_sample_n`.

```
one_sample <- airbnb |>
  rep_sample_n(40)
```

We can create a histogram to visualize the distribution of observations in the sample (Figure 10.4), and calculate the mean of our sample.

```
sample_distribution <- ggplot(one_sample, aes(price)) +
  geom_histogram(fill = "dodgerblue3", color = "lightgrey") +
  labs(x = "Price per night (Canadian dollars)", y = "Count") +
  theme(text = element_text(size = 12))

sample_distribution
```

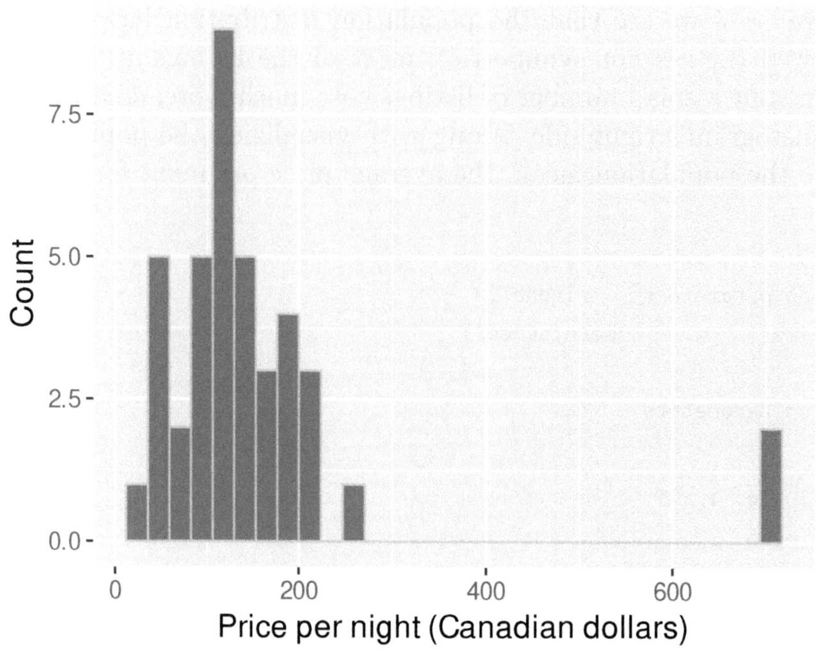

FIGURE 10.4: Distribution of price per night (Canadian dollars) for sample of 40 Airbnb listings.

```
estimates <- one_sample |>
  summarize(sample_mean = mean(price))

estimates
```

```
## # A tibble: 1 x 2
##   replicate sample_mean
##       <int>       <dbl>
## 1         1      155.80
```

The average value of the sample of size 40 is \$155.8. This number is a point estimate for the mean of the full population. Recall that the population mean was \$154.51. So our estimate was fairly close to the population parameter: the mean was about 0.8% off. Note that we usually cannot compute the estimate's accuracy in practice since we do not have access to the population parameter; if we did, we wouldn't need to estimate it!

Also, recall from the previous section that the point estimate can vary; if we took another random sample from the population, our estimate's value might change. So then, did we just get lucky with our point estimate above? How much does our estimate vary across different samples of size 40 in this example? Again, since we have access to the population, we can take many samples and plot the sampling distribution of sample means for samples of size 40 to get a sense for this variation. In this case, we'll use 20,000 samples of size 40.

```
samples <- rep_sample_n(airbnb, size = 40, reps = 20000)
samples
```

```
## # A tibble: 800,000 x 9
## # Groups:   replicate [20,000]
##    replicate    id neighbourhood room_type accommodates bathrooms bedrooms  beds
##        <int> <dbl> <chr>         <chr>            <dbl> <chr>        <dbl> <dbl>
##  1         1  1177 Downtown      Entire h~            4 2 baths          2     2
##  2         1  4063 Downtown      Entire h~            2 1 bath           1     1
##  3         1  2641 Kitsilano     Private ~            1 1 shared~        1     1
##  4         1  1941 West End      Entire h~            2 1 bath           1     1
##  5         1  2431 Mount Pleasa~ Entire h~            2 1 bath           1     1
##  6         1  1871 Arbutus Ridge Entire h~            4 1 bath           2     2
##  7         1  2557 Marpole       Private ~            3 1 privat~        1     2
##  8         1  3534 Downtown      Entire h~            2 1 bath           1     1
##  9         1  4379 Downtown      Entire h~            4 1 bath           1     0
## 10         1  2161 Downtown      Entire h~            4 2 baths          2     2
```

```
## # ... with 799,990 more rows, and 1 more variable: price <dbl>
```

Now we can calculate the sample mean for each replicate and plot the sampling distribution of sample means for samples of size 40.

```
sample_estimates <- samples |>
  group_by(replicate) |>
  summarize(sample_mean = mean(price))

sample_estimates
```

```
## # A tibble: 20,000 x 2
##    replicate sample_mean
##        <int>       <dbl>
##  1         1      160.06
##  2         2      173.18
##  3         3      131.20
##  4         4      176.96
##  5         5      125.65
##  6         6      148.84
##  7         7      134.82
##  8         8      137.26
##  9         9      166.11
## 10        10      157.81
## # ... with 19,990 more rows
```

```
sampling_distribution_40 <- ggplot(sample_estimates, aes(x = sample_mean)) +
  geom_histogram(fill = "dodgerblue3", color = "lightgrey") +
  labs(x = "Sample mean price per night (Canadian dollars)", y = "Count") +
  theme(text = element_text(size = 12))

sampling_distribution_40
```

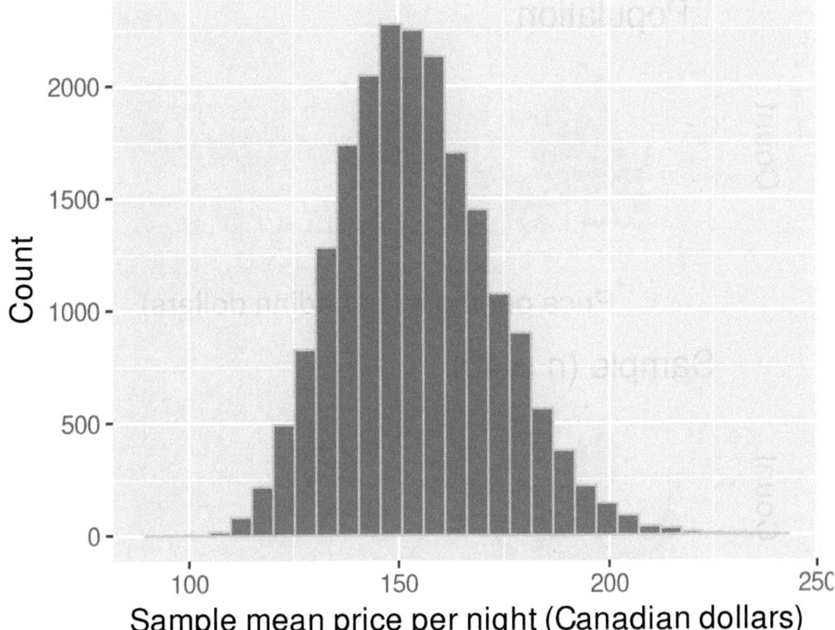

FIGURE 10.5: Sampling distribution of the sample means for sample size of 40.

In Figure 10.5, the sampling distribution of the mean has one peak and is bell-shaped. Most of the estimates are between about \$140 and \$170; but there are a good fraction of cases outside this range (i.e., where the point estimate was not close to the population parameter). So it does indeed look like we were quite lucky when we estimated the population mean with only 0.8% error.

Let's visualize the population distribution, distribution of the sample, and the sampling distribution on one plot to compare them in Figure 10.6. Comparing these three distributions, the centers of the distributions are all around the same price (around \$150). The original population distribution has a long right tail, and the sample distribution has a similar shape to that of the population distribution. However, the sampling distribution is not shaped like the population or sample distribution. Instead, it has a bell shape, and it has a lower spread than the population or sample distributions. The sample means vary less than the individual observations because there will be some high values and some small values in any random sample, which will keep the average from being too extreme.

Given that there is quite a bit of variation in the sampling distribution of the sample mean—i.e., the point estimate that we obtain is not very reliable—is there any way to improve the estimate? One way to improve a point estimate is to take a *larger* sample. To illustrate what effect this has, we will take many

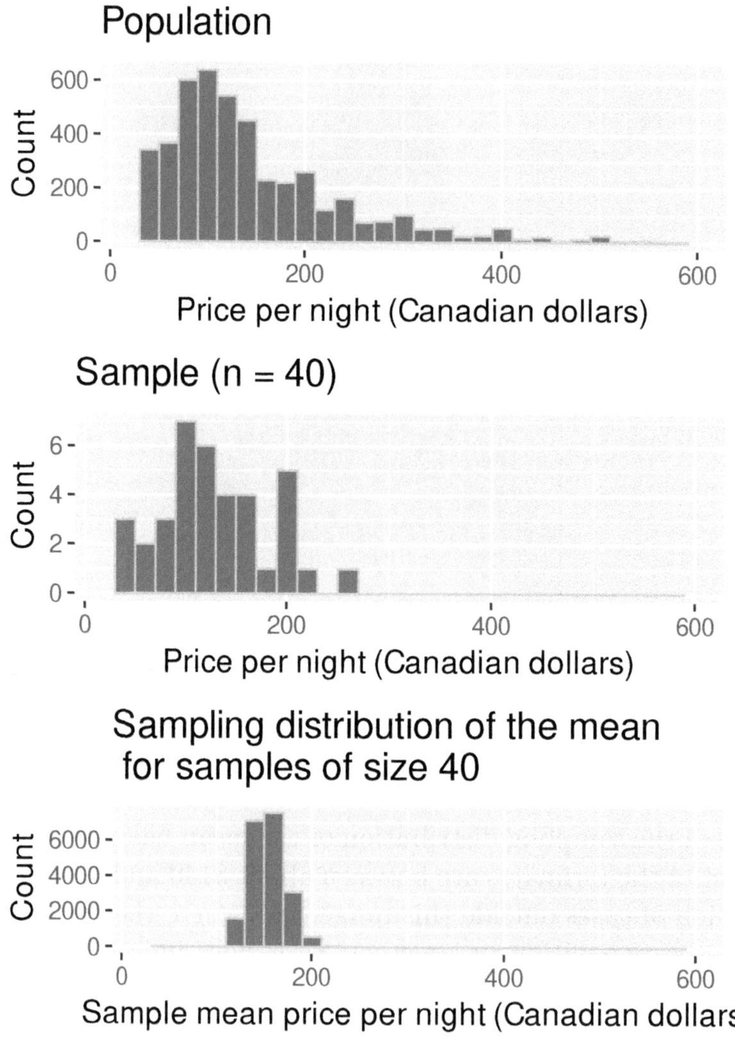

FIGURE 10.6: Comparison of population distribution, sample distribution, and sampling distribution.

samples of size 20, 50, 100, and 500, and plot the sampling distribution of the sample mean. We indicate the mean of the sampling distribution with a red vertical line.

Based on the visualization in Figure 10.7, three points about the sample mean become clear. First, the mean of the sample mean (across samples) is equal to the population mean. In other words, the sampling distribution is centered at the population mean. Second, increasing the size of the sample decreases the spread (i.e., the variability) of the sampling distribution. Therefore, a larger sample size results in a more reliable point estimate of the population parameter. And third, the distribution of the sample mean is roughly bell-shaped.

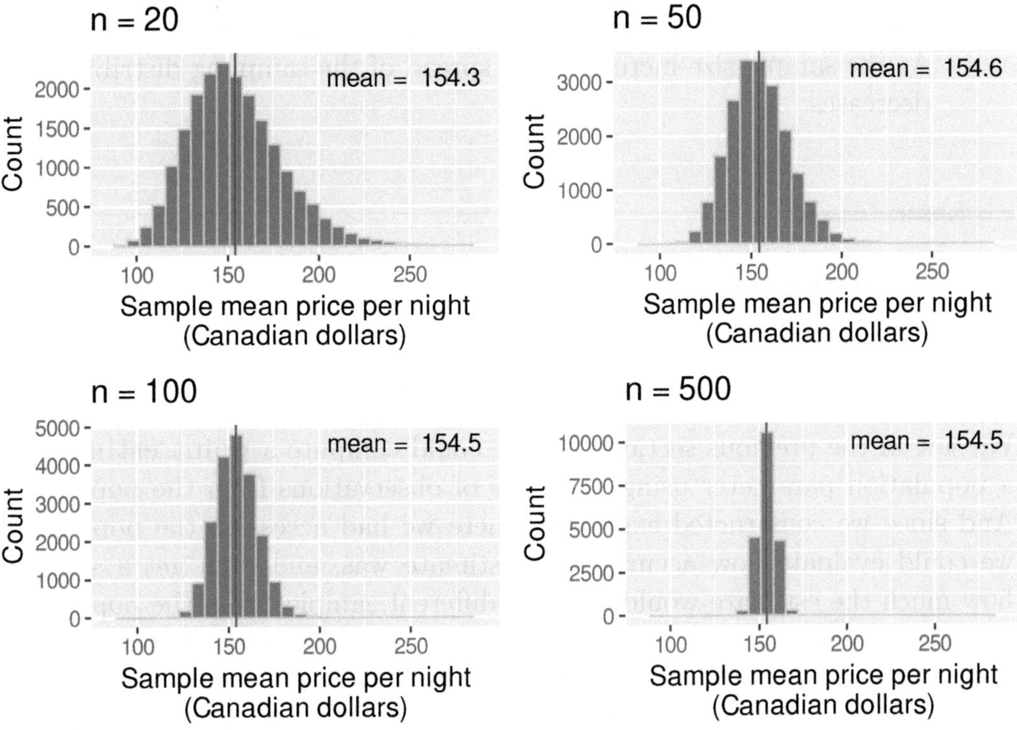

FIGURE 10.7: Comparison of sampling distributions, with mean highlighted as a vertical red line.

Note: You might notice that in the `n = 20` case in Figure 10.7, the distribution is not *quite* bell-shaped. There is a bit of skew towards the right! You might also notice that in the `n = 50` case and larger, that skew seems to disappear. In general, the sampling distribution—for both means and proportions—only becomes bell-shaped *once the sample size is large enough*. How large is "large enough?" Unfortunately, it depends entirely on the problem at hand. But as a rule of thumb, often a sample size of at least 20 will suffice.

10.4.3 Summary

1. A point estimate is a single value computed using a sample from a population (e.g., a mean or proportion).
2. The sampling distribution of an estimate is the distribution of the estimate for all possible samples of a fixed size from the same population.
3. The shape of the sampling distribution is usually bell-shaped with one peak and centered at the population mean or proportion.

4. The spread of the sampling distribution is related to the sample size. As the sample size increases, the spread of the sampling distribution decreases.

10.5 Bootstrapping

10.5.1 Overview

Why all this emphasis on sampling distributions?

We saw in the previous section that we could compute a **point estimate** of a population parameter using a sample of observations from the population. And since we constructed examples where we had access to the population, we could evaluate how accurate the estimate was, and even get a sense of how much the estimate would vary for different samples from the population. But in real data analysis settings, we usually have *just one sample* from our population and do not have access to the population itself. Therefore we cannot construct the sampling distribution as we did in the previous section. And as we saw, our sample estimate's value can vary significantly from the population parameter. So reporting the point estimate from a single sample alone may not be enough. We also need to report some notion of *uncertainty* in the value of the point estimate.

Unfortunately, we cannot construct the exact sampling distribution without full access to the population. However, if we could somehow *approximate* what the sampling distribution would look like for a sample, we could use that approximation to then report how uncertain our sample point estimate is (as we did above with the *exact* sampling distribution). There are several methods to accomplish this; in this book, we will use the *bootstrap*. We will discuss **interval estimation** and construct **confidence intervals** using just a single sample from a population. A confidence interval is a range of plausible values for our population parameter.

Here is the key idea. First, if you take a big enough sample, it *looks like* the population. Notice the histograms' shapes for samples of different sizes taken from the population in Figure 10.8. We see that the sample's distribution looks like that of the population for a large enough sample.

In the previous section, we took many samples of the same size *from our population* to get a sense of the variability of a sample estimate. But if our sample is big enough that it looks like our population, we can pretend that our sample *is* the population, and take more samples (with replacement) of the

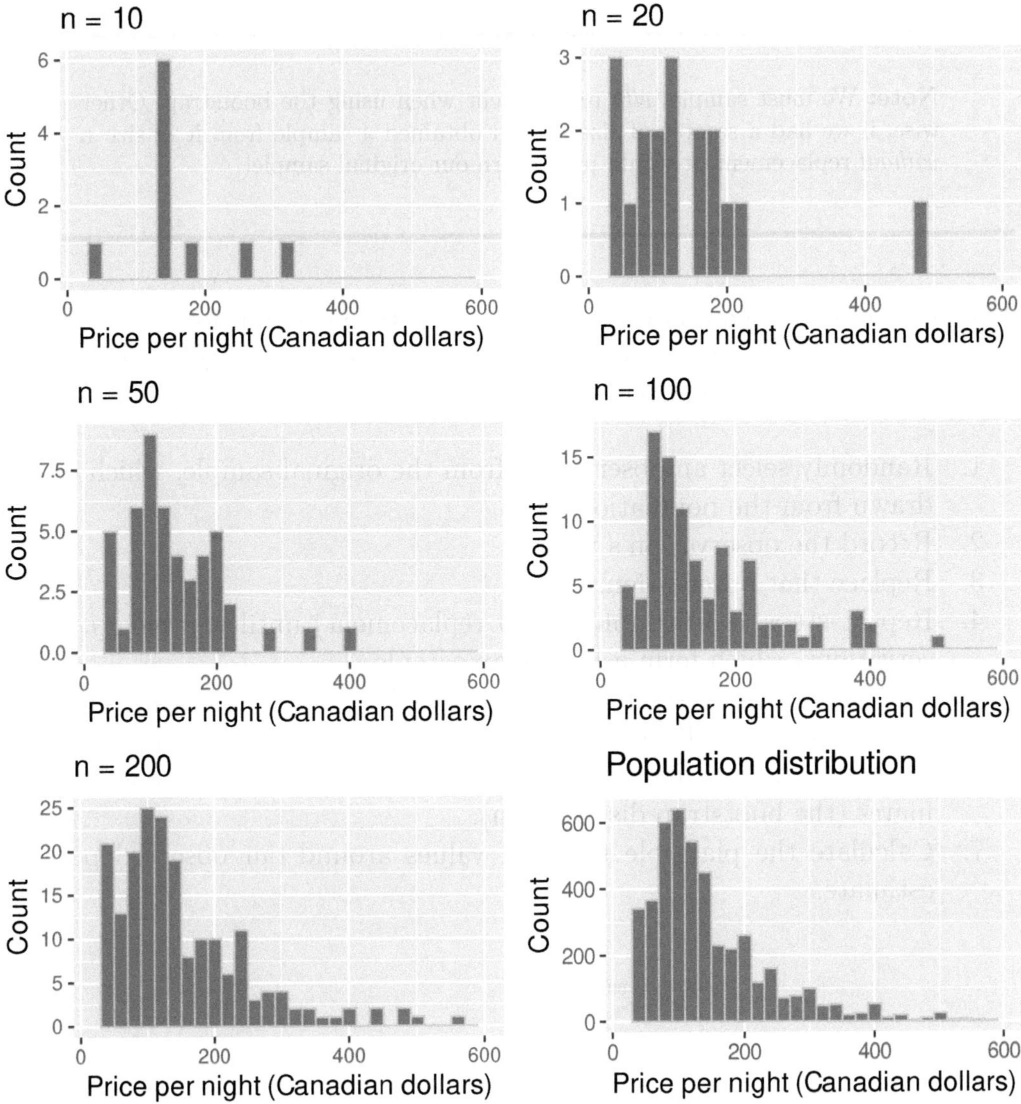

FIGURE 10.8: Comparison of samples of different sizes from the population.

same size from it instead! This very clever technique is called **the bootstrap**. Note that by taking many samples from our single, observed sample, we do not obtain the true sampling distribution, but rather an approximation that we call **the bootstrap distribution**.

Note: We must sample *with* replacement when using the bootstrap. Otherwise, if we had a sample of size n, and obtained a sample from it of size n *without* replacement, it would just return our original sample!

This section will explore how to create a bootstrap distribution from a single sample using R. The process is visualized in Figure 10.9. For a sample of size n, you would do the following:

1. Randomly select an observation from the original sample, which was drawn from the population.
2. Record the observation's value.
3. Replace that observation.
4. Repeat steps 1–3 (sampling *with* replacement) until you have n observations, which form a bootstrap sample.
5. Calculate the bootstrap point estimate (e.g., mean, median, proportion, slope, etc.) of the n observations in your bootstrap sample.
6. Repeat steps 1–5 many times to create a distribution of point estimates (the bootstrap distribution).
7. Calculate the plausible range of values around our observed point estimate.

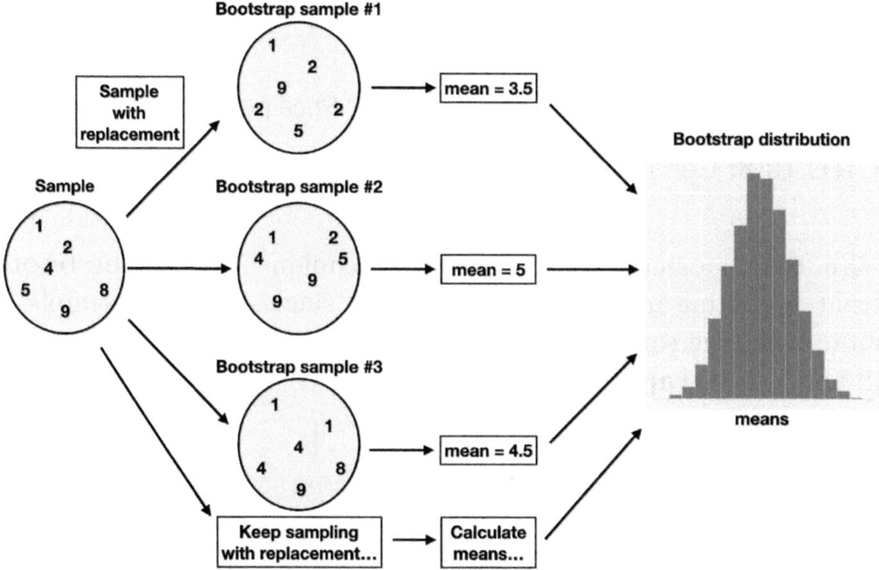

FIGURE 10.9: Overview of the bootstrap process.

10.5.2 Bootstrapping in R

Let's continue working with our Airbnb example to illustrate how we might create and use a bootstrap distribution using just a single sample from the population. Once again, suppose we are interested in estimating the population mean price per night of all Airbnb listings in Vancouver, Canada, using a single sample size of 40. Recall our point estimate was \$155.8. The histogram of prices in the sample is displayed in Figure 10.10.

```
one_sample
```

```
## # A tibble: 40 x 8
##       id neighbourhood     room_type  accommodates bathrooms bedrooms  beds price
##    <dbl> <chr>             <chr>             <dbl> <chr>        <dbl> <dbl> <dbl>
##  1  3928 Marpole           Private r~            2 1 shared~        1     1    58
##  2  3013 Kensington-Ceda~ Entire ho~            4 1 bath           2     2   112
##  3  3156 Downtown          Entire ho~            6 2 baths          2     2   151
##  4  3873 Dunbar Southlan~ Private r~            5 1 bath           2     3   700
##  5  3632 Downtown Eastsi~ Entire ho~            6 2 baths          3     3   157
##  6   296 Kitsilano         Private r~            1 1 shared~        1     1   100
##  7  3514 West End          Entire ho~            2 1 bath           1     1   110
##  8   594 Sunset            Entire ho~            5 1 bath           3     3   105
##  9  3305 Dunbar Southlan~ Entire ho~            4 1 bath           1     2   196
## 10   938 Downtown          Entire ho~            7 2 baths          2     3   269
## # ... with 30 more rows
```

```
one_sample_dist <- ggplot(one_sample, aes(price)) +
  geom_histogram(fill = "dodgerblue3", color = "lightgrey") +
  labs(x = "Price per night (Canadian dollars)", y = "Count") +
  theme(text = element_text(size = 12))

one_sample_dist
```

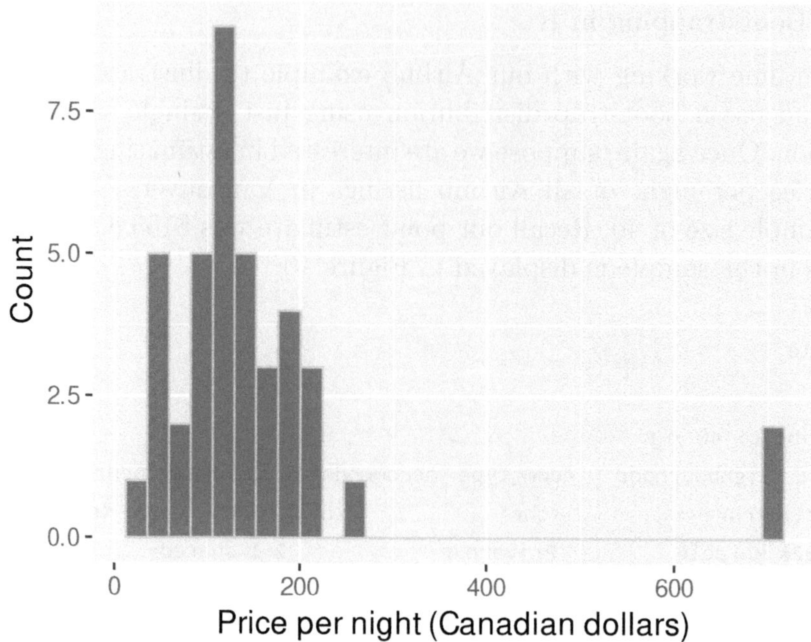

FIGURE 10.10: Histogram of price per night (Canadian dollars) for one sample of size 40.

The histogram for the sample is skewed, with a few observations out to the right. The mean of the sample is $155.8. Remember, in practice, we usually only have this one sample from the population. So this sample and estimate are the only data we can work with.

We now perform steps 1–5 listed above to generate a single bootstrap sample in R and calculate a point estimate from that bootstrap sample. We will use the `rep_sample_n` function as we did when we were creating our sampling distribution. But critically, note that we now pass `one_sample`—our single sample of size 40—as the first argument. And since we need to sample with replacement, we change the argument for `replace` from its default value of `FALSE` to `TRUE`.

```
boot1 <- one_sample |>
  rep_sample_n(size = 40, replace = TRUE, reps = 1)
boot1_dist <- ggplot(boot1, aes(price)) +
  geom_histogram(fill = "dodgerblue3", color = "lightgrey") +
  labs(x = "Price per night (Canadian dollars)", y =  "Count") +
  theme(text = element_text(size = 12))

boot1_dist
```

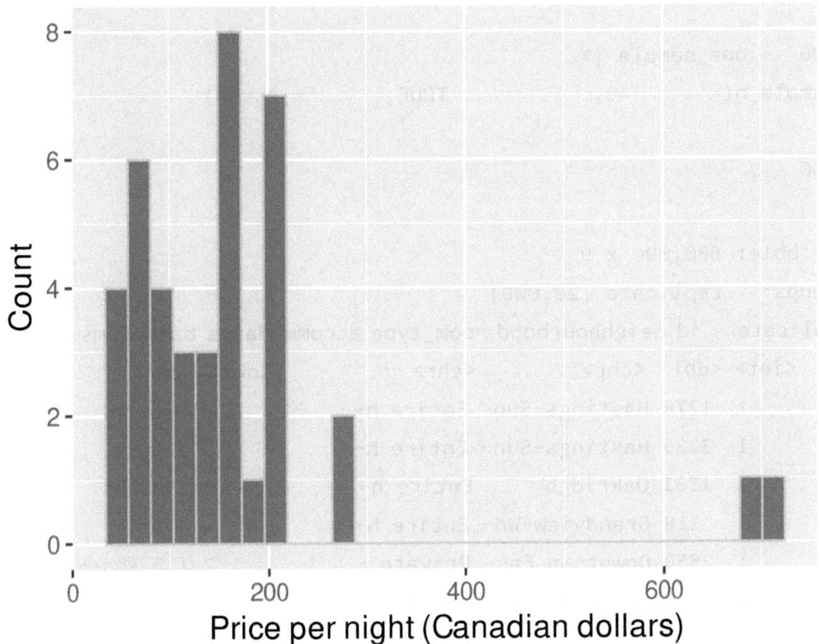

FIGURE 10.11: Bootstrap distribution.

```
summarize(boot1, mean = mean(price))
```

```
## # A tibble: 1 x 2
##   replicate   mean
##       <int>  <dbl>
## 1         1 164.20
```

Notice in Figure 10.11 that the histogram of our bootstrap sample has a similar shape to the original sample histogram. Though the shapes of the distributions are similar, they are not identical. You'll also notice that the original sample mean and the bootstrap sample mean differ. How might that happen? Remember that we are sampling with replacement from the original sample, so we don't end up with the same sample values again. We are *pretending* that our single sample is close to the population, and we are trying to mimic drawing another sample from the population by drawing one from our original sample.

Let's now take 20,000 bootstrap samples from the original sample (`one_sample`) using `rep_sample_n`, and calculate the means for each of those replicates. Recall that this assumes that `one_sample` *looks like* our original population; but since we do not have access to the population itself, this is often the best we can do.

```
boot20000 <- one_sample |>
  rep_sample_n(size = 40, replace = TRUE, reps = 20000)

boot20000
```

```
## # A tibble: 800,000 x 9
## # Groups:   replicate [20,000]
##    replicate    id neighbourhood room_type accommodates bathrooms bedrooms  beds
##        <int> <dbl> <chr>         <chr>            <dbl> <chr>        <dbl> <dbl>
##  1         1  1276 Hastings-Sun~ Entire h~            2 1 bath           1     1
##  2         1  3235 Hastings-Sun~ Entire h~            2 1 bath           1     1
##  3         1  1301 Oakridge      Entire h~           12 2 baths          2    12
##  4         1   118 Grandview-Wo~ Entire h~            4 1 bath           2     2
##  5         1  2550 Downtown Eas~ Private ~            2 1.5 shar~        1     1
##  6         1  1006 Grandview-Wo~ Entire h~            5 1 bath           3     4
##  7         1  3632 Downtown Eas~ Entire h~            6 2 baths          3     3
##  8         1  1923 West End      Entire h~            4 2 baths          2     2
##  9         1  3873 Dunbar South~ Private ~            5 1 bath           2     3
## 10         1  2349 Kerrisdale    Private ~            2 1 shared~        1     1
## # ... with 799,990 more rows, and 1 more variable: price <dbl>
```

```
tail(boot20000)
```

```
## # A tibble: 6 x 9
## # Groups:   replicate [1]
##   replicate    id neighbourhood  room_type accommodates bathrooms bedrooms  beds
##       <int> <dbl> <chr>          <chr>            <dbl> <chr>        <dbl> <dbl>
## 1     20000  1949 Kitsilano      Entire h~            3 1 bath           1     1
## 2     20000  1025 Kensington-Ce~ Entire h~            3 1 bath           1     1
## 3     20000  3013 Kensington-Ce~ Entire h~            4 1 bath           2     2
## 4     20000  2868 Downtown       Entire h~            2 1 bath           1     1
## 5     20000  3156 Downtown       Entire h~            6 2 baths          2     2
## 6     20000  1923 West End       Entire h~            4 2 baths          2     2
## # ... with 1 more variable: price <dbl>
```

Let's take a look at histograms of the first six replicates of our bootstrap samples.

```
six_bootstrap_samples <- boot20000 |>
  filter(replicate <= 6)
```

```
ggplot(six_bootstrap_samples, aes(price)) +
  geom_histogram(fill = "dodgerblue3", color = "lightgrey") +
  labs(x = "Price per night (Canadian dollars)", y = "Count") +
  facet_wrap(~replicate) +
  theme(text = element_text(size = 12))
```

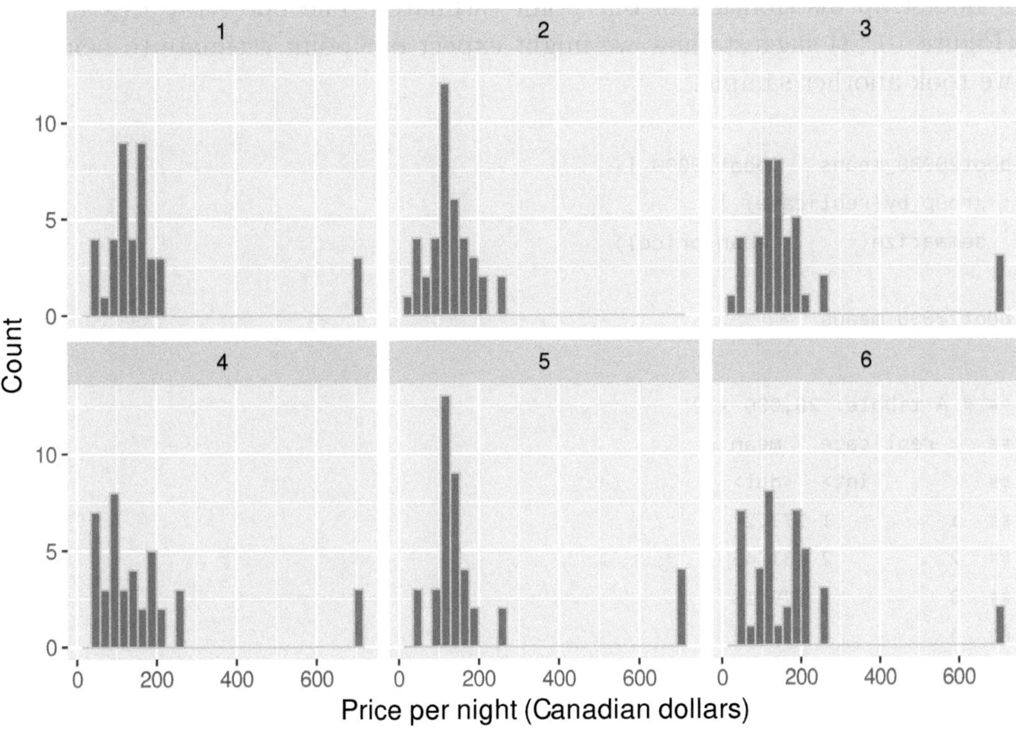

FIGURE 10.12: Histograms of first six replicates of bootstrap samples.

We see in Figure 10.12 how the bootstrap samples differ. We can also calculate the sample mean for each of these six replicates.

```
six_bootstrap_samples |>
  group_by(replicate) |>
  summarize(mean = mean(price))
```

```
## # A tibble: 6 x 2
##   replicate   mean
##       <int>  <dbl>
## 1         1 177.2
## 2         2 131.45
## 3         3 179.10
```

```
## 4          4 171.35
## 5          5 191.32
## 6          6 170.05
```

We can see that the bootstrap sample distributions and the sample means are different. They are different because we are sampling *with replacement*. We will now calculate point estimates for our 20,000 bootstrap samples and generate a bootstrap distribution of our point estimates. The bootstrap distribution (Figure 10.13) suggests how we might expect our point estimate to behave if we took another sample.

```
boot20000_means <- boot20000 |>
  group_by(replicate) |>
  summarize(mean = mean(price))

boot20000_means
```

```
## # A tibble: 20,000 x 2
##    replicate   mean
##        <int>  <dbl>
##  1         1 177.2
##  2         2 131.45
##  3         3 179.10
##  4         4 171.35
##  5         5 191.32
##  6         6 170.05
##  7         7 178.83
##  8         8 154.78
##  9         9 163.85
## 10        10 209.28
## # ... with 19,990 more rows
```

```
tail(boot20000_means)
```

```
## # A tibble: 6 x 2
##   replicate   mean
##       <int>  <dbl>
## 1     19995 130.40
## 2     19996 189.18
## 3     19997 168.98
## 4     19998 168.23
## 5     19999 155.73
```

```
## 6     20000 136.95
```

```
boot_est_dist <- ggplot(boot20000_means, aes(x = mean)) +
  geom_histogram(fill = "dodgerblue3", color = "lightgrey") +
  labs(x = "Sample mean price per night \n (Canadian dollars)", y = "Count") +
  theme(text = element_text(size = 12))

boot_est_dist
```

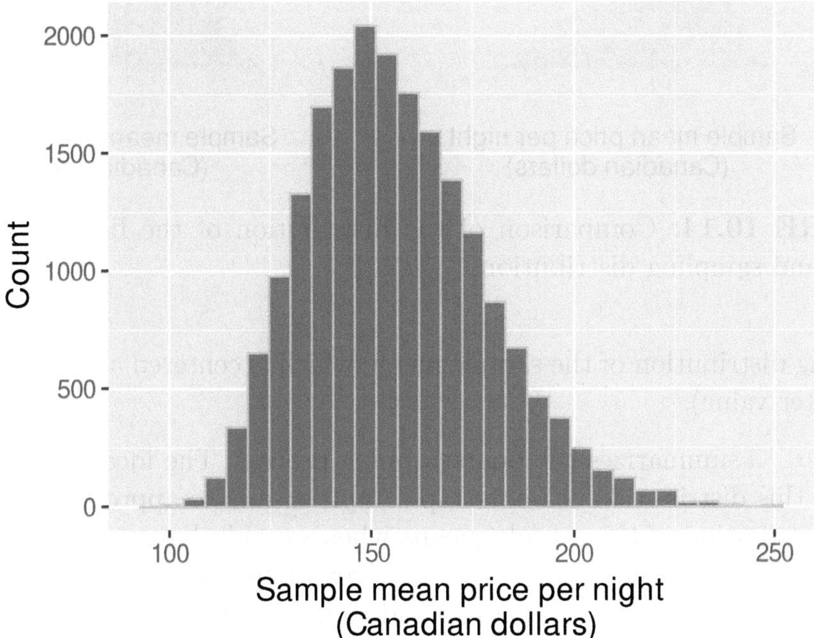

FIGURE 10.13: Distribution of the bootstrap sample means.

Let's compare the bootstrap distribution—which we construct by taking many samples from our original sample of size 40—with the true sampling distribution—which corresponds to taking many samples from the population.

There are two essential points that we can take away from Figure 10.14. First, the shape and spread of the true sampling distribution and the bootstrap distribution are similar; the bootstrap distribution lets us get a sense of the point estimate's variability. The second important point is that the means of these two distributions are different. The sampling distribution is centered at \$154.51, the population mean value. However, the bootstrap distribution is centered at the original sample's mean price per night, \$155.87. Because we are resampling from the original sample repeatedly, we see that the bootstrap distribution is centered at the original sample's mean value (unlike the

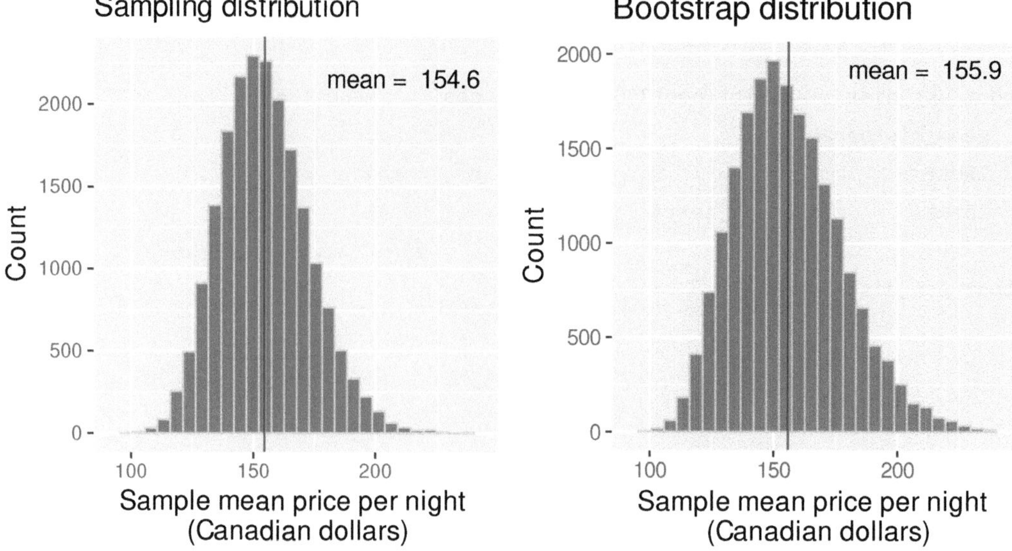

FIGURE 10.14: Comparison of the distribution of the bootstrap sample means and sampling distribution.

sampling distribution of the sample mean, which is centered at the population parameter value).

Figure 10.15 summarizes the bootstrapping process. The idea here is that we can use this distribution of bootstrap sample means to approximate the sampling distribution of the sample means when we only have one sample. Since the bootstrap distribution pretty well approximates the sampling distribution spread, we can use the bootstrap spread to help us develop a plausible range for our population parameter along with our estimate!

10.5.3 Using the bootstrap to calculate a plausible range

Now that we have constructed our bootstrap distribution, let's use it to create an approximate 95% percentile bootstrap confidence interval. A **confidence interval** is a range of plausible values for the population parameter. We will find the range of values covering the middle 95% of the bootstrap distribution, giving us a 95% confidence interval. You may be wondering, what does "95% confidence" mean? If we took 100 random samples and calculated 100 95% confidence intervals, then about 95% of the ranges would capture the population parameter's value. Note there's nothing special about 95%. We could have used other levels, such as 90% or 99%. There is a balance between our level of confidence and precision. A higher confidence level corresponds to a wider range of the interval, and a lower confidence level corresponds to a narrower range. Therefore the level we choose is based on what chance we are

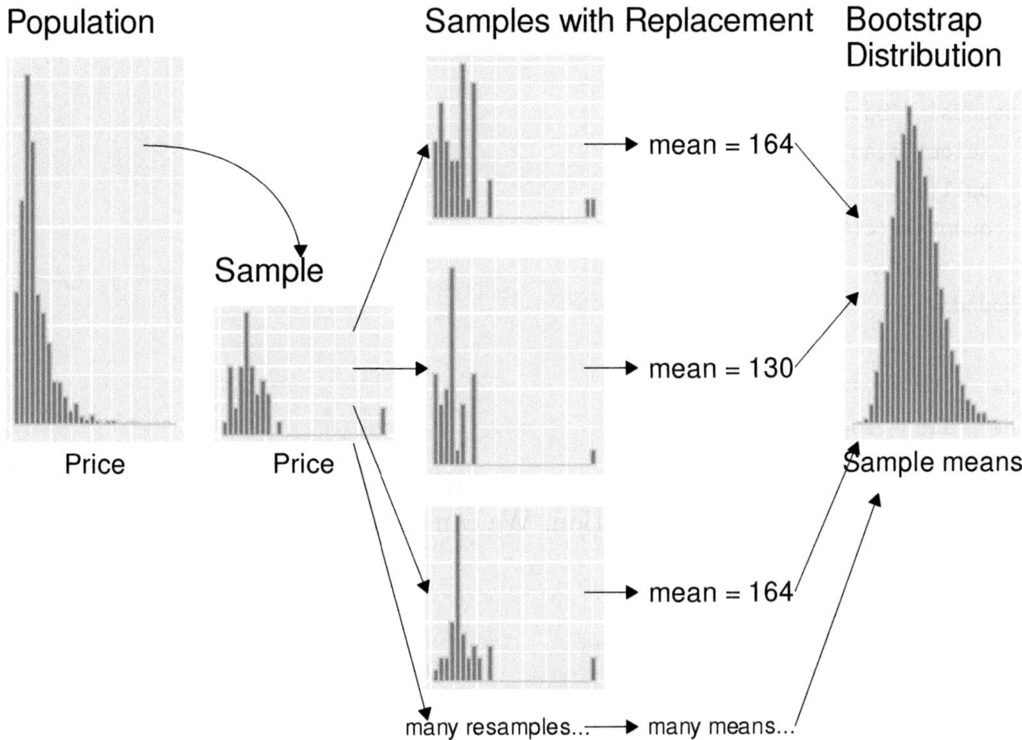

FIGURE 10.15: Summary of bootstrapping process.

willing to take of being wrong based on the implications of being wrong for our application. In general, we choose confidence levels to be comfortable with our level of uncertainty but not so strict that the interval is unhelpful. For instance, if our decision impacts human life and the implications of being wrong are deadly, we may want to be very confident and choose a higher confidence level.

To calculate a 95% percentile bootstrap confidence interval, we will do the following:

1. Arrange the observations in the bootstrap distribution in ascending order.
2. Find the value such that 2.5% of observations fall below it (the 2.5% percentile). Use that value as the lower bound of the interval.
3. Find the value such that 97.5% of observations fall below it (the 97.5% percentile). Use that value as the upper bound of the interval.

To do this in R, we can use the `quantile()` function:

```
bounds <- boot20000_means |>
  select(mean) |>
  pull() |>
  quantile(c(0.025, 0.975))
```

```
bounds
```

```
##  2.5% 97.5%
##   119   204
```

Our interval, \$119.28 to \$203.63, captures the middle 95% of the sample mean prices in the bootstrap distribution. We can visualize the interval on our distribution in Figure 10.16.

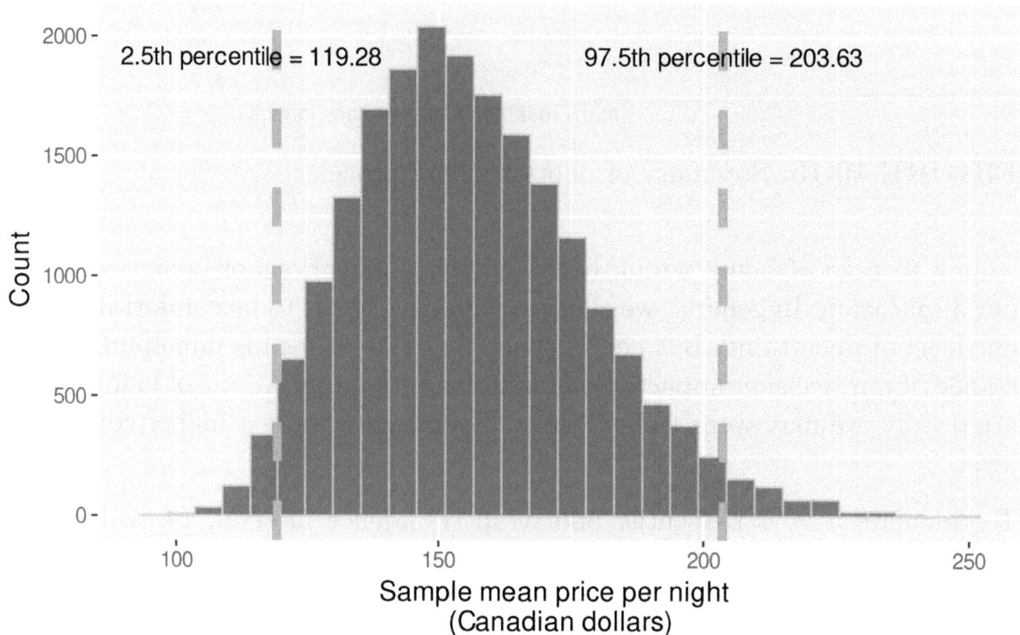

FIGURE 10.16: Distribution of the bootstrap sample means with percentile lower and upper bounds.

To finish our estimation of the population parameter, we would report the point estimate and our confidence interval's lower and upper bounds. Here the sample mean price per night of 40 Airbnb listings was \$155.8, and we are 95% "confident" that the true population mean price per night for all Airbnb listings in Vancouver is between \$(119.28, 203.63). Notice that our interval does indeed contain the true population mean value, \$154.51! However, in practice, we would not know whether our interval captured the population parameter or not because we usually only have a single sample, not the entire population. This is the best we can do when we only have one sample!

This chapter is only the beginning of the journey into statistical inference. We can extend the concepts learned here to do much more than report point estimates and confidence intervals, such as testing for real differences between populations, tests for associations between variables, and so much more. We have just scratched the surface of statistical inference; however, the material presented here will serve as the foundation for more advanced statistical techniques you may learn about in the future!

10.6 Exercises

Practice exercises for the material covered in this chapter can be found in the accompanying worksheets repository[2] in the two "Statistical inference" rows. You can launch an interactive version of each worksheet in your browser by clicking the "launch binder" button. You can also preview a non-interactive version of each worksheet by clicking "view worksheet." If you instead decide to download the worksheets and run them on your own machine, make sure to follow the instructions for computer setup found in Chapter 13. This will ensure that the automated feedback and guidance that the worksheets provide will function as intended.

10.7 Additional resources

- Chapters 7 to 10 of *Modern Dive* [Ismay and Kim, 2020] provide a great next step in learning about inference. In particular, Chapters 7 and 8 cover sampling and bootstrapping using `tidyverse` and `infer` in a slightly more in-depth manner than the present chapter. Chapters 9 and 10 take the next step beyond the scope of this chapter and begin to provide some of the initial mathematical underpinnings of inference and more advanced applications of the concept of inference in testing hypotheses and performing regression. This material offers a great starting point for getting more into the technical side of statistics.
- Chapters 4 to 7 of *OpenIntro Statistics* [Diez et al., 2019] provide a good next step after *Modern Dive*. Although it is still certainly an introductory text,

[2] https://github.com/UBC-DSCI/data-science-a-first-intro-worksheets#readme

things get a bit more mathematical here. Depending on your background, you may actually want to start going through Chapters 1 to 3 first, where you will learn some fundamental concepts in probability theory. Although it may seem like a diversion, probability theory is *the language of statistics*; if you have a solid grasp of probability, more advanced statistics will come naturally to you!

11

Combining code and text with Jupyter

11.1 Overview

A typical data analysis involves not only writing and executing code, but also writing text and displaying images that help tell the story of the analysis. In fact, ideally, we would like to *interleave* these three media, with the text and images serving as narration for the code and its output. In this chapter we will show you how to accomplish this using Jupyter notebooks, a common coding platform in data science. Jupyter notebooks do precisely what we need: they let you combine text, images, and (executable!) code in a single document. In this chapter, we will focus on the *use* of Jupyter notebooks to program in R and write text via a web interface. These skills are essential to getting your analysis running; think of it like getting dressed in the morning! Note that we assume that you already have Jupyter set up and ready to use. If that is not the case, please first read Chapter 13 to learn how to install and configure Jupyter on your own computer.

11.2 Chapter learning objectives

By the end of the chapter, readers will be able to do the following:

- Create new Jupyter notebooks.
- Write, edit, and execute R code in a Jupyter notebook.
- Write, edit, and view text in a Jupyter notebook.
- Open and view plain text data files in Jupyter.
- Export Jupyter notebooks to other standard file types (e.g., `.html`, `.pdf`).

11.3 Jupyter

Jupyter is a web-based interactive development environment for creating, editing, and executing documents called Jupyter notebooks. Jupyter notebooks are documents that contain a mix of computer code (and its output) and formattable text. Given that they combine these two analysis artifacts in a single document—code is not separate from the output or written report—notebooks are one of the leading tools to create reproducible data analyses. Reproducible data analysis is one where you can reliably and easily re-create the same results when analyzing the same data. Although this sounds like something that should always be true of any data analysis, in reality, this is not often the case; one needs to make a conscious effort to perform data analysis in a reproducible manner. An example of what a Jupyter notebook looks like is shown in Figure 11.1.

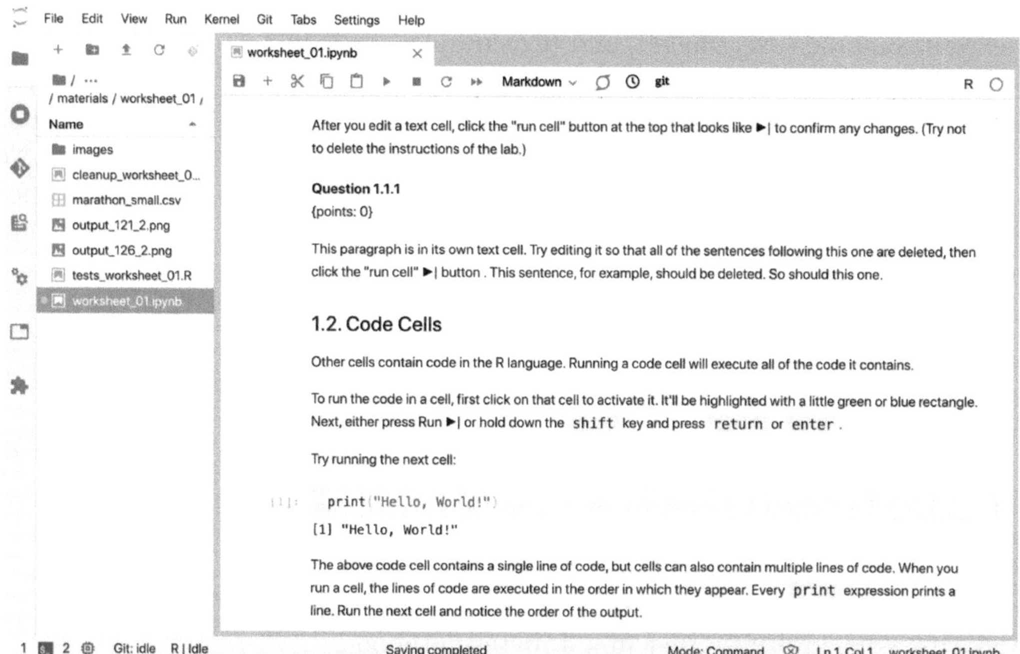

FIGURE 11.1: A screenshot of a Jupyter Notebook.

11.3.1 Accessing Jupyter

One of the easiest ways to start working with Jupyter is to use a web-based platform called JupyterHub. JupyterHubs often have Jupyter, R, a number of R packages, and collaboration tools installed, configured and ready to use. JupyterHubs are usually created and provisioned by organizations, and require authentication to gain access. For example, if you are reading this book as part of a course, your instructor may have a JupyterHub already set up for you to use! Jupyter can also be installed on your own computer; see Chapter 13 for instructions.

11.4 Code cells

The sections of a Jupyter notebook that contain code are referred to as code cells. A code cell that has not yet been executed has no number inside the square brackets to the left of the cell (Figure 11.2). Running a code cell will execute all of the code it contains, and the output (if any exists) will be displayed directly underneath the code that generated it. Outputs may include printed text or numbers, data frames and data visualizations. Cells that have been executed also have a number inside the square brackets to the left of the cell. This number indicates the order in which the cells were run (Figure 11.3).

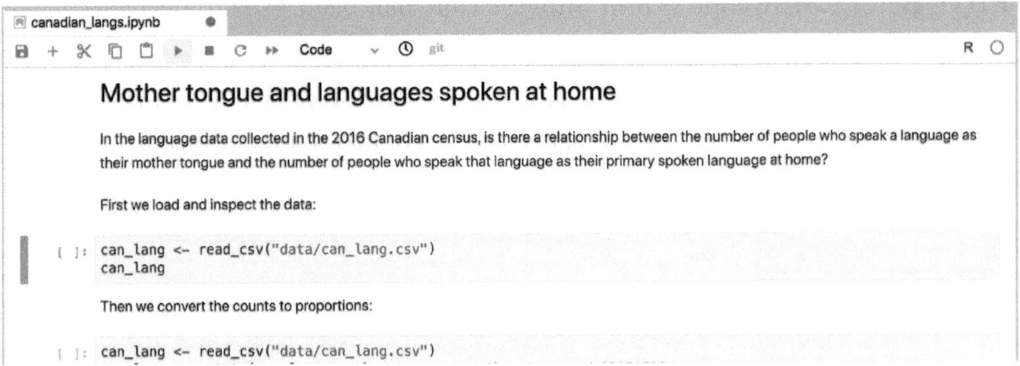

FIGURE 11.2: A code cell in Jupyter that has not yet been executed.

11.4.1 Executing code cells

Code cells can be run independently or as part of executing the entire notebook using one of the "**Run all**" commands found in the **Run** or **Kernel** menus in Jupyter. Running a single code cell independently is a workflow typically used when editing or writing your own R code. Executing an entire notebook is a workflow typically used to ensure that your analysis runs in its entirety before sharing it with others, and when using a notebook as part of an automated process.

To run a code cell independently, the cell needs to first be activated. This is done by clicking on it with the cursor. Jupyter will indicate a cell has been activated by highlighting it with a blue rectangle to its left. After the cell has

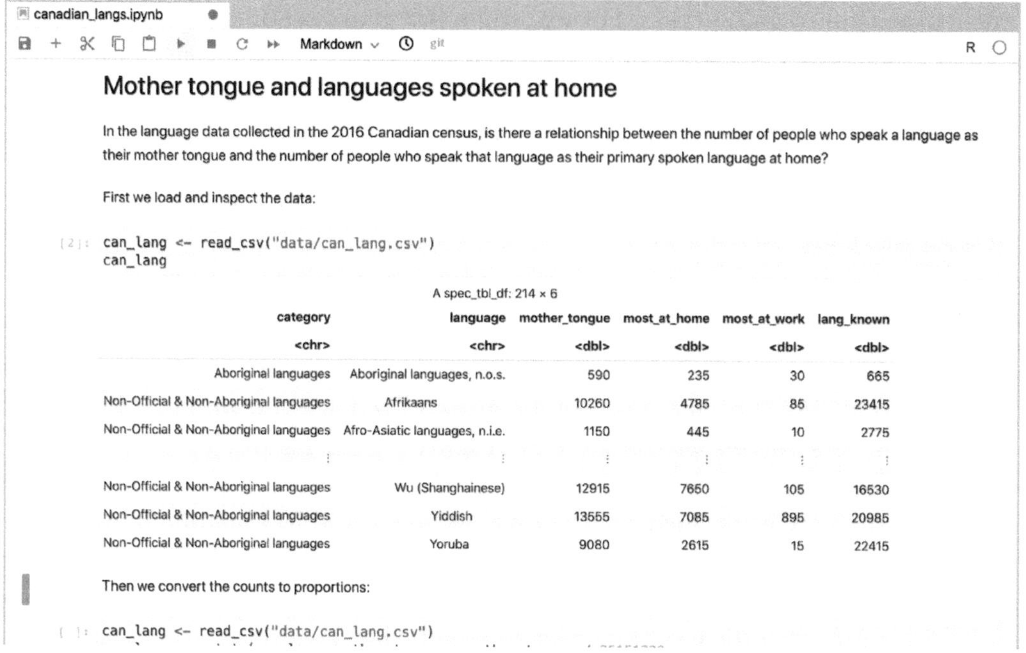

FIGURE 11.3: A code cell in Jupyter that has been executed.

been activated (Figure 11.4), the cell can be run by either pressing the **Run** (▶) button in the toolbar, or by using the keyboard shortcut `Shift + Enter`.

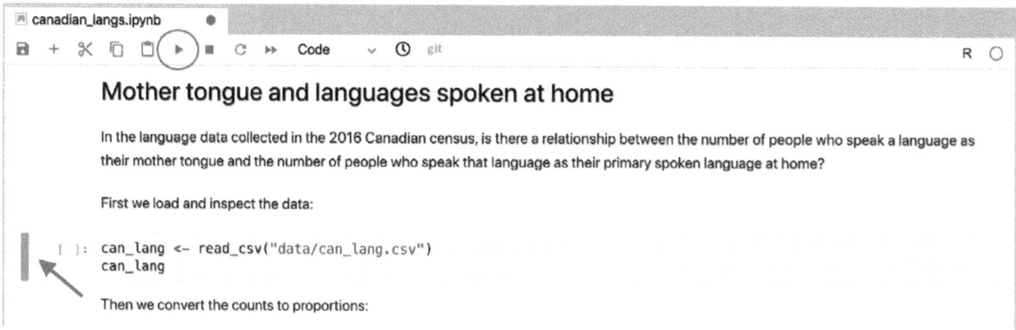

FIGURE 11.4: An activated cell that is ready to be run. The blue rectangle to the cell's left (annotated by a red arrow) indicates that it is ready to be run. The cell can be run by clicking the run button (circled in red).

To execute all of the code cells in an entire notebook, you have three options:

1. Select **Run** » **Run All Cells** from the menu.
2. Select **Kernel** » **Restart Kernel and Run All Cells...** from the menu (Figure 11.5).

3. Click the (⏭) button in the tool bar.

All of these commands result in all of the code cells in a notebook being run. However, there is a slight difference between them. In particular, only options 2 and 3 above will restart the R session before running all of the cells; option 1 will not restart the session. Restarting the R session means that all previous objects that were created from running cells before this command was run will be deleted. In other words, restarting the session and then running all cells (options 2 or 3) emulates how your notebook code would run if you completely restarted Jupyter before executing your entire notebook.

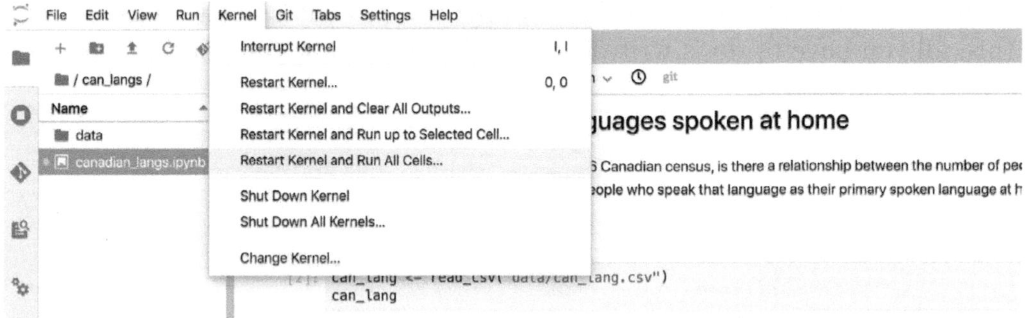

FIGURE 11.5: Restarting the R session can be accomplished by clicking Restart Kernel and Run All Cells...

11.4.2 The Kernel

The kernel is a program that executes the code inside your notebook and outputs the results. Kernels for many different programming languages have been created for Jupyter, which means that Jupyter can interpret and execute the code of many different programming languages. To run R code, your notebook will need an R kernel. In the top right of your window, you can see a circle that indicates the status of your kernel. If the circle is empty (○), the kernel is idle and ready to execute code. If the circle is filled in (●), the kernel is busy running some code.

You may run into problems where your kernel is stuck for an excessive amount of time, your notebook is very slow and unresponsive, or your kernel loses its connection. If this happens, try the following steps:

1. At the top of your screen, click **Kernel**, then **Interrupt Kernel**.
2. If that doesn't help, click **Kernel**, then **Restart Kernel...** If you do this, you will have to run your code cells from the start of your notebook up until where you paused your work.
3. If that still doesn't help, restart Jupyter. First, save your work by

clicking **File** at the top left of your screen, then **Save Notebook**. Next, if you are accessing Jupyter using a JupyterHub server, from the **File** menu click **Hub Control Panel**. Choose **Stop My Server** to shut it down, then the **My Server** button to start it back up. If you are running Jupyter on your own computer, from the **File** menu click **Shut Down**, then start Jupyter again. Finally, navigate back to the notebook you were working on.

11.4.3 Creating new code cells

To create a new code cell in Jupyter (Figure 11.6), click the + button in the toolbar. By default, all new cells in Jupyter start out as code cells, so after this, all you have to do is write R code within the new cell you just created!

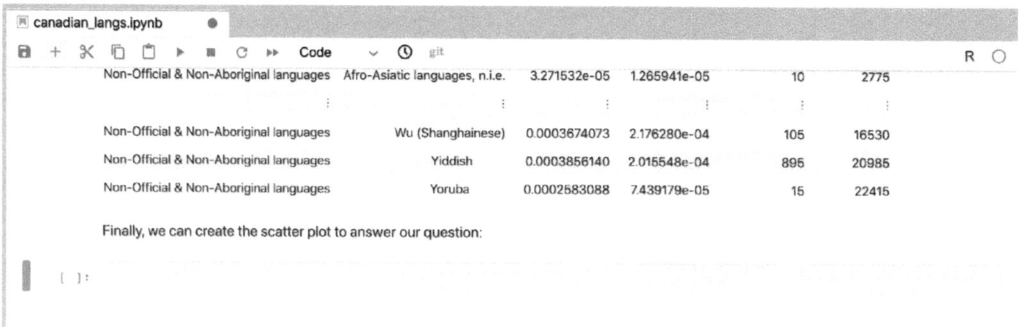

FIGURE 11.6: New cells can be created by clicking the + button, and are by default code cells.

11.5 Markdown cells

Text cells inside a Jupyter notebook are called Markdown cells. Markdown cells are rich formatted text cells, which means you can **bold** and *italicize* text, create subject headers, create bullet and numbered lists, and more. These cells are given the name "Markdown" because they use *Markdown language* to specify the rich text formatting. You do not need to learn Markdown to write text in the Markdown cells in Jupyter; plain text will work just fine. However, you might want to learn a bit about Markdown eventually to enable you to create nicely formatted analyses. See the additional resources at the end of this chapter to find out where you can start learning Markdown.

11.5.1 Editing Markdown cells

To edit a Markdown cell in Jupyter, you need to double click on the cell. Once you do this, the unformatted (or *unrendered*) version of the text will be shown (Figure 11.7). You can then use your keyboard to edit the text. To view the formatted (or *rendered*) text (Figure 11.8), click the **Run** (▶) button in the toolbar, or use the `Shift + Enter` keyboard shortcut.

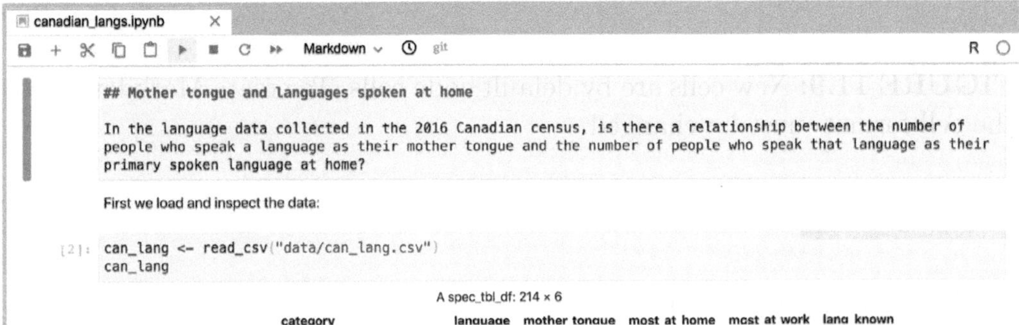

FIGURE 11.7: A Markdown cell in Jupyter that has not yet been rendered and can be edited.

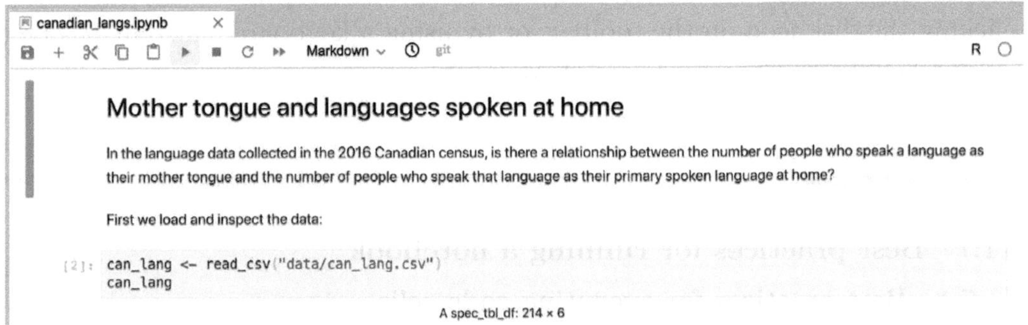

FIGURE 11.8: A Markdown cell in Jupyter that has been rendered and exhibits rich text formatting.

11.5.2 Creating new Markdown cells

To create a new Markdown cell in Jupyter, click the + button in the toolbar. By default, all new cells in Jupyter start as code cells, so the cell format needs to be changed to be recognized and rendered as a Markdown cell. To do this, click on the cell with your cursor to ensure it is activated. Then click on the drop-down box on the toolbar that says "Code" (it is next to the ⏭ button), and change it from "**Code**" to "**Markdown**" (Figure 11.9).

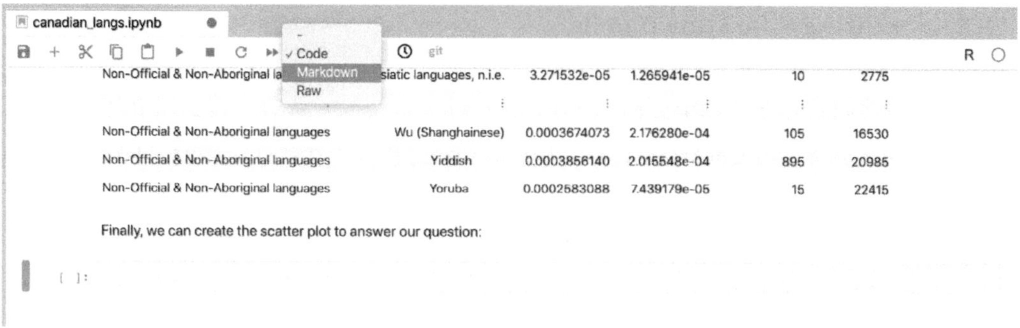

FIGURE 11.9: New cells are by default code cells. To create Markdown cells, the cell format must be changed.

11.6 Saving your work

As with any file you work on, it is critical to save your work often so you don't lose your progress! Jupyter has an autosave feature, where open files are saved periodically. The default for this is every two minutes. You can also manually save a Jupyter notebook by selecting **Save Notebook** from the **File** menu, by clicking the disk icon on the toolbar, or by using a keyboard shortcut (`Control + S` for Windows, or `Command + S` for Mac OS).

11.7 Best practices for running a notebook

11.7.1 Best practices for executing code cells

As you might know (or at least imagine) by now, Jupyter notebooks are great for interactively editing, writing and running R code; this is what they were designed for! Consequently, Jupyter notebooks are flexible in regards to code cell execution order. This flexibility means that code cells can be run in any arbitrary order using the **Run** (▶) button. But this flexibility has a downside: it can lead to Jupyter notebooks whose code cannot be executed in a linear order (from top to bottom of the notebook). A nonlinear notebook is problematic because a linear order is the conventional way code documents are run, and others will have this expectation when running your notebook. Finally, if the code is used in some automated process, it will need to run in a linear order, from top to bottom of the notebook.

The most common way to inadvertently create a nonlinear notebook is to rely solely on using the ▶ button to execute cells. For example, suppose you write some R code that creates an R object, say a variable named `y`. When you execute that cell and create `y`, it will continue to exist until it is deliberately deleted with R code, or when the Jupyter notebook R session (*i.e.*, kernel) is stopped or restarted. It can also be referenced in another distinct code cell (Figure 11.10). Together, this means that you could then write a code cell further above in the notebook that references `y` and execute it without error in the current session (Figure 11.11). This could also be done successfully in future sessions if, and only if, you run the cells in the same unconventional order. However, it is difficult to remember this unconventional order, and it is not the order that others would expect your code to be executed in. Thus, in the future, this would lead to errors when the notebook is run in the conventional linear order (Figure 11.12).

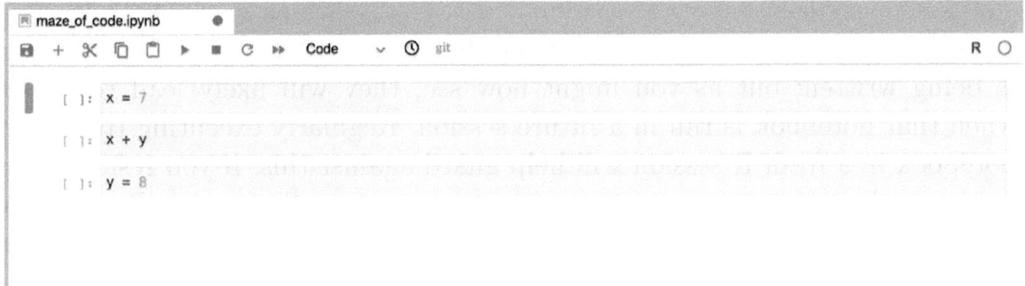

FIGURE 11.10: Code that was written out of order, but not yet executed.

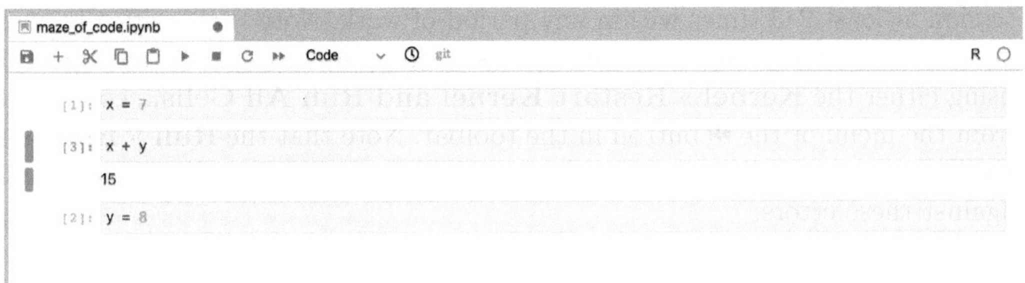

FIGURE 11.11: Code that was written out of order, and was executed using the run button in a nonlinear order without error. The order of execution can be traced by following the numbers to the left of the code cells; their order indicates the order in which the cells were executed.

You can also accidentally create a nonfunctioning notebook by creating an object in a cell that later gets deleted. In such a scenario, that object only exists for that one particular R session and will not exist once the notebook is restarted and run again. If that object was referenced in another cell in that notebook, an error would occur when the notebook was run again in a new session.

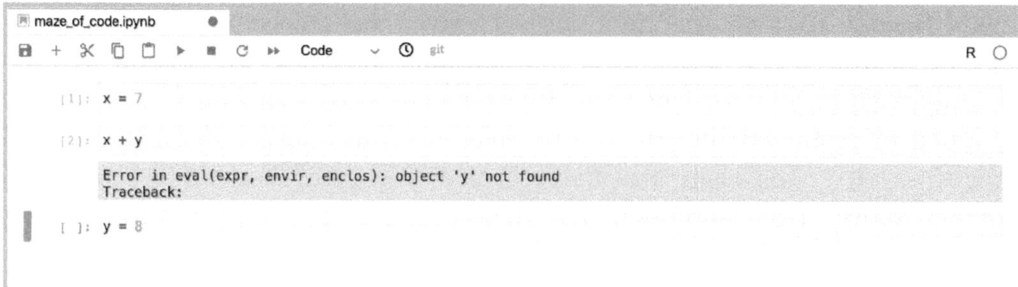

FIGURE 11.12: Code that was written out of order, and was executed in a linear order using "Restart Kernel and Run All Cells..." This resulted in an error at the execution of the second code cell and it failed to run all code cells in the notebook.

These events may not negatively affect the current R session when the code is being written; but as you might now see, they will likely lead to errors when that notebook is run in a future session. Regularly executing the entire notebook in a fresh R session will help guard against this. If you restart your session and new errors seem to pop up when you run all of your cells in linear order, you can at least be aware that there is an issue. Knowing this sooner rather than later will allow you to fix the issue and ensure your notebook can be run linearly from start to finish.

We recommend as a best practice to run the entire notebook in a fresh R session at least 2–3 times within any period of work. Note that, critically, you *must do this in a fresh R session* by restarting your kernel. We recommend using either the **Kernel** » **Restart Kernel and Run All Cells...** command from the menu or the ⏭ button in the toolbar. Note that the **Run** » **Run All Cells** menu item will not restart the kernel, and so it is not sufficient to guard against these errors.

11.7.2 Best practices for including R packages in notebooks

Most data analyses these days depend on functions from external R packages that are not built into R. One example is the `tidyverse` metapackage that we heavily rely on in this book. This package provides us access to functions like `read_csv` for reading data, `select` for subsetting columns, and `ggplot` for creating high-quality graphics.

As mentioned earlier in the book, external R packages need to be loaded before the functions they contain can be used. Our recommended way to do this is via `library(package_name)`. But where should this line of code be written in a Jupyter notebook? One idea could be to load the library right before the function is used in the notebook. However, although this technically works, this causes hidden, or at least non-obvious, R package dependencies when others view or try to run the notebook. These hidden dependencies can lead to errors when the notebook is executed on another computer if the needed R packages are not installed. Additionally, if the data analysis code takes a long time to run, uncovering the hidden dependencies that need to be installed so that the analysis can run without error can take a great deal of time to uncover.

Therefore, we recommend you load all R packages in a code cell near the top of the Jupyter notebook. Loading all your packages at the start ensures that all packages are loaded before their functions are called, assuming the notebook is run in a linear order from top to bottom as recommended above. It also makes it easy for others viewing or running the notebook to see what external R packages are used in the analysis, and hence, what packages they should install on their computer to run the analysis successfully.

11.7.3 Summary of best practices for running a notebook

1. Write code so that it can be executed in a linear order.
2. As you write code in a Jupyter notebook, run the notebook in a linear order and in its entirety often (2–3 times every work session) via the **Kernel » Restart Kernel and Run All Cells...** command from the Jupyter menu or the ⏭ button in the toolbar.
3. Write the code that loads external R packages near the top of the Jupyter notebook.

11.8 Exploring data files

It is essential to preview data files before you try to read them into R to see whether or not there are column names, what the delimiters are, and if there are lines you need to skip. In Jupyter, you preview data files stored as plain text files (e.g., comma- and tab-separated files) in their plain text format (Figure 11.14) by right-clicking on the file's name in the Jupyter file explorer, selecting **Open with**, and then selecting **Editor** (Figure 11.13). Suppose you do not specify to open the data file with an editor. In that case, Jupyter will render a nice table for you, and you will not be able to see the column delimiters, and therefore you will not know which function to use, nor which arguments to use and values to specify for them.

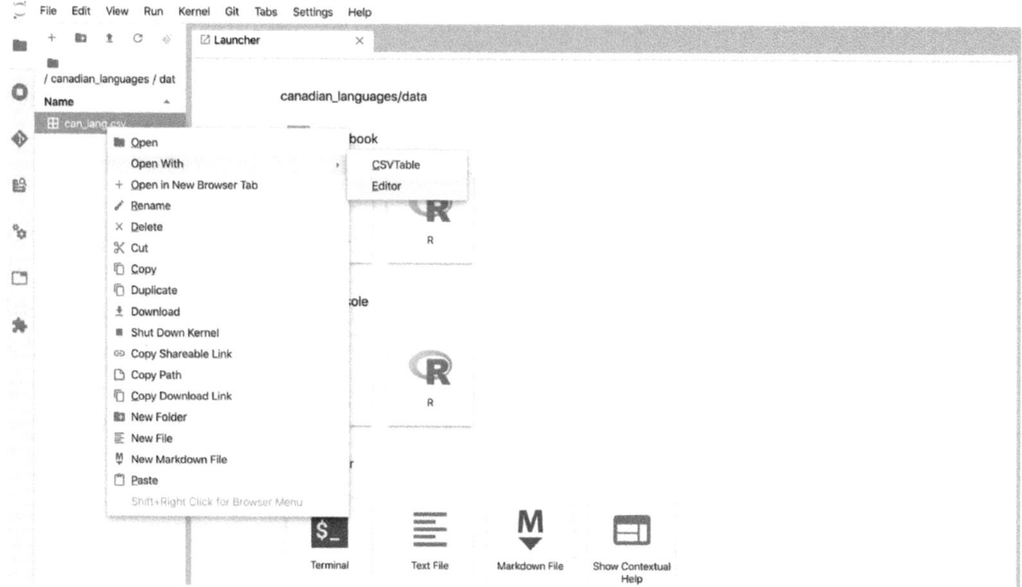

FIGURE 11.13: Opening data files with an editor in Jupyter.

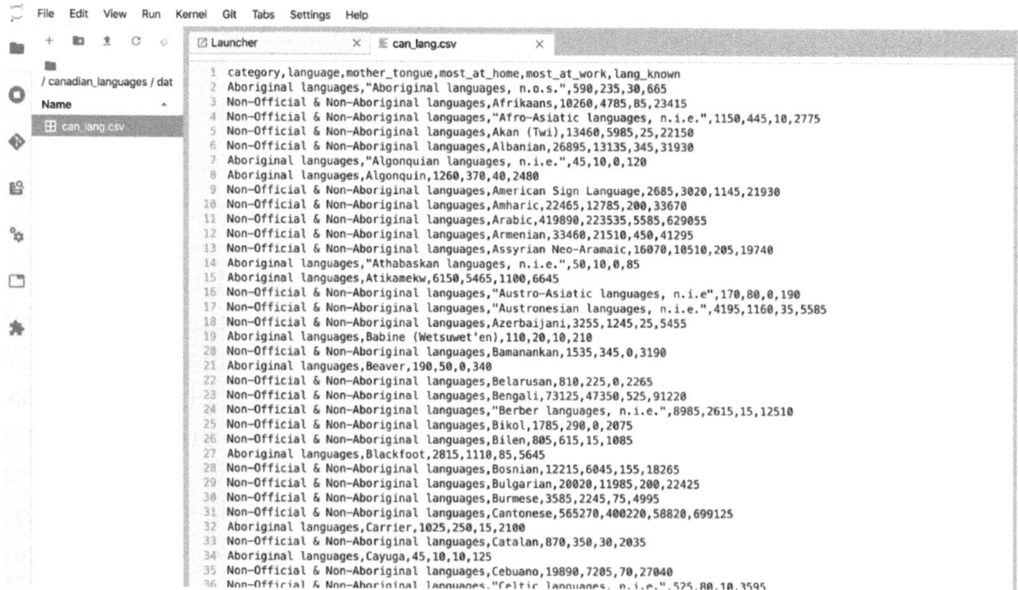

FIGURE 11.14: A data file as viewed in an editor in Jupyter.

11.9 Exporting to a different file format

In Jupyter, viewing, editing and running R code is done in the Jupyter notebook file format with file extension `.ipynb`. This file format is not easy to open and view outside of Jupyter. Thus, to share your analysis with people who do not commonly use Jupyter, it is recommended that you export your executed analysis as a more common file type, such as an `.html` file, or a `.pdf`. We recommend exporting the Jupyter notebook after executing the analysis so that you can also share the outputs of your code. Note, however, that your audience will not be able to *run* your analysis using a `.html` or `.pdf` file. If you want your audience to be able to reproduce the analysis, you must provide them with the `.ipynb` Jupyter notebook file.

11.9.1 Exporting to HTML

Exporting to `.html` will result in a shareable file that anyone can open using a web browser (e.g., Firefox, Safari, Chrome, or Edge). The `.html` output will produce a document that is visually similar to what the Jupyter notebook looked like inside Jupyter. One point of caution here is that if there are images in your Jupyter notebook, you will need to share the image files and the `.html` file to see them.

11.9.2 Exporting to PDF

Exporting to `.pdf` will result in a shareable file that anyone can open using many programs, including Adobe Acrobat, Preview, web browsers and many more. The benefit of exporting to PDF is that it is a standalone document, even if the Jupyter notebook included references to image files. Unfortunately, the default settings will result in a document that visually looks quite different from what the Jupyter notebook looked like. The font, page margins, and other details will appear different in the `.pdf` output.

11.10 Creating a new Jupyter notebook

At some point, you will want to create a new, fresh Jupyter notebook for your own project instead of viewing, running or editing a notebook that was started by someone else. To do this, navigate to the **Launcher** tab, and click on the R icon under the **Notebook** heading. If no **Launcher** tab is visible, you can get a new one via clicking the + button at the top of the Jupyter file explorer (Figure 11.15).

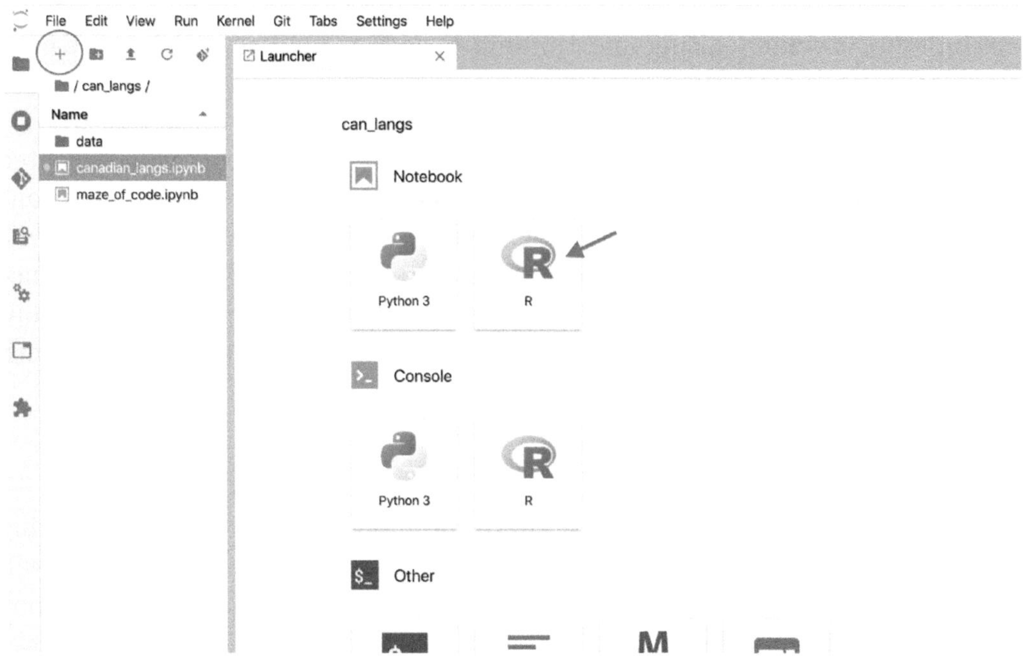

FIGURE 11.15: Clicking on the R icon under the Notebook heading will create a new Jupyter notebook with an R kernel.

Once you have created a new Jupyter notebook, be sure to give it a descriptive name, as the default file name is `Untitled.ipynb`. You can rename files by first right-clicking on the file name of the notebook you just created, and then clicking **Rename**. This will make the file name editable. Use your keyboard to change the name. Pressing `Enter` or clicking anywhere else in the Jupyter interface will save the changed file name.

We recommend not using white space or non-standard characters in file names. Doing so will not prevent you from using that file in Jupyter. However, these sorts of things become troublesome as you start to do more advanced data science projects that involve repetition and automation. We recommend naming files using lower case characters and separating words by a dash (-) or an underscore (_).

11.11 Additional resources

- The JupyterLab Documentation[1] is a good next place to look for more information about working in Jupyter notebooks. This documentation goes into significantly more detail about all of the topics we covered in this chapter, and covers more advanced topics as well.
- If you are keen to learn about the Markdown language for rich text formatting, two good places to start are CommonMark's Markdown cheatsheet[2] and Markdown tutorial[3].

[1] https://jupyterlab.readthedocs.io/en/latest/

[2] https://commonmark.org/help/

[3] https://commonmark.org/help/tutorial/

12

Collaboration with version control

> *You mostly collaborate with yourself, and me-from-two-months-ago never responds to email.*
>
> –Mark T. Holder

12.1 Overview

This chapter will introduce the concept of using version control systems to track changes to a project over its lifespan, to share and edit code in a collaborative team, and to distribute the finished project to its intended audience. This chapter will also introduce how to use the two most common version control tools: Git for local version control, and GitHub for remote version control. We will focus on the most common version control operations used day-to-day in a standard data science project. There are many user interfaces for Git; in this chapter we will cover the Jupyter Git interface.

12.2 Chapter learning objectives

By the end of the chapter, readers will be able to do the following:

- Describe what version control is and why data analysis projects can benefit from it.
- Create a remote version control repository on GitHub.
- Use Jupyter's Git version control tools for project versioning and collaboration:
  - Clone a remote version control repository to create a local repository.
  - Commit changes to a local version control repository.

  - Push local changes to a remote version control repository.
  - Pull changes from a remote version control repository to a local version control repository.
  - Resolve merge conflicts.
- Give collaborators access to a remote GitHub repository.
- Communicate with collaborators using GitHub issues.
- Use best practices when collaborating on a project with others.

12.3 What is version control, and why should I use it?

Data analysis projects often require iteration and revision to move from an initial idea to a finished product ready for the intended audience. Without deliberate and conscious effort towards tracking changes made to the analysis, projects tend to become messy. This mess can have serious, negative repercussions on an analysis project, including interesting results files that your code cannot reproduce, temporary files with snippets of ideas that are forgotten or not easy to find, mind-boggling file names that make it unclear which is the current working version of the file (e.g., `document_final_draft_final.txt`, `to_hand_in_final_v2.txt`, etc.), and more.

Additionally, the iterative nature of data analysis projects means that most of the time, the final version of the analysis that is shared with the audience is only a fraction of what was explored during the development of that analysis. Changes in data visualizations and modeling approaches, as well as some negative results, are often not observable from reviewing only the final, polished analysis. The lack of observability of these parts of the analysis development can lead to others repeating things that did not work well, instead of seeing what did not work well, and using that as a springboard to new, more fruitful approaches.

Finally, data analyses are typically completed by a team of people rather than a single person. This means that files need to be shared across multiple computers, and multiple people often end up editing the project simultaneously. In such a situation, determining who has the latest version of the project—and how to resolve conflicting edits—can be a real challenge.

Version control helps solve these challenges. Version control is the process of keeping a record of changes to documents, including when the changes were made and who made them, throughout the history of their development. It also provides the means both to view earlier versions of the project and to revert changes. Version control is most commonly used in software development, but can be used for any electronic files for any type of project, including data analyses. Being able to record and view the history of a data analysis project is important for understanding how and why decisions to use one method or another were made, among other things. Version control also facilitates collaboration via tools to share edits with others and resolve conflicting edits. But even if you're working on a project alone, you should still use version control. It helps you keep track of what you've done, when you did it, and what you're planning to do next!

To version control a project, you generally need two things: a *version control system* and a *repository hosting service*. The version control system is the software responsible for tracking changes, sharing changes you make with others, obtaining changes from others, and resolving conflicting edits. The repository hosting service is responsible for storing a copy of the version-controlled project online (a *repository*), where you and your collaborators can access it remotely, discuss issues and bugs, and distribute your final product. For both of these items, there is a wide variety of choices. In this textbook we'll use Git for version control, and GitHub for repository hosting, because both are currently the most widely used platforms. In the additional resources section at the end of the chapter, we list many of the common version control systems and repository hosting services in use today.

Note: Technically you don't *have to* use a repository hosting service. You can, for example, version control a project that is stored only in a folder on your computer—never sharing it on a repository hosting service. But using a repository hosting service provides a few big benefits, including managing collaborator access permissions, tools to discuss and track bugs, and the ability to have external collaborators contribute work, not to mention the safety of having your work backed up in the cloud. Since most repository hosting services now offer free accounts, there are not many situations in which you wouldn't want to use one for your project.

12.4 Version control repositories

Typically, when we put a data analysis project under version control, we create two copies of the repository (Figure 12.1). One copy we use as our primary workspace where we create, edit, and delete files. This copy is commonly referred to as the **local repository**. The local repository most commonly exists on our computer or laptop, but can also exist within a workspace on a server (e.g., JupyterHub). The other copy is typically stored in a repository hosting service (e.g., GitHub), where we can easily share it with our collaborators. This copy is commonly referred to as the **remote repository**.

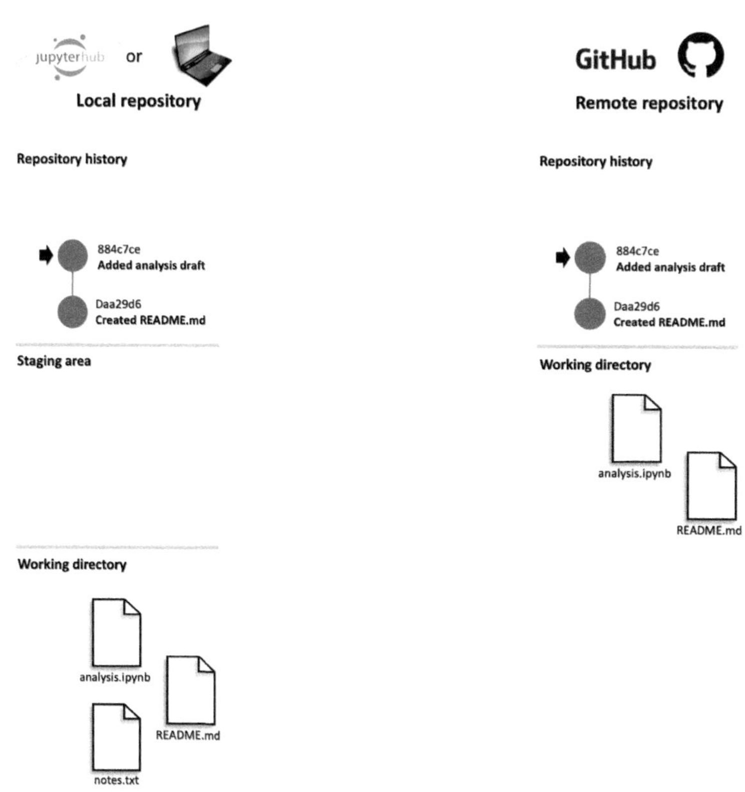

FIGURE 12.1: Schematic of local and remote version control repositories.

Both copies of the repository have a **working directory** where you can create, store, edit, and delete files (e.g., `analysis.ipynb` in Figure 12.1). Both copies of the repository also maintain a full project history (Figure 12.1). This history is a record of all versions of the project files that have been created. The repository history is not automatically generated; Git must be explicitly told when to record a version of the project. These records are called **commits**. They are a snapshot of the file contents as well metadata about the repository at that time the record was created (who made the commit, when it was made, etc.). In the local and remote repositories shown in Figure 12.1, there are two commits represented as gray circles. Each commit can be identified by a human-readable **message**, which you write when you make a commit, and a **commit hash** that Git automatically adds for you.

The purpose of the message is to contain a brief, rich description of what work was done since the last commit. Messages act as a very useful narrative of the changes to a project over its lifespan. If you ever want to view or revert to an earlier version of the project, the message can help you identify which commit to view or revert to. In Figure 12.1, you can see two such messages, one for each commit: `Created README.md` and `Added analysis draft`.

The hash is a string of characters consisting of about 40 letters and numbers. The purpose of the hash is to serve as a unique identifier for the commit, and is used by Git to index project history. Although hashes are quite long—imagine having to type out 40 precise characters to view an old project version!—Git is able to work with shorter versions of hashes. In Figure 12.1, you can see two of these shortened hashes, one for each commit: `Daa29d6` and `884c7ce`.

12.5 Version control workflows

When you work in a local version-controlled repository, there are generally three additional steps you must take as part of your regular workflow. In addition to just working on files—creating, editing, and deleting files as you normally would—you must:

1. Tell Git when to make a commit of your own changes in the local repository.
2. Tell Git when to send your new commits to the remote GitHub repository.
3. Tell Git when to retrieve any new changes (that others made) from the remote GitHub repository.

In this section we will discuss all three of these steps in detail.

12.5.1 Committing changes to a local repository

When working on files in your local version control repository (e.g., using Jupyter) and saving your work, these changes will only initially exist in the working directory of the local repository (Figure 12.2).

Once you reach a point that you want Git to keep a record of the current version of your work, you need to commit (i.e., snapshot) your changes. A prerequisite to this is telling Git which files should be included in that

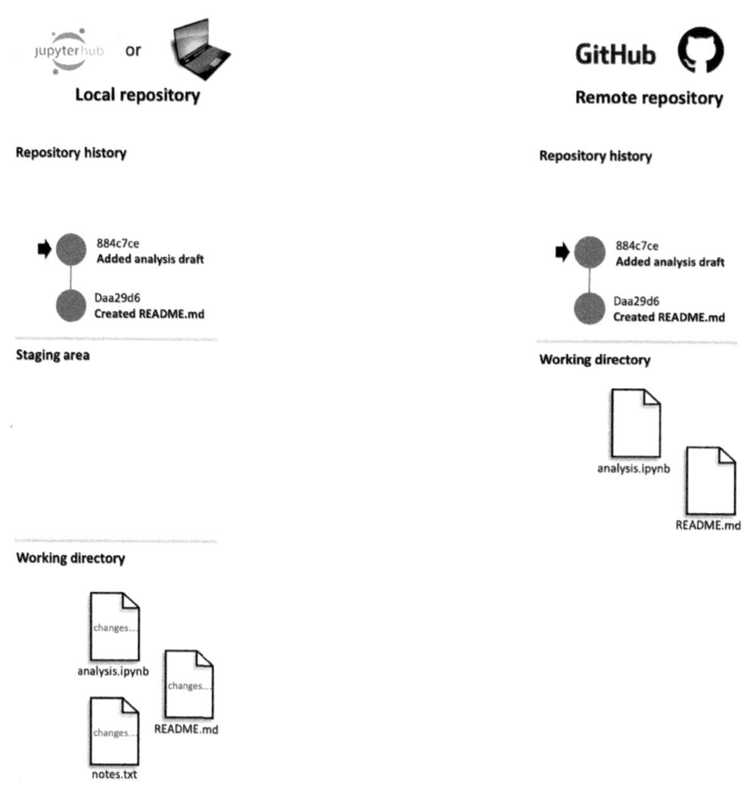

FIGURE 12.2: Local repository with changes to files.

snapshot. We call this step **adding** the files to the **staging area**. Note that the staging area is not a real physical location on your computer; it is instead a conceptual placeholder for these files until they are committed. The benefit of the Git version control system using a staging area is that you can choose to commit changes in only certain files. For example, in Figure 12.3, we add only the two files that are important to the analysis project (`analysis.ipynb` and `README.md`) and not our personal scratch notes for the project (`notes.txt`).

Once the files we wish to commit have been added to the staging area, we can then commit those files to the repository history (Figure 12.4). When we do this, we are required to include a helpful *commit message* to tell collaborators (which often includes future you!) about the changes that were made. In Figure 12.4, the message is `Message about changes...`; in your work you should make sure to replace this with an informative message about what changed. It is also important to note here that these changes are only being committed to the

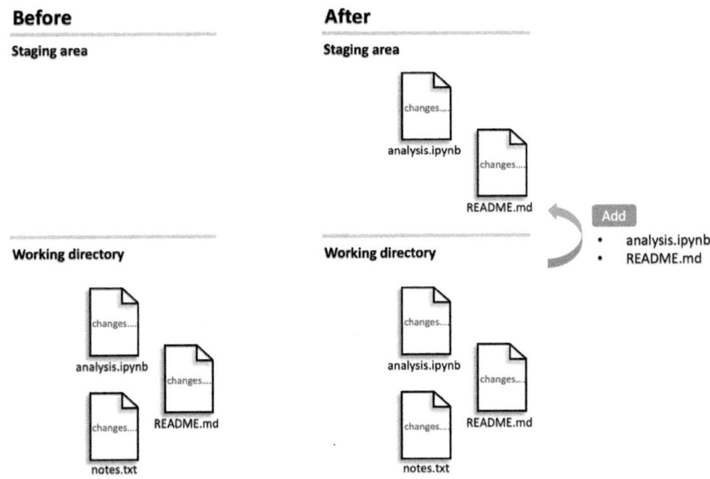

FIGURE 12.3: Adding modified files to the staging area in the local repository.

local repository's history. The remote repository on GitHub has not changed, and collaborators are not yet able to see your new changes.

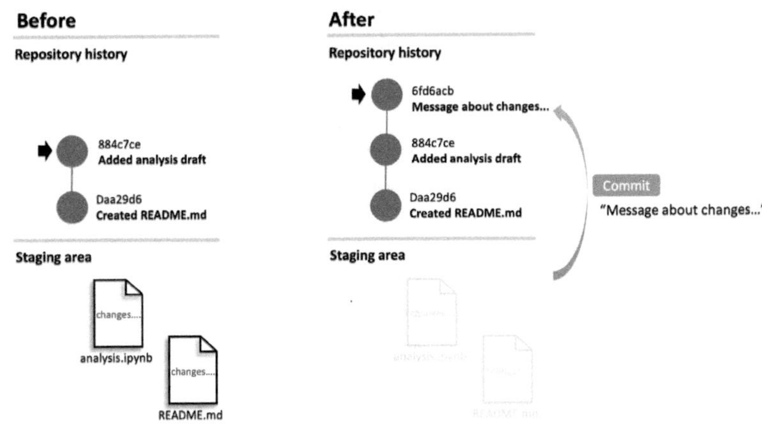

FIGURE 12.4: Committing the modified files in the staging area to the local repository history, with an informative message about what changed.

12.5.2 Pushing changes to a remote repository

Once you have made one or more commits that you want to share with your collaborators, you need to **push** (i.e., send) those commits back to GitHub

(Figure 12.5). This updates the history in the remote repository (i.e., GitHub) to match what you have in your local repository. Now when collaborators interact with the remote repository, they will be able to see the changes you made. And you can also take comfort in the fact that your work is now backed up in the cloud!

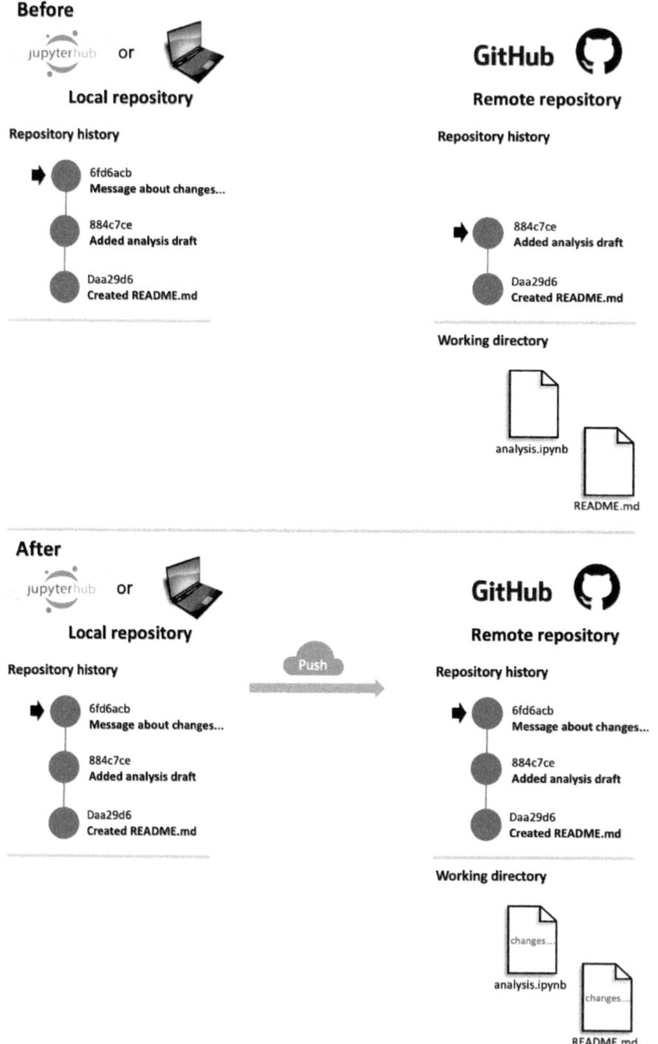

FIGURE 12.5: Pushing the commit to send the changes to the remote repository on GitHub.

12.5.3 Pulling changes from a remote repository

If you are working on a project with collaborators, they will also be making changes to files (e.g., to the analysis code in a Jupyter notebook and the project's README file), committing them to their own local repository, and pushing their commits to the remote GitHub repository to share them with you. When they push their changes, those changes will only initially exist in the remote GitHub repository and not in your local repository (Figure 12.6).

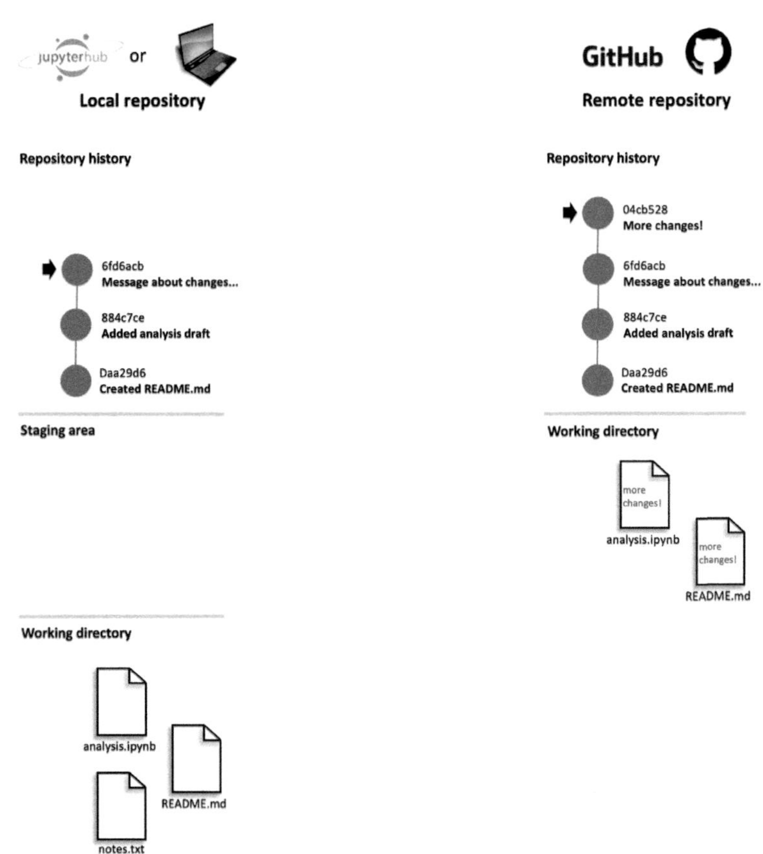

FIGURE 12.6: Changes pushed by collaborators, or created directly on GitHub will not be automatically sent to your local repository.

To obtain the new changes from the remote repository on GitHub, you will need to **pull** those changes to your own local repository. By pulling changes, you synchronize your local repository to what is present on GitHub (Figure 12.7). Additionally, until you pull changes from the remote repository, you will not be able to push any more changes yourself (though you will still be able to work and make commits in your own local repository).

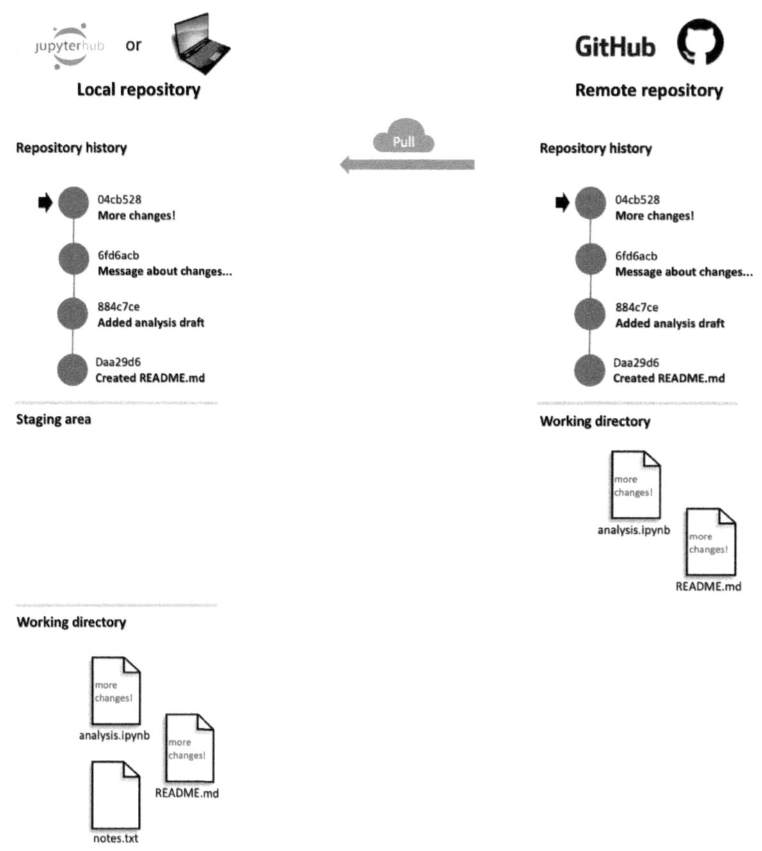

FIGURE 12.7: Pulling changes from the remote GitHub repository to synchronize your local repository.

12.6 Working with remote repositories using GitHub

Now that you have been introduced to some of the key general concepts and workflows of Git version control, we will walk through the practical steps. There are several different ways to start using version control with a new project. For simplicity and ease of setup, we recommend creating a remote repository first. This section covers how to both create and edit a remote repository on GitHub. Once you have a remote repository set up, we recommend **cloning** (or copying) that repository to create a local repository in which you primarily work. You can clone the repository either on your own computer or in a workspace on a server (e.g., a JupyterHub server). Section 12.7 below will cover this second step in detail.

12.6.1 Creating a remote repository on GitHub

Before you can create remote repositories on GitHub, you will need a GitHub account; you can sign up for a free account at `https://github.com/`. Once you have logged into your account, you can create a new repository to host your project by clicking on the "+" icon in the upper right-hand corner, and then on "New Repository," as shown in Figure 12.8.

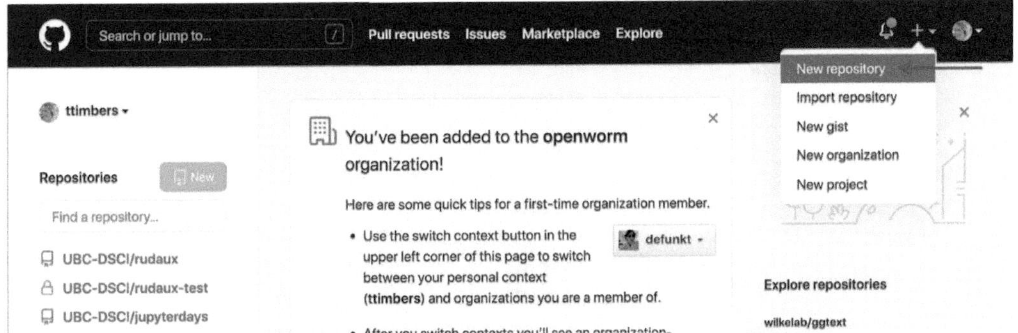

FIGURE 12.8: New repositories on GitHub can be created by clicking on "New Repository" from the + menu.

Repositories can be set up with a variety of configurations, including a name, optional description, and the inclusion (or not) of several template files. One of the most important configuration items to choose is the visibility to the outside world, either public or private. *Public* repositories can be viewed by anyone. *Private* repositories can be viewed by only you. Both public and private repositories are only editable by you, but you can change that by giving access to other collaborators.

To get started with a *public* repository having a template `README.md` file, take the following steps shown in Figure 12.9:

1. Enter the name of your project repository. In the example below, we use `canadian_languages`. Most repositories follow a similar naming convention involving only lowercase letter words separated by either underscores or hyphens.
2. Choose an option for the privacy of your repository.
3. Select "Add a README file." This creates a template `README.md` file in your repository's root folder.
4. When you are happy with your repository name and configuration, click on the green "Create Repository" button.

A newly created public repository with a `README.md` template file should look something like what is shown in Figure 12.10.

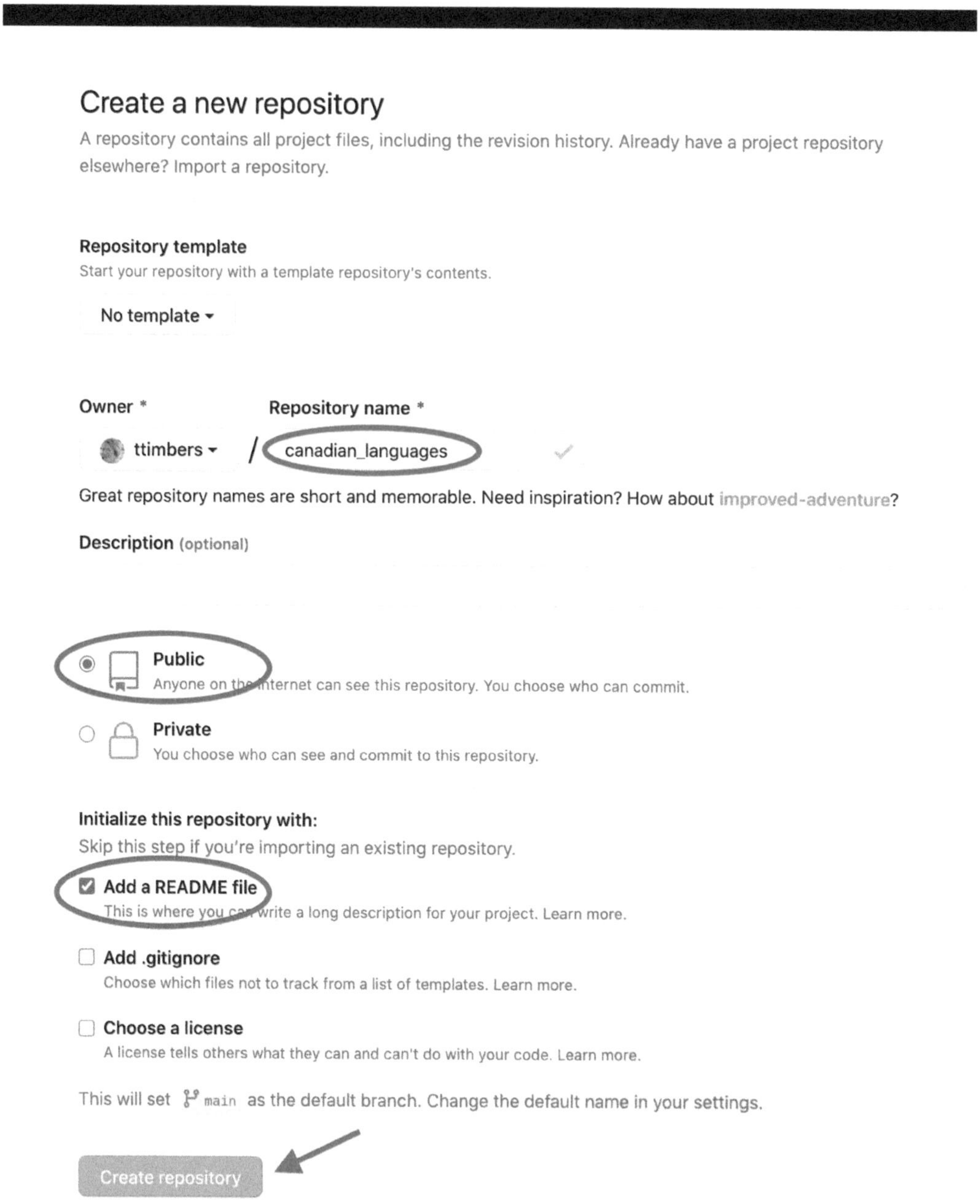

FIGURE 12.9: Repository configuration for a project that is public and initialized with a README.md template file.

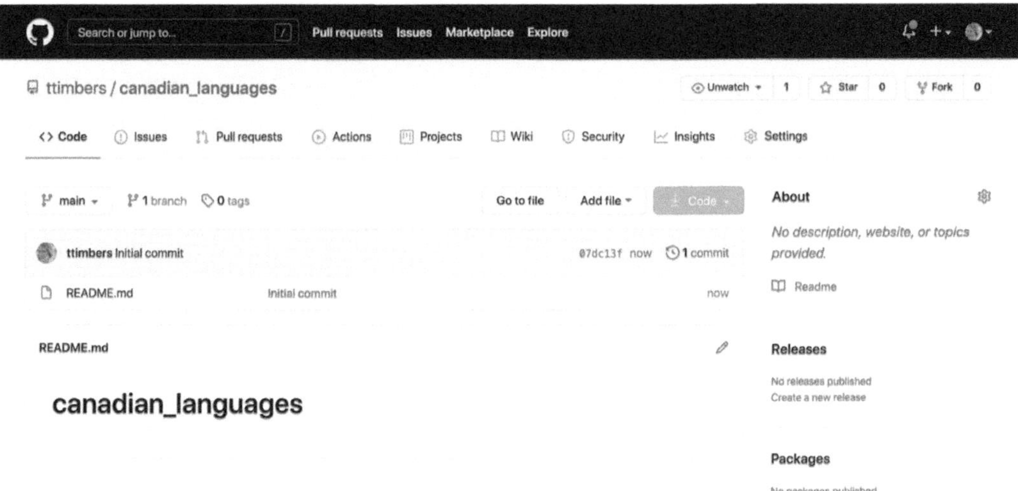

FIGURE 12.10: Respository configuration for a project that is public and initialized with a README.md template file.

12.6.2 Editing files on GitHub with the pen tool

The pen tool can be used to edit existing plain text files. When you click on the pen tool, the file will be opened in a text box where you can use your keyboard to make changes (Figures 12.11 and 12.12).

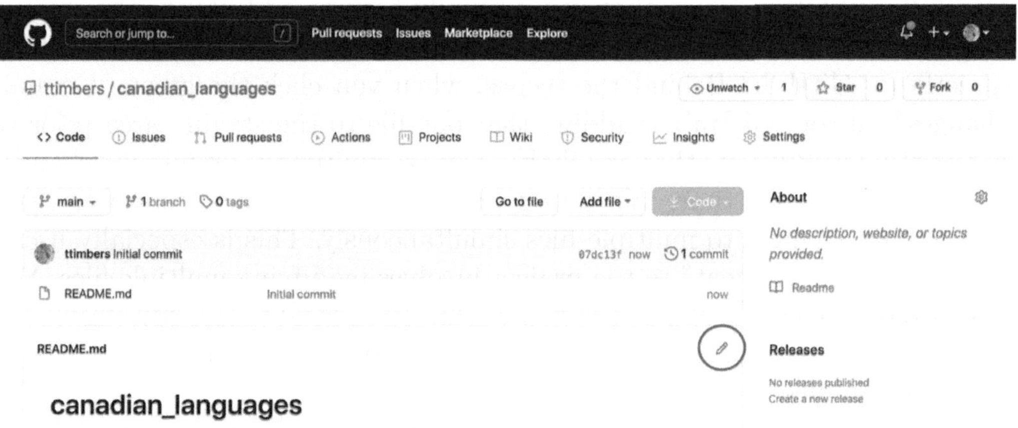

FIGURE 12.11: Clicking on the pen tool opens a text box for editing plain text files.

After you are done with your edits, they can be "saved" by *committing* your changes. When you *commit a file* in a repository, the version control system takes a snapshot of what the file looks like. As you continue working on the project, over time you will possibly make many commits to a single file; this generates a useful version history for that file. On GitHub, if you click the

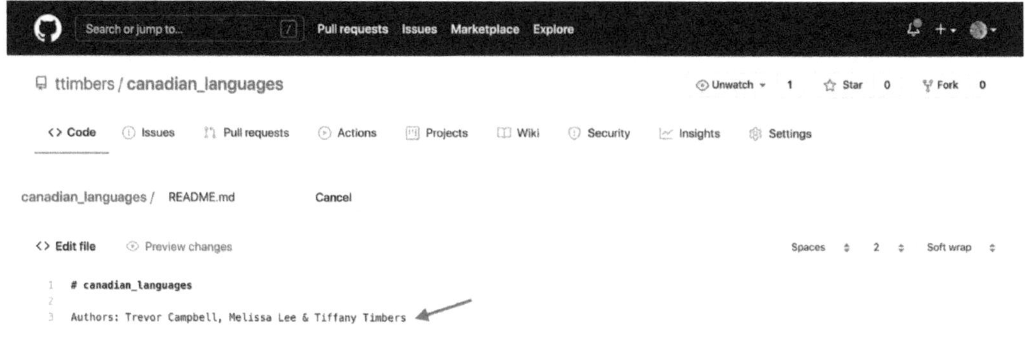

FIGURE 12.12: The text box where edits can be made after clicking on the pen tool.

green "Commit changes" button, it will save the file and then make a commit (Figure 12.13).

Recall from Section 12.5.1 that you normally have to add files to the staging area before committing them. Why don't we have to do that when we work directly on GitHub? Behind the scenes, when you click the green "Commit changes" button, GitHub *is* adding that one file to the staging area prior to committing it. But note that on GitHub you are limited to committing changes to only one file at a time. When you work in your own local repository, you can commit changes to multiple files simultaneously. This is especially useful when one "improvement" to the project involves modifying multiple files. You can also do things like run code when working in a local repository, which you cannot do on GitHub. In general, editing on GitHub is reserved for small edits to plain text files.

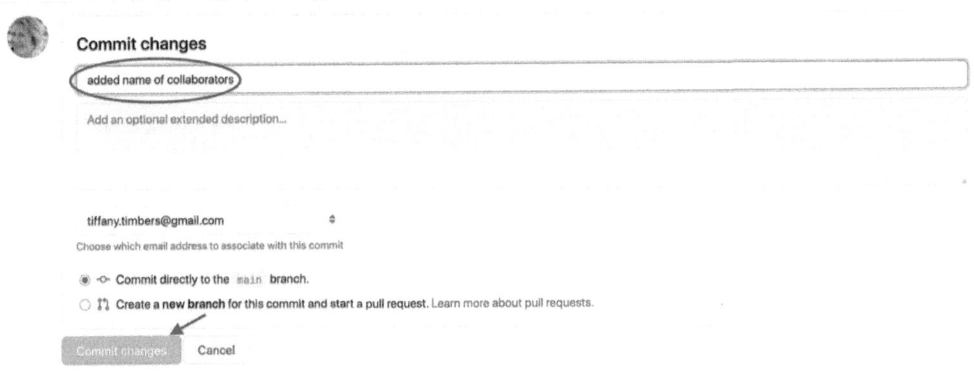

FIGURE 12.13: Saving changes using the pen tool requires committing those changes, and an associated commit message.

12.6.3 Creating files on GitHub with the "Add file" menu

The "Add file" menu can be used to create new plain text files and upload files from your computer. To create a new plain text file, click the "Add file" drop-down menu and select the "Create new file" option (Figure 12.14).

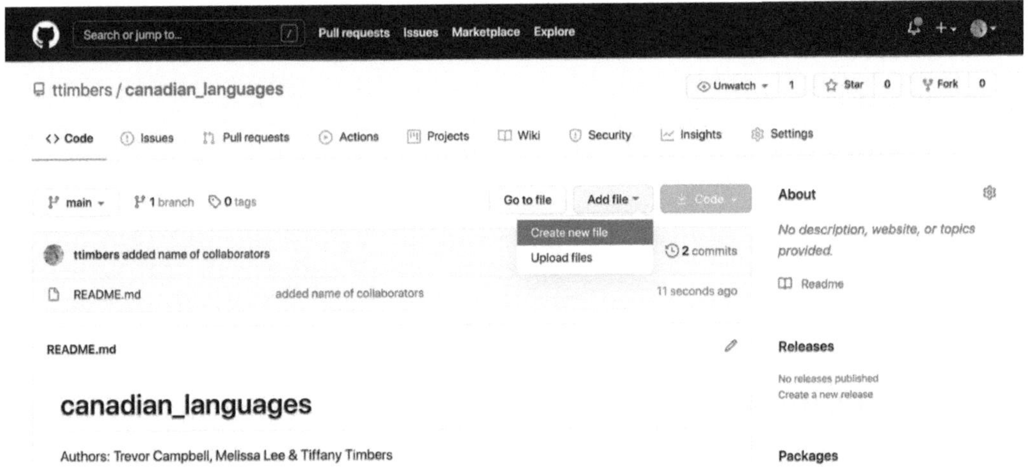

FIGURE 12.14: New plain text files can be created directly on GitHub.

A page will open with a small text box for the file name to be entered, and a larger text box where the desired file content text can be entered. Note the two tabs, "Edit new file" and "Preview". Toggling between them lets you enter and edit text and view what the text will look like when rendered, respectively (Figure 12.15). Note that GitHub understands and renders `.md` files using a markdown syntax[1] very similar to Jupyter notebooks, so the "Preview" tab is especially helpful for checking markdown code correctness.

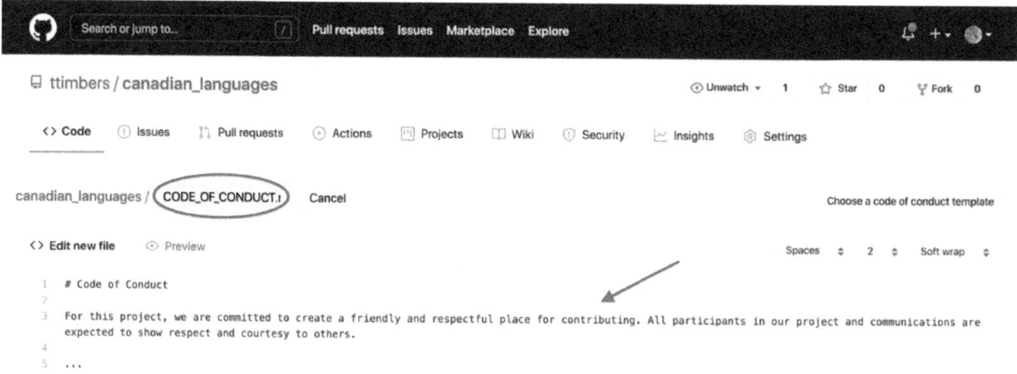

FIGURE 12.15: New plain text files require a file name in the text box circled in red, and file content entered in the larger text box (red arrow).

Save and commit your changes by clicking the green "Commit changes" button at the bottom of the page (Figure 12.16).

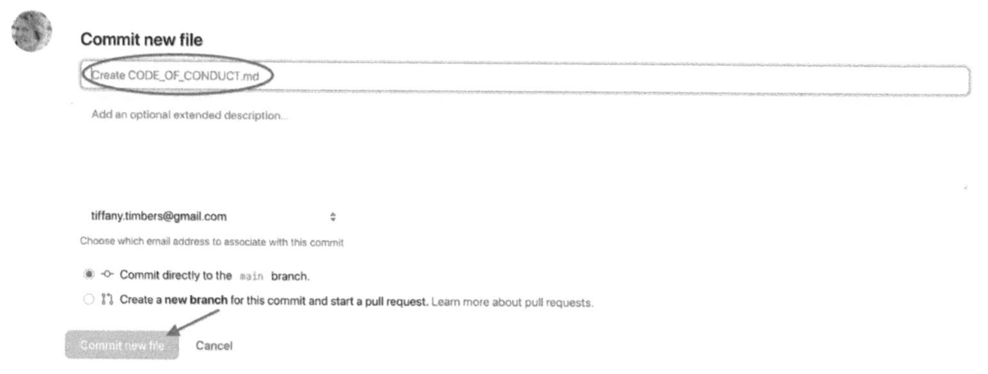

FIGURE 12.16: To be saved, newly created files are required to be committed along with an associated commit message.

You can also upload files that you have created on your local machine by using the "Add file" drop-down menu and selecting "Upload files" (Figure 12.17). To select the files from your local computer to upload, you can either drag and drop them into the gray box area shown below, or click the "choose your files" link to access a file browser dialog. Once the files you want to upload have been selected, click the green "Commit changes" button at the bottom of the page (Figure 12.18).

[1] https://guides.github.com/pdfs/markdown-cheatsheet-online.pdf

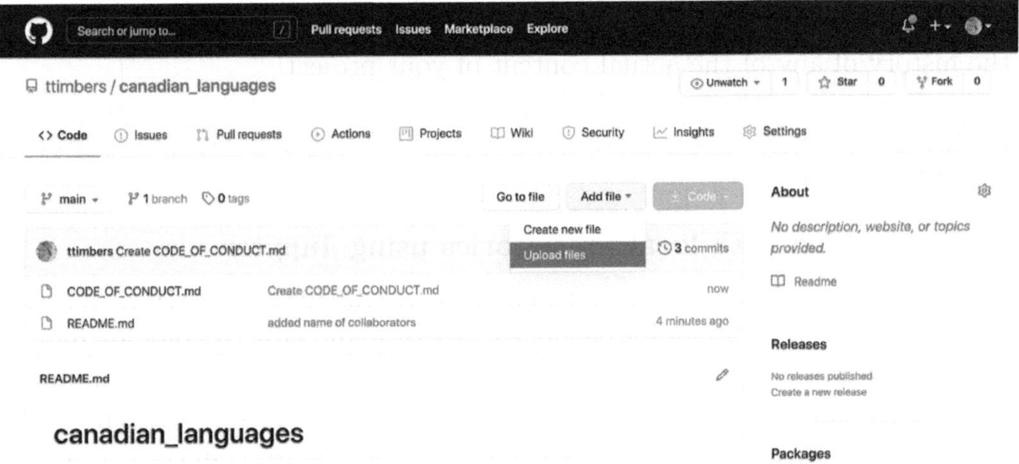

FIGURE 12.17: New files of any type can be uploaded to GitHub.

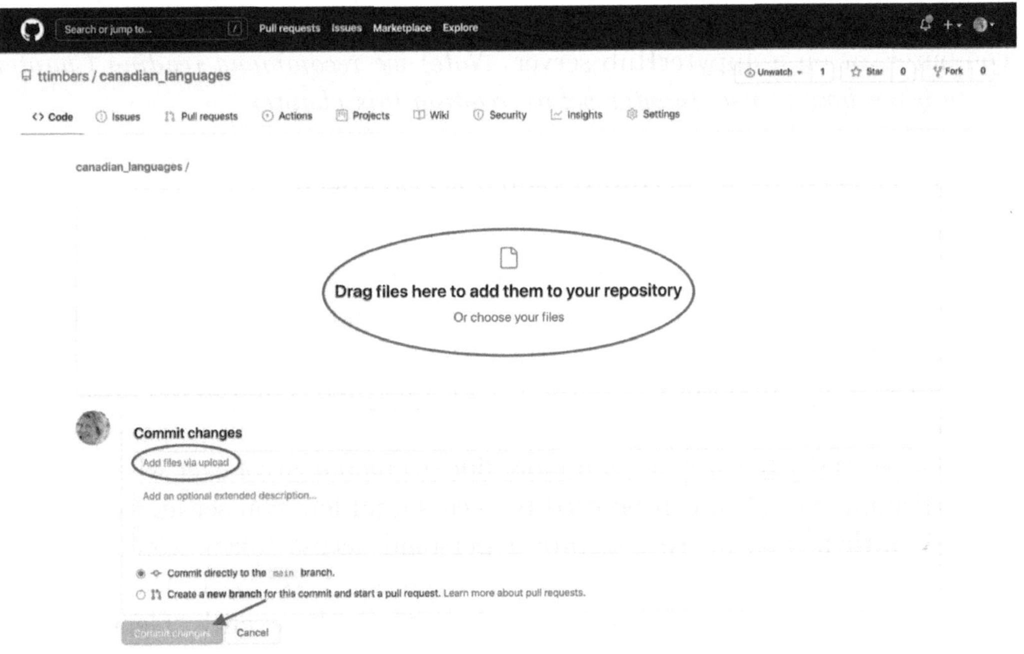

FIGURE 12.18: Specify files to upload by dragging them into the GitHub website (red circle) or by clicking on "choose your files." Uploaded files are also required to be committed along with an associated commit message.

Note that Git and GitHub are designed to track changes in individual files. **Do not** upload your whole project in an archive file (e.g., `.zip`). If you do, then Git can only keep track of changes to the entire `.zip` file, which will not be human-readable. Committing one big archive defeats the whole purpose of using version control: you won't be able to see, interpret, or find changes in the history of any of the actual content of your project!

12.7 Working with local repositories using Jupyter

Although there are several ways to create and edit files on GitHub, they are not quite powerful enough for efficiently creating and editing complex files, or files that need to be executed to assess whether they work (e.g., files containing code). For example, you wouldn't be able to run an analysis written with R code directly on GitHub. Thus, it is useful to be able to connect the remote repository that was created on GitHub to a local coding environment. This can be done by creating and working in a local copy of the repository. In this chapter, we focus on interacting with Git via Jupyter using the Jupyter Git extension. The Jupyter Git extension can be run by Jupyter on your local computer, or on a JupyterHub server. *Note: we recommend reading Chapter 11 to learn how to use Jupyter before reading this chapter.*

12.7.1 Generating a GitHub personal access token

To send and retrieve work between your local repository and the remote repository on GitHub, you will frequently need to authenticate with GitHub to prove you have the required permission. There are several methods to do this, but for beginners we recommend using the HTTPS method because it is easier and requires less setup. In order to use the HTTPS method, GitHub requires you to provide a *personal access token*. A personal access token is like a password—so keep it a secret!—but it gives you more fine-grained control over what parts of your account the token can be used to access, and lets you set an expiry date for the authentication. To generate a personal access token, you must first visit `https://github.com/settings/tokens`, which will take you to the "Personal access tokens" page in your account settings. Once there, click "Generate new token" (Figure 12.19). Note that you may be asked to re-authenticate with your username and password to proceed.

You will be asked to add a note to describe the purpose for your personal access token. Next, you need to select permissions for the token; this is where you can control what parts of your account the token can be used to access. Make sure to choose only those permissions that you absolutely require. In Figure 12.20, we tick only the "repo" box, which gives the token access to our repositories (so that we can push and pull) but none of our other GitHub account features. Finally, to generate the token, scroll to the bottom of that page and click the green "Generate token" button (Figure 12.20).

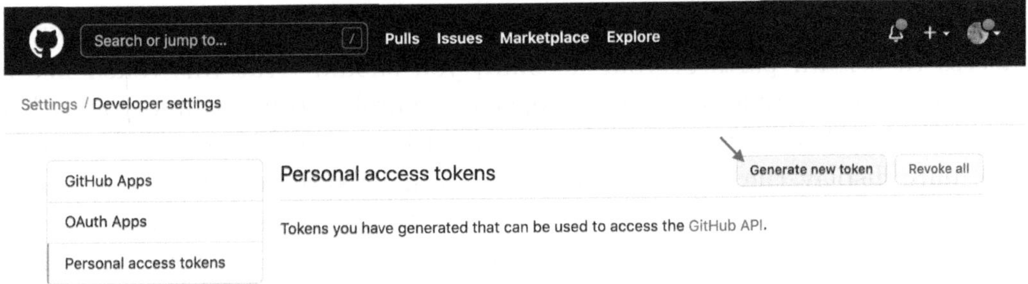

FIGURE 12.19: The "Generate new token" button used to initiate the creation of a new personal access token. It is found in the "Personal access tokens" section of the "Developer settings" page in your account settings.

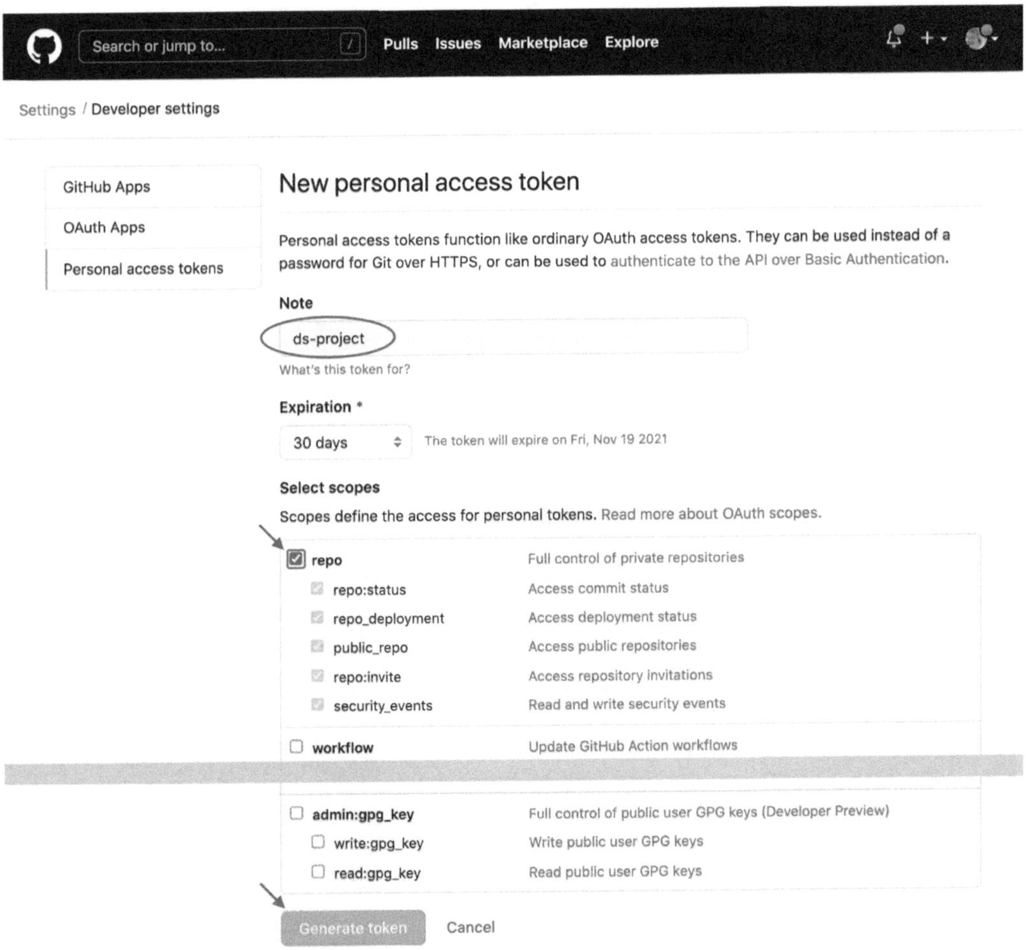

FIGURE 12.20: Webpage for creating a new personal access token.

Finally, you will be taken to a page where you will be able to see and copy the personal access token you just generated (Figure 12.21). Since it provides access to certain parts of your account, you should treat this token like a password; for example, you should consider securely storing it (and your other passwords and tokens, too!) using a password manager. Note that this page will only display the token to you once, so make sure you store it in a safe place right away. If you accidentally forget to store it, though, do not fret—you can delete that token by clicking the "Delete" button next to your token, and generate a new one from scratch. To learn more about GitHub authentication, see the additional resources section at the end of this chapter.

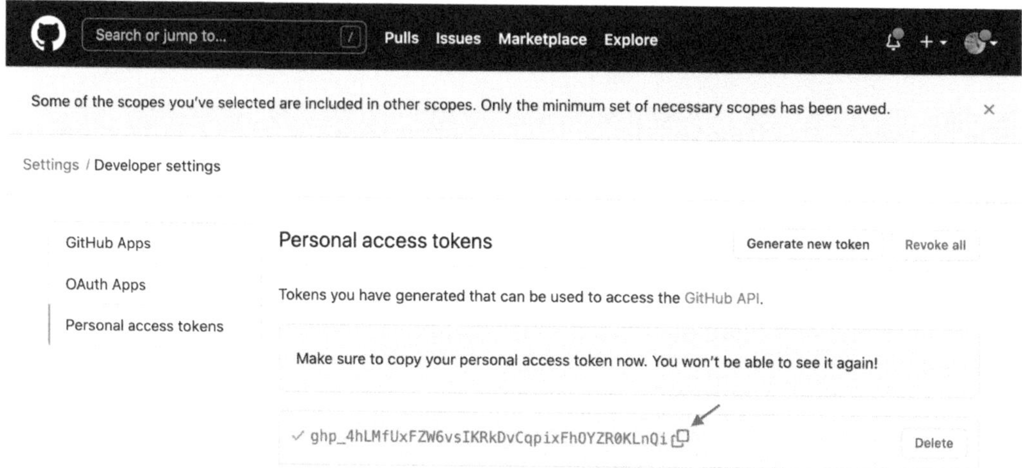

FIGURE 12.21: Display of the newly generated personal access token.

12.7.2 Cloning a repository using Jupyter

Cloning a remote repository from GitHub to create a local repository results in a copy that knows where it was obtained from so that it knows where to send/receive new committed edits. In order to do this, first copy the URL from the HTTPS tab of the Code drop-down menu on GitHub (Figure 12.22).

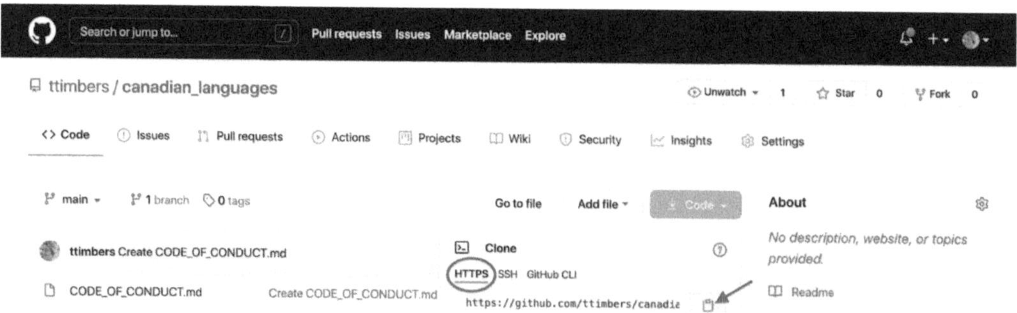

FIGURE 12.22: The green "Code" drop-down menu contains the remote address (URL) corresponding to the location of the remote GitHub repository.

Open Jupyter, and click the Git+ icon on the file browser tab (Figure 12.23).

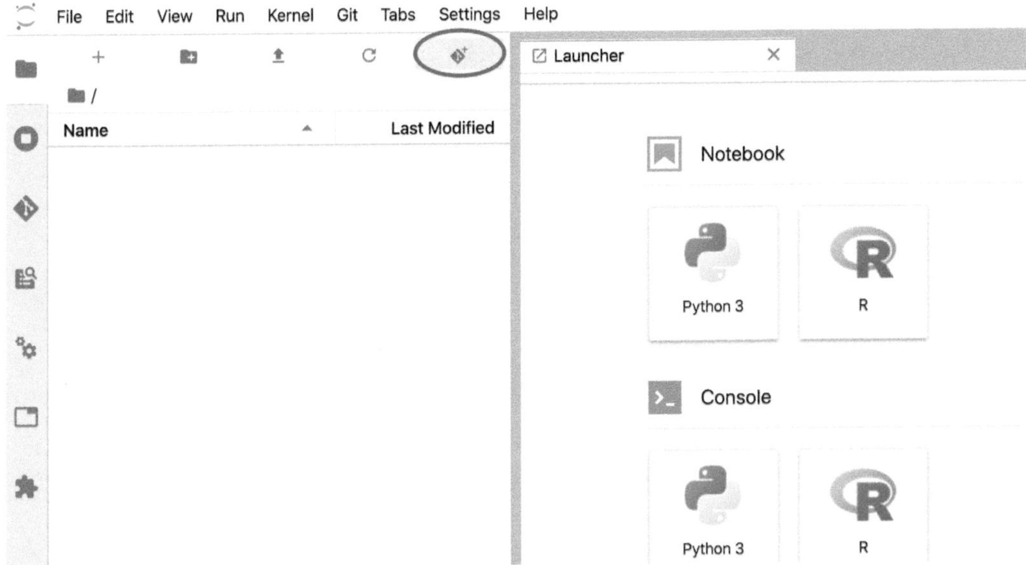

FIGURE 12.23: The Jupyter Git Clone icon (red circle).

Paste the URL of the GitHub project repository you created and click the blue "CLONE" button (Figure 12.24).

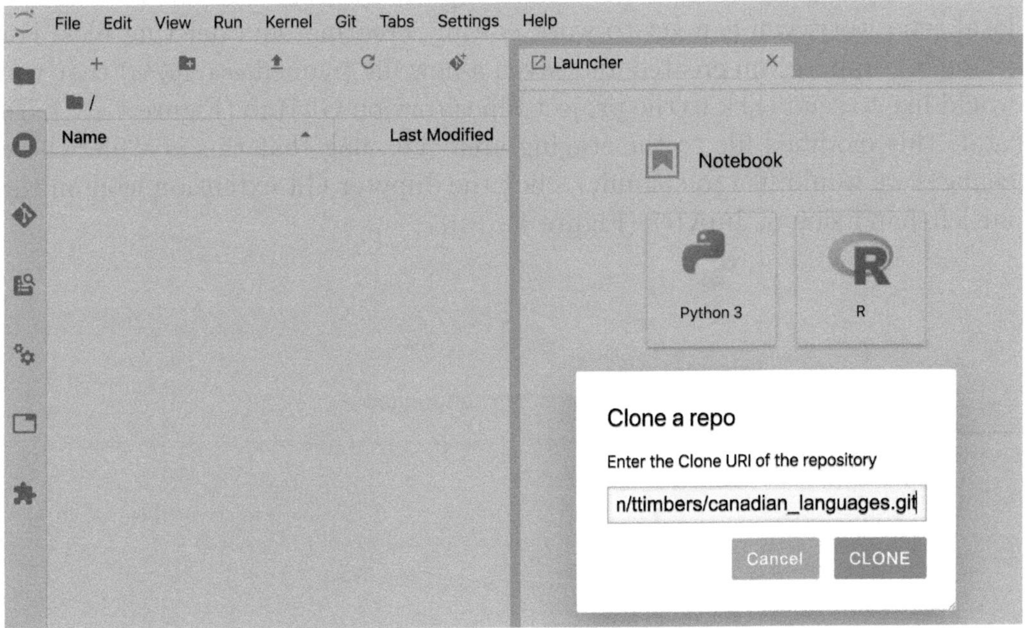

FIGURE 12.24: Prompt where the remote address (URL) corresponding to the location of the GitHub repository needs to be input in Jupyter.

On the file browser tab, you will now see a folder for the repository. Inside this folder will be all the files that existed on GitHub (Figure 12.25).

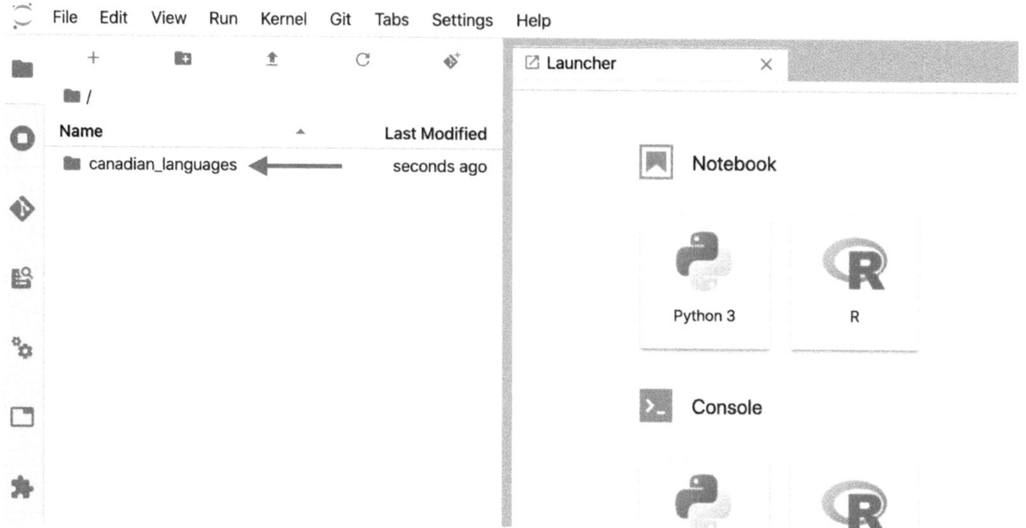

FIGURE 12.25: Cloned GitHub repositories can been seen and accessed via the Jupyter file browser.

12.7.3 Specifying files to commit

Now that you have cloned the remote repository from GitHub to create a local repository, you can get to work editing, creating, and deleting files. For example, suppose you created and saved a new file (named `eda.ipynb`) that you would like to send back to the project repository on GitHub (Figure 12.26). To "add" this modified file to the staging area (i.e., flag that this is a file whose changes we would like to commit), click the Jupyter Git extension icon on the far left-hand side of Jupyter (Figure 12.26).

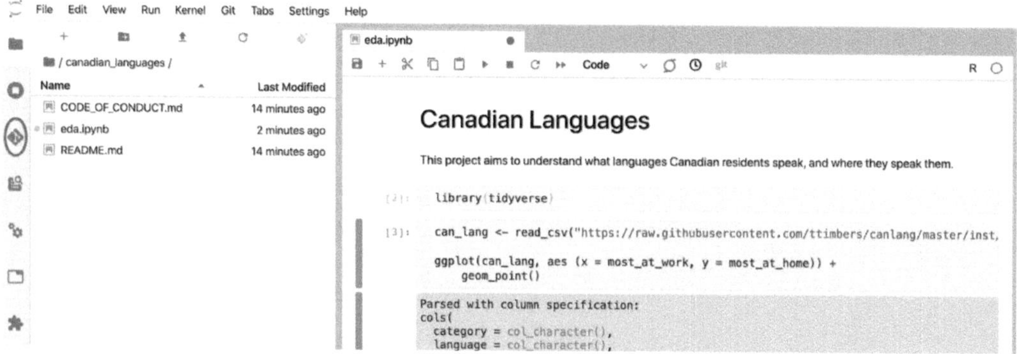

FIGURE 12.26: Jupyter Git extension icon (circled in red).

This opens the Jupyter Git graphical user interface pane. Next, click the plus sign (+) beside the file(s) that you want to "add" (Figure 12.27). Note that because this is the first change for this file, it falls under the "Untracked" heading. However, next time you edit this file and want to add the changes, you will find it under the "Changed" heading.

You will also see an `eda-checkpoint.ipynb` file under the "Untracked" heading. This is a temporary "checkpoint file" created by Jupyter when you work on `eda.ipynb`. You generally do not want to add auto-generated files to Git repositories; only add the files you directly create and edit.

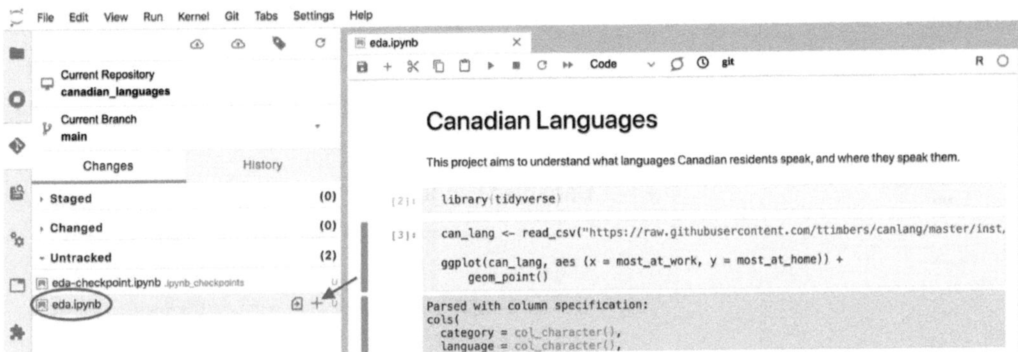

FIGURE 12.27: `eda.ipynb` is added to the staging area via the plus sign (+).

Clicking the plus sign (+) moves the file from the "Untracked" heading to the "Staged" heading, so that Git knows you want a snapshot of its current state as a commit (Figure 12.28). Now you are ready to "commit" the changes. Make sure to include a (clear and helpful!) message about what was changed so that your collaborators (and future you) know what happened in this commit.

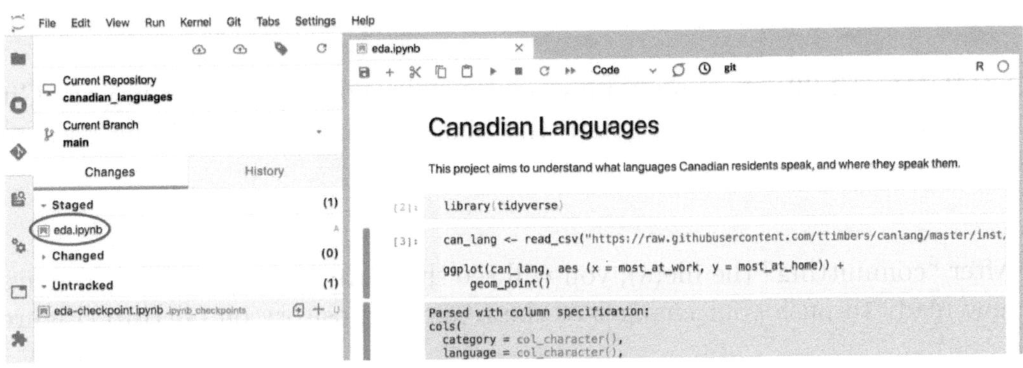

FIGURE 12.28: Adding `eda.ipynb` makes it visible in the staging area.

12.7.4 Making the commit

To snapshot the changes with an associated commit message, you must put a message in the text box at the bottom of the Git pane and click on the blue "Commit" button (Figure 12.29). It is highly recommended to write useful and meaningful messages about what was changed. These commit messages, and the datetime stamp for a given commit, are the primary means to navigate through the project's history in the event that you need to view or retrieve a past version of a file, or revert your project to an earlier state. When you click the "Commit" button for the first time, you will be prompted to enter your name and email. This only needs to be done once for each machine you use Git on.

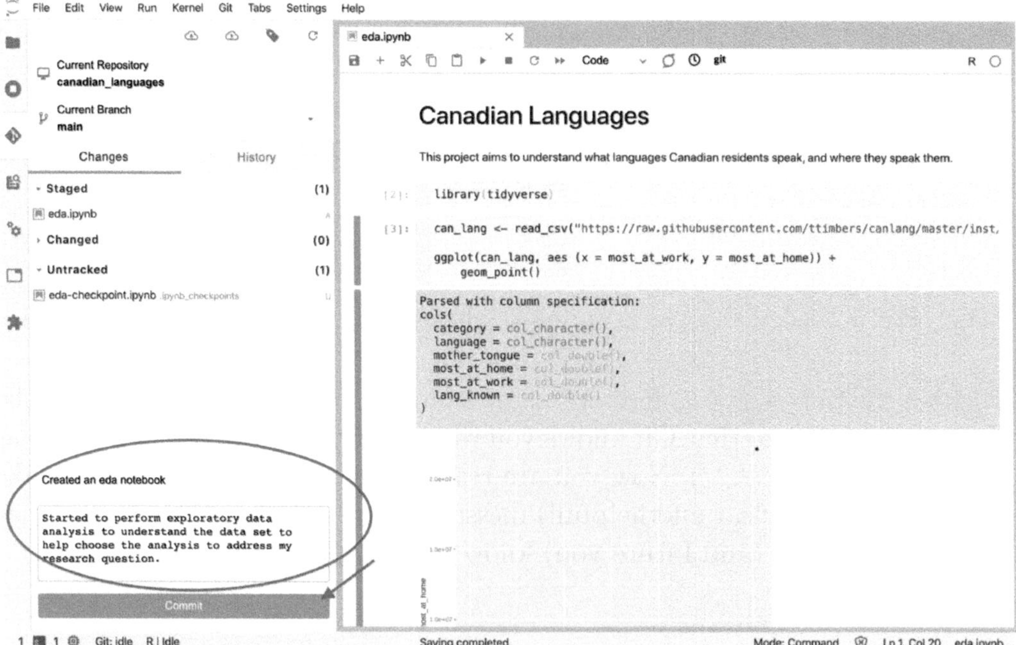

FIGURE 12.29: A commit message must be added into the Jupyter Git extension commit text box before the blue Commit button can be used to record the commit.

After "committing" the file(s), you will see there are 0 "Staged" files. You are now ready to push your changes to the remote repository on GitHub (Figure 12.30).

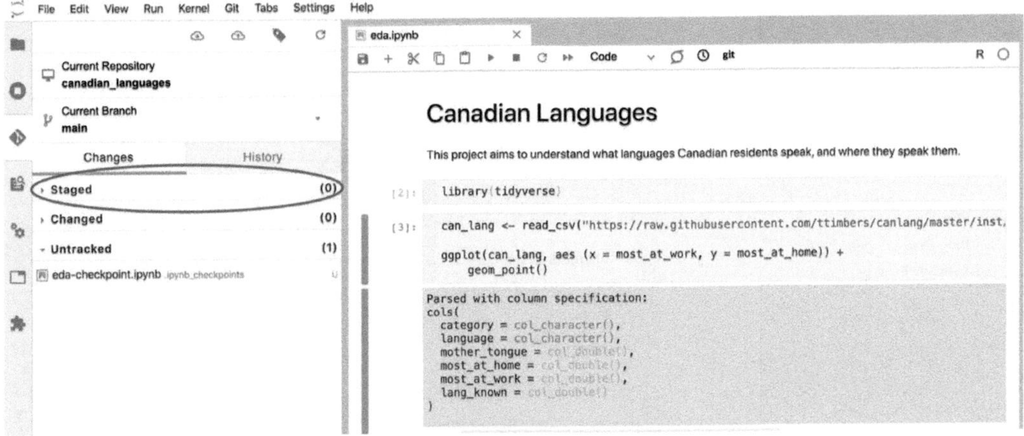

FIGURE 12.30: After recording a commit, the staging area should be empty.

12.7.5 Pushing the commits to GitHub

To send the committed changes back to the remote repository on GitHub, you need to *push* them. To do this, click on the cloud icon with the up arrow on the Jupyter Git tab (Figure 12.31).

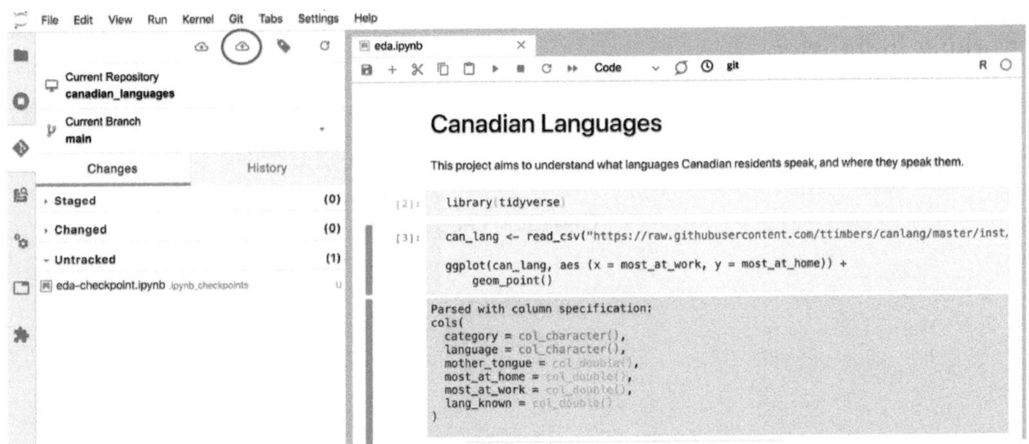

FIGURE 12.31: The Jupyter Git extension "push" button (circled in red).

You will then be prompted to enter your GitHub username and the personal access token that you generated earlier (not your account password!). Click the blue "OK" button to initiate the push (Figure 12.32).

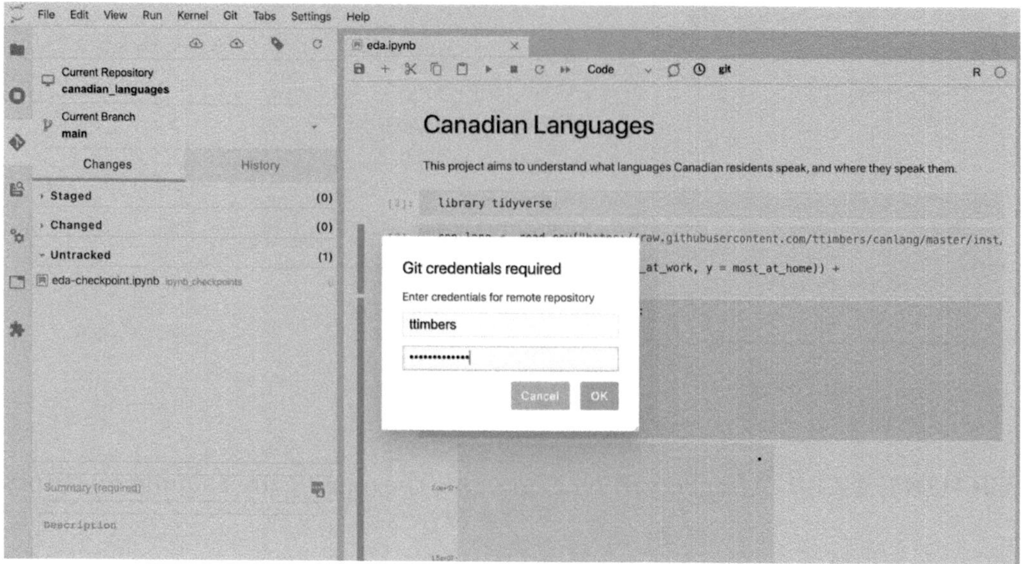

FIGURE 12.32: Enter your Git credentials to authorize the push to the remote repository.

If the files were successfully pushed to the project repository on GitHub, you will be shown a success message (Figure 12.33). Click "Dismiss" to continue working in Jupyter.

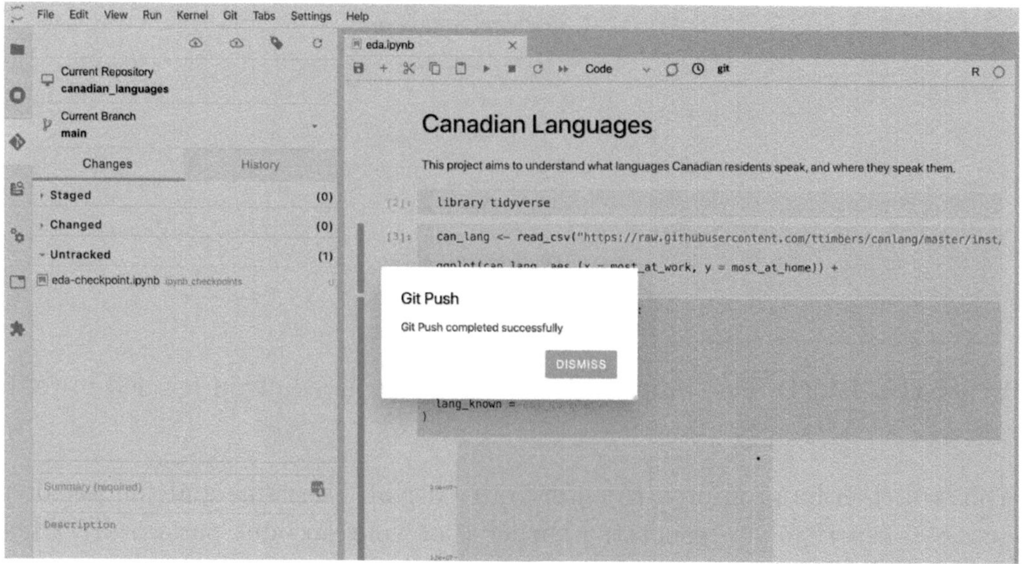

FIGURE 12.33: The prompt that the push was successful.

If you visit the remote repository on GitHub, you will see that the changes now exist there too (Figure 12.34)!

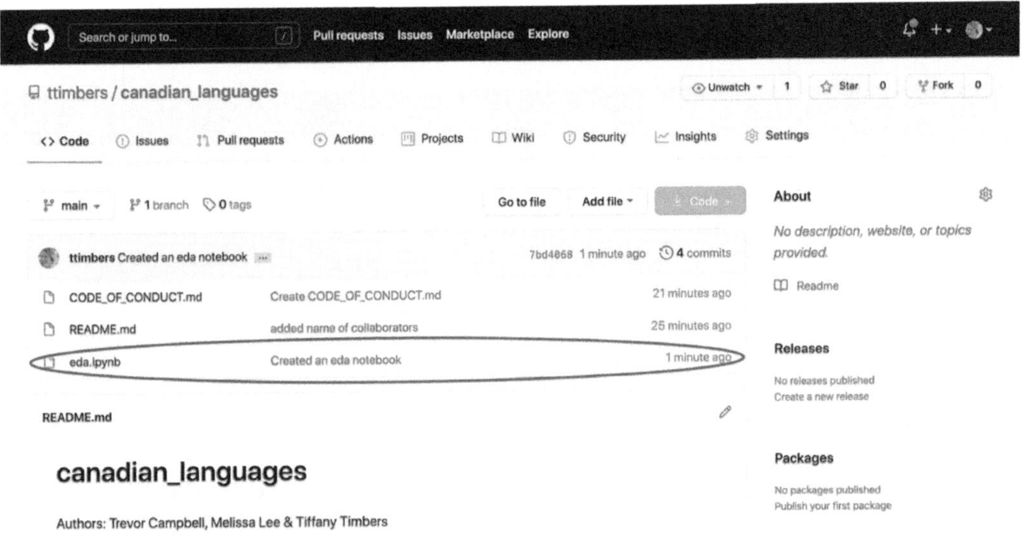

FIGURE 12.34: The GitHub web interface shows a preview of the commit message, and the time of the most recently pushed commit for each file.

12.8 Collaboration

12.8.1 Giving collaborators access to your project

As mentioned earlier, GitHub allows you to control who has access to your project. The default of both public and private projects are that only the person who created the GitHub repository has permissions to create, edit and delete files (*write access*). To give your collaborators write access to the projects, navigate to the "Settings" tab (Figure 12.35).

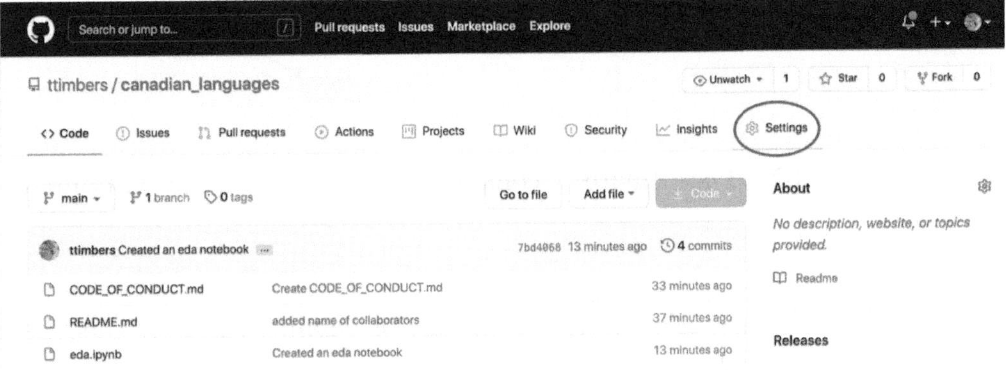

FIGURE 12.35: The "Settings" tab on the GitHub web interface.

Then click "Manage access" (Figure 12.36).

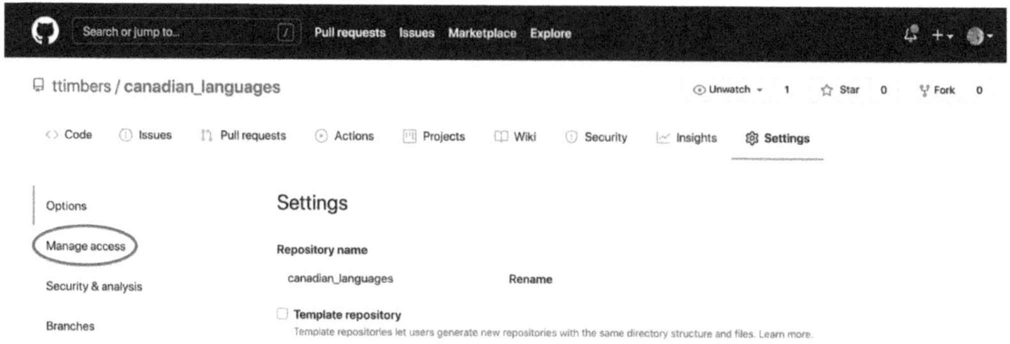

FIGURE 12.36: The "Manage access" tab on the GitHub web interface.

(Figure 12.37).

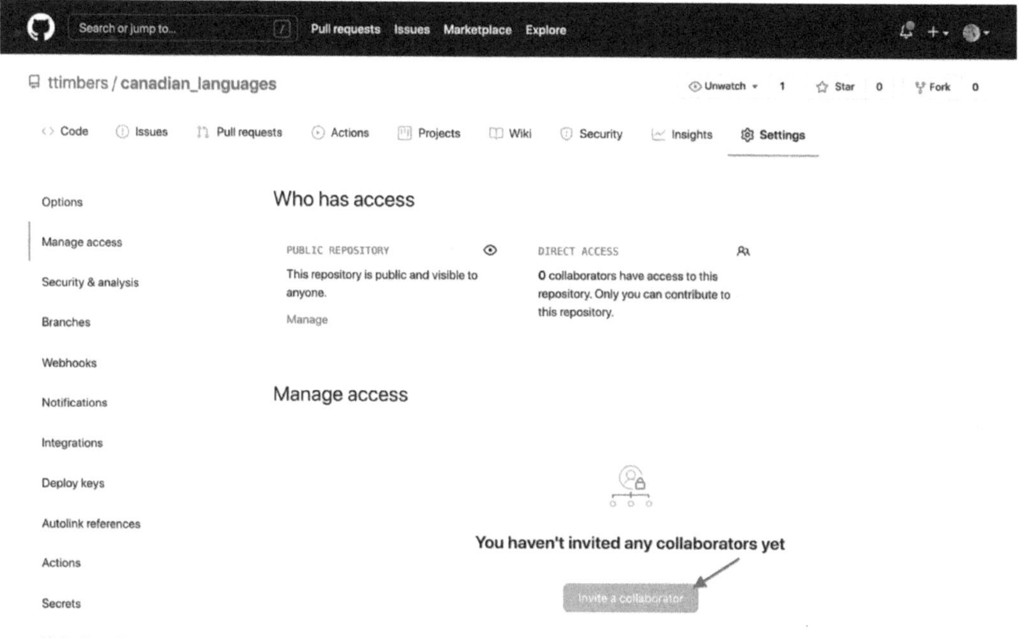

FIGURE 12.37: The "Invite a collaborator" button on the GitHub web interface.

Type in the collaborator's GitHub username or email, and select their name when it appears (Figure 12.38).

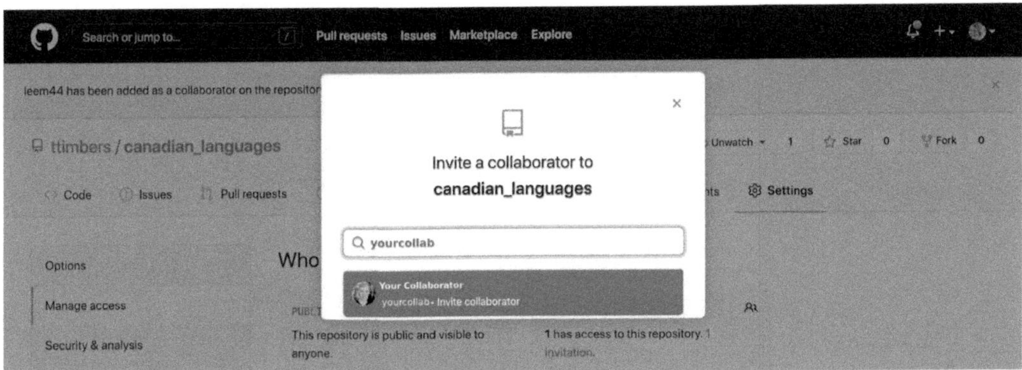

FIGURE 12.38: The text box where a collaborator's GitHub username or email can be entered.

Finally, click the green "Add to this repository" button (Figure 12.39).

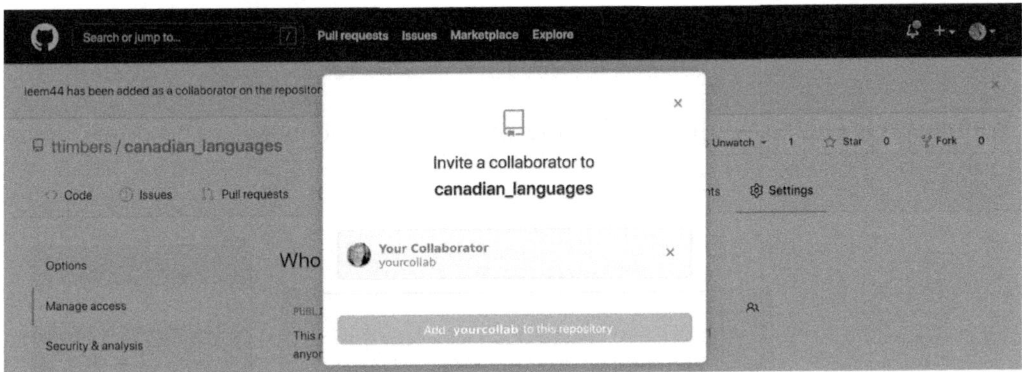

FIGURE 12.39: The confirmation button for adding a collaborator to a repository on the GitHub web interface.

After this, you should see your newly added collaborator listed under the "Manage access" tab. They should receive an email invitation to join the GitHub repository as a collaborator. They need to accept this invitation to enable write access.

12.8.2 Pulling changes from GitHub using Jupyter

We will now walk through how to use the Jupyter Git extension tool to pull changes to our `eda.ipynb` analysis file that were made by a collaborator (Figure 12.40).

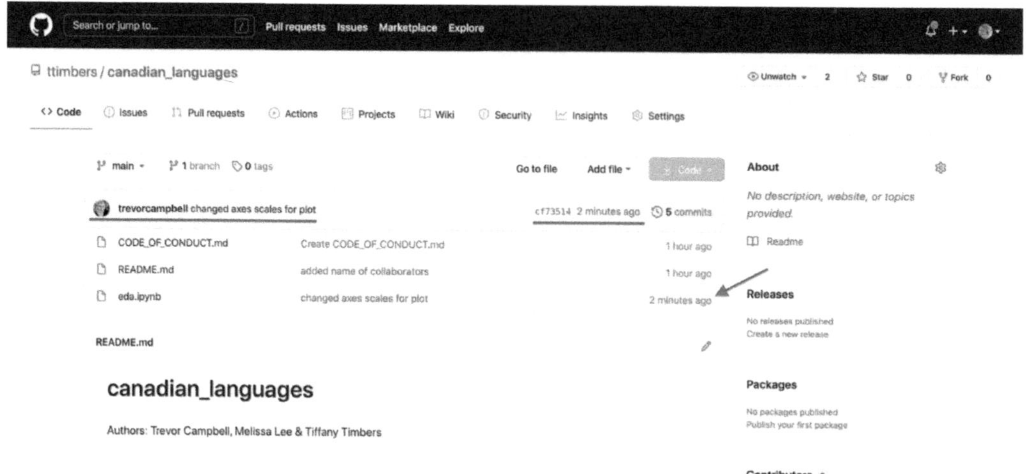

FIGURE 12.40: The GitHub interface indicates the name of the last person to push a commit to the remote repository, a preview of the associated commit message, the unique commit identifier, and how long ago the commit was snapshotted.

You can tell Git to "pull" by clicking on the cloud icon with the down arrow in Jupyter (Figure 12.41).

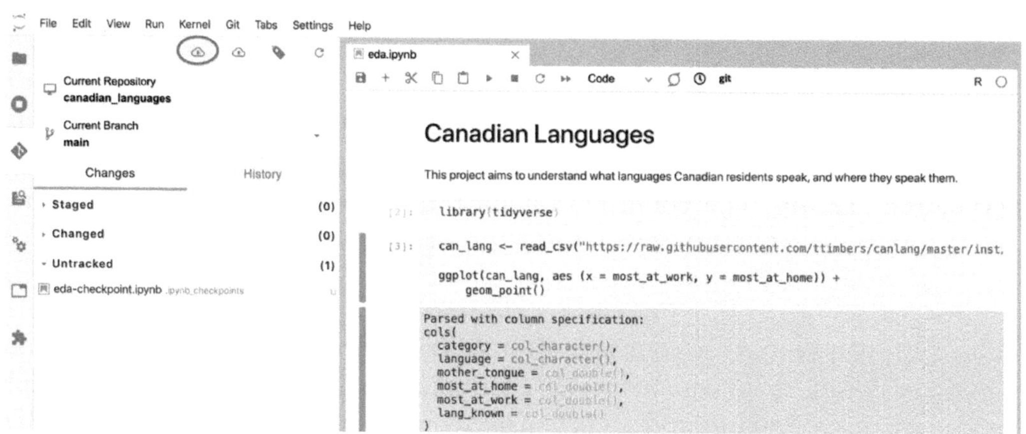

FIGURE 12.41: The Jupyter Git extension clone button.

Once the files are successfully pulled from GitHub, you need to click "Dismiss" to keep working (Figure 12.42).

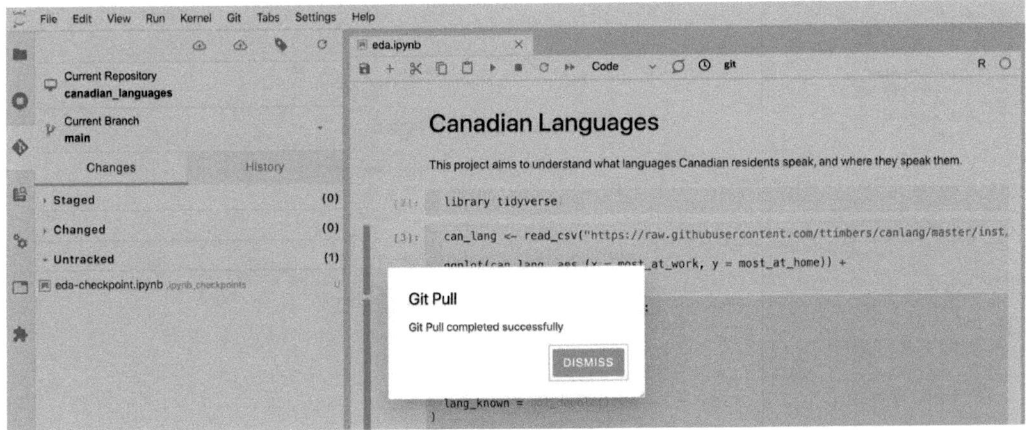

FIGURE 12.42: The prompt after changes have been successfully pulled from a remote repository.

And then when you open (or refresh) the files whose changes you just pulled, you should be able to see them (Figure 12.43).

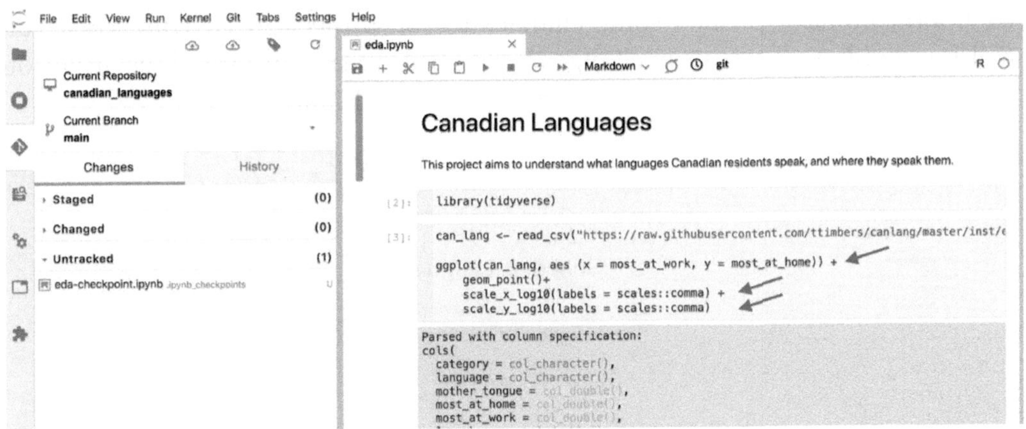

FIGURE 12.43: Changes made by the collaborator to `eda.ipynb` (code highlighted by red arrows).

It can be very useful to review the history of the changes to your project. You can do this directly in Jupyter by clicking "History" in the Git tab (Figure 12.44).

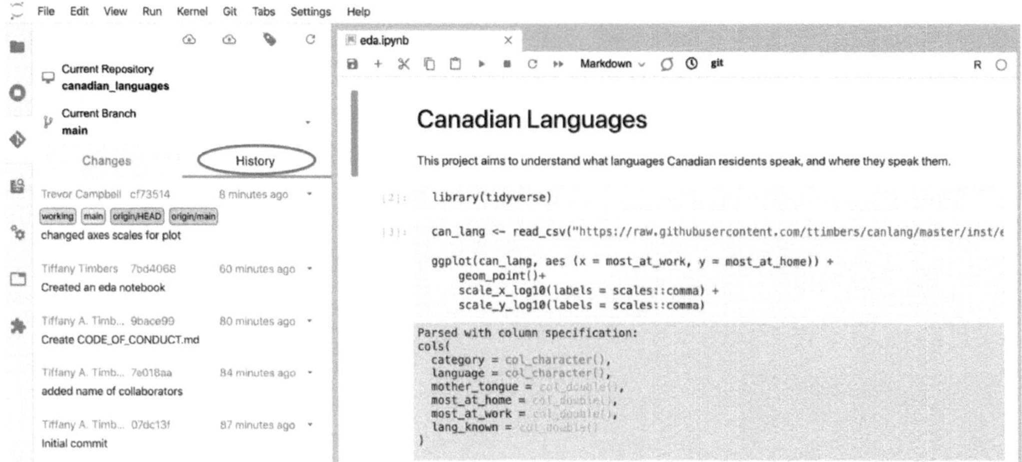

FIGURE 12.44: Version control repository history viewed using the Jupyter Git extension.

It is good practice to pull any changes at the start of *every* work session before you start working on your local copy. If you do not do this, and your collaborators have pushed some changes to the project to GitHub, then you will be unable to push your changes to GitHub until you pull. This situation can be recognized by the error message shown in Figure 12.45.

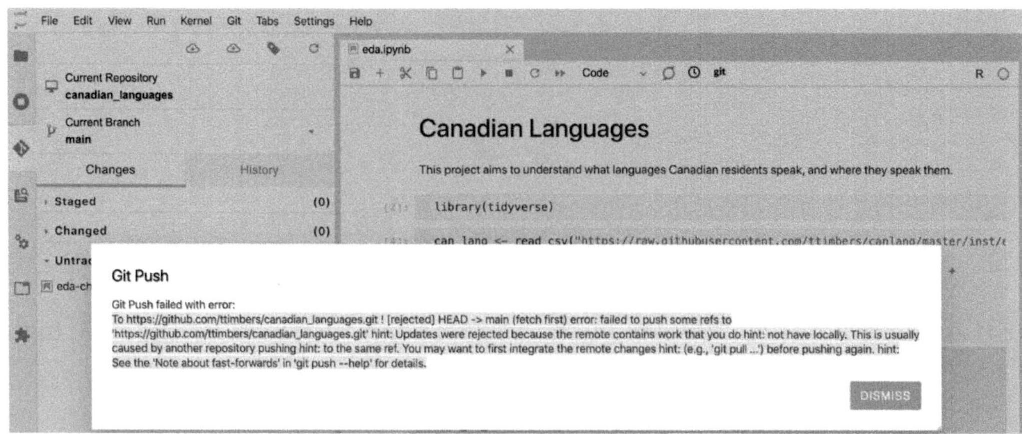

FIGURE 12.45: Error message that indicates that there are changes on the remote repository that you do not have locally.

Usually, getting out of this situation is not too troublesome. First you need to pull the changes that exist on GitHub that you do not yet have in the local repository. Usually when this happens, Git can automatically merge the changes for you, even if you and your collaborators were working on different parts of the same file!

If, however, you and your collaborators made changes to the same line of the same file, Git will not be able to automatically merge the changes—it will not know whether to keep your version of the line(s), your collaborators version of the line(s), or some blend of the two. When this happens, Git will tell you that you have a merge conflict in certain file(s) (Figure 12.46).

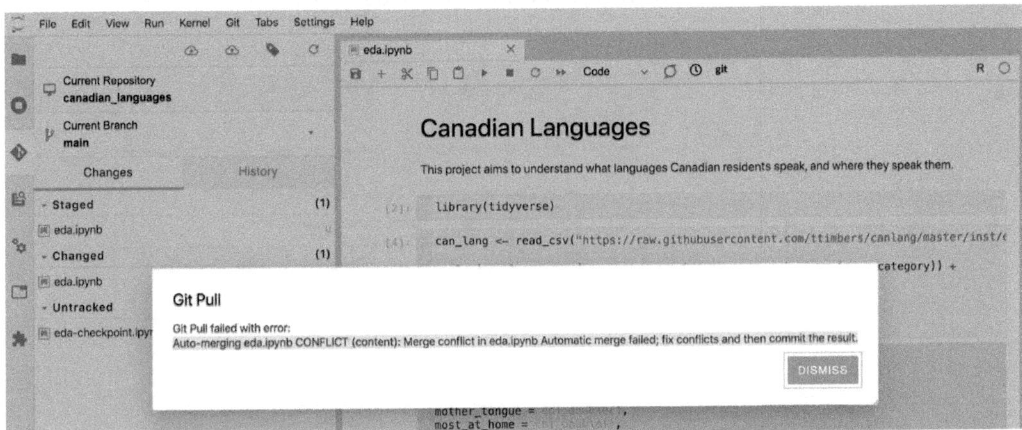

FIGURE 12.46: Error message that indicates you and your collaborators made changes to the same line of the same file and that Git will not be able to automatically merge the changes.

12.8.3 Handling merge conflicts

To fix the merge conflict, you need to open the offending file in a plain text editor and look for special marks that Git puts in the file to tell you where the merge conflict occurred (Figure 12.47).

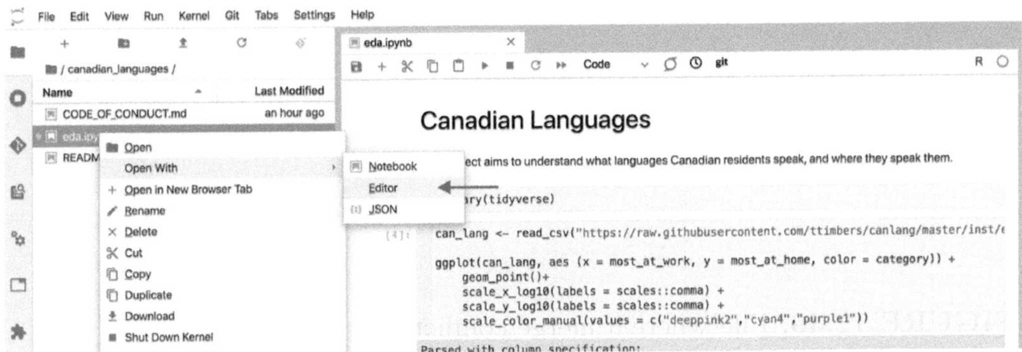

FIGURE 12.47: How to open a Jupyter notebook as a plain text file view in Jupyter.

The beginning of the merge conflict is preceded by `<<<<<<< HEAD` and the end of the merge conflict is marked by `>>>>>>>`. Between these markings, Git also inserts a separator (`=======`). The version of the change before the separator is your change, and the version that follows the separator was the change that existed on GitHub. In Figure 12.48, you can see that in your local repository there is a line of code that calls `scale_color_manual` with three color values (`deeppink2`, `cyan4`, and `purple1`). It looks like your collaborator made an edit to that line too, except with different colors (to `blue3`, `red3`, and `black`)!

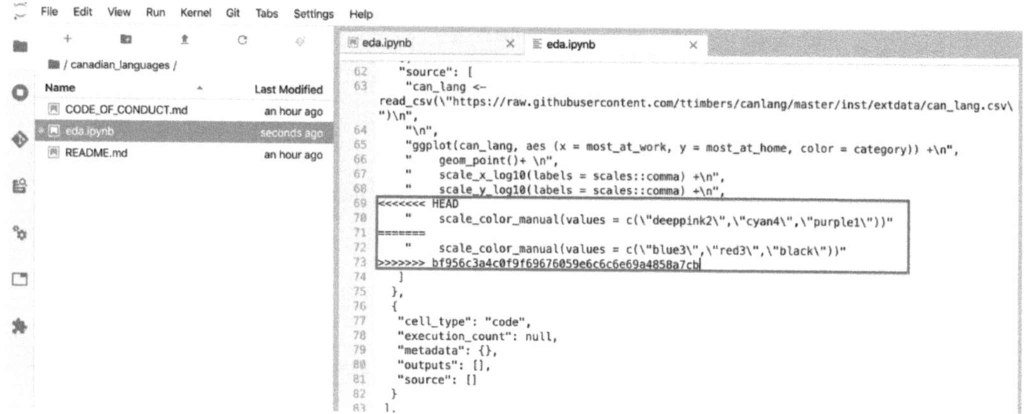

FIGURE 12.48: Merge conflict identifiers (highlighted in red).

Once you have decided which version of the change (or what combination!) to keep, you need to use the plain text editor to remove the special marks that Git added (Figure 12.49).

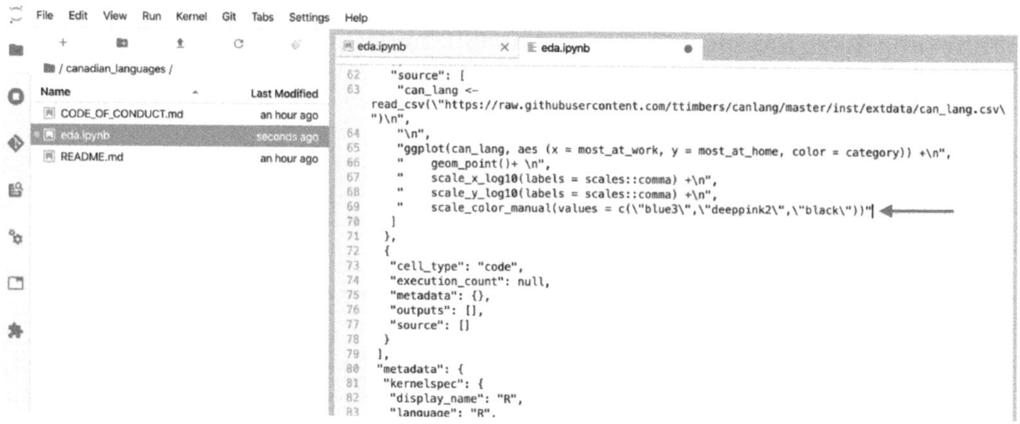

FIGURE 12.49: File where a merge conflict has been resolved.

The file must be saved, added to the staging area, and then committed before you will be able to push your changes to GitHub.

12.8.4 Communicating using GitHub issues

When working on a project in a team, you don't just want a historical record of who changed what file and when in the project—you also want a record of decisions that were made, ideas that were floated, problems that were identified and addressed, and all other communication surrounding the project. Email and messaging apps are both very popular for general communication, but are not designed for project-specific communication: they both generally do not have facilities for organizing conversations by project subtopics, searching for conversations related to particular bugs or software versions, etc.

GitHub *issues* are an alternative written communication medium to email and messaging apps, and were designed specifically to facilitate project-specific communication. Issues are *opened* from the "Issues" tab on the project's GitHub page, and they persist there even after the conversation is over and the issue is *closed* (in contrast to email, issues are not usually deleted). One issue thread is usually created per topic, and they are easily searchable using GitHub's search tools. All issues are accessible to all project collaborators, so no one is left out of the conversation. Finally, issues can be set up so that team members get email notifications when a new issue is created or a new post is made in an issue thread. Replying to issues from email is also possible. Given all of these advantages, we highly recommend the use of issues for project-related communication.

To open a GitHub issue, first click on the "Issues" tab (Figure 12.50).

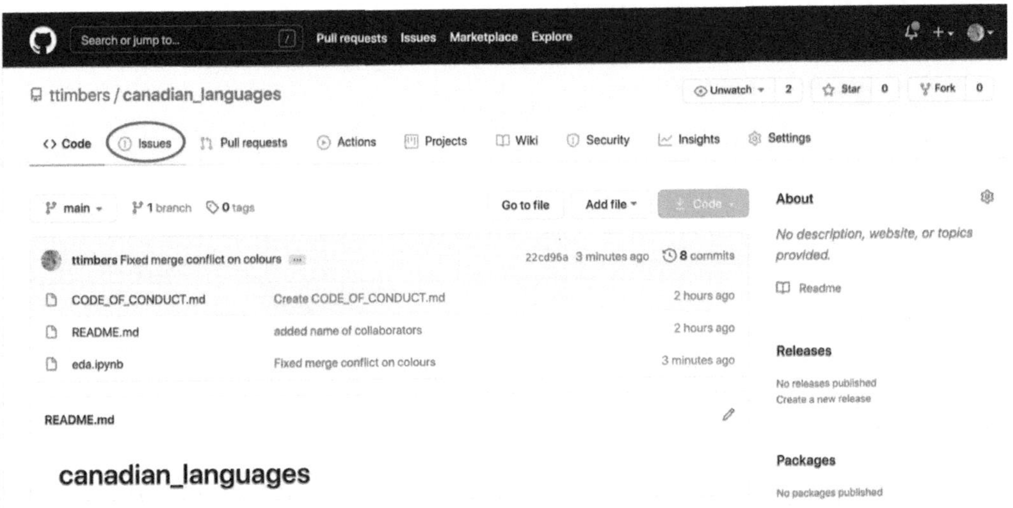

FIGURE 12.50: The "Issues" tab on the GitHub web interface.

Next click the "New issue" button (Figure 12.51).

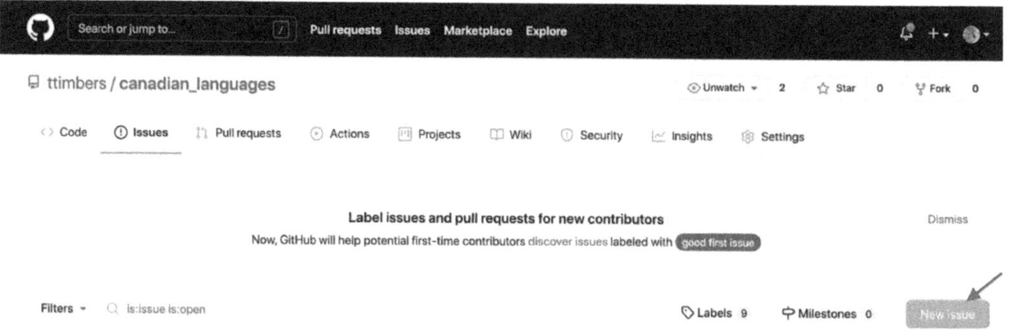

FIGURE 12.51: The "New issue" button on the GitHub web interface.

Add an issue title (which acts like an email subject line), and then put the body of the message in the larger text box. Finally, click "Submit new issue" to post the issue to share with others (Figure 12.52).

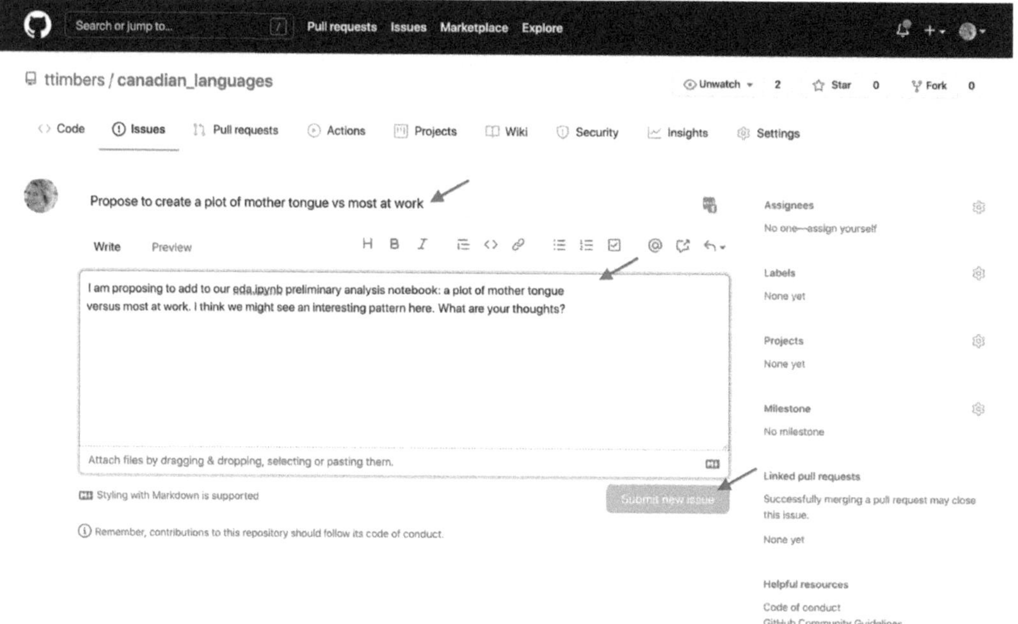

FIGURE 12.52: Dialog boxes and submission button for creating new GitHub issues.

You can reply to an issue that someone opened by adding your written response to the large text box and clicking comment (Figure 12.53).

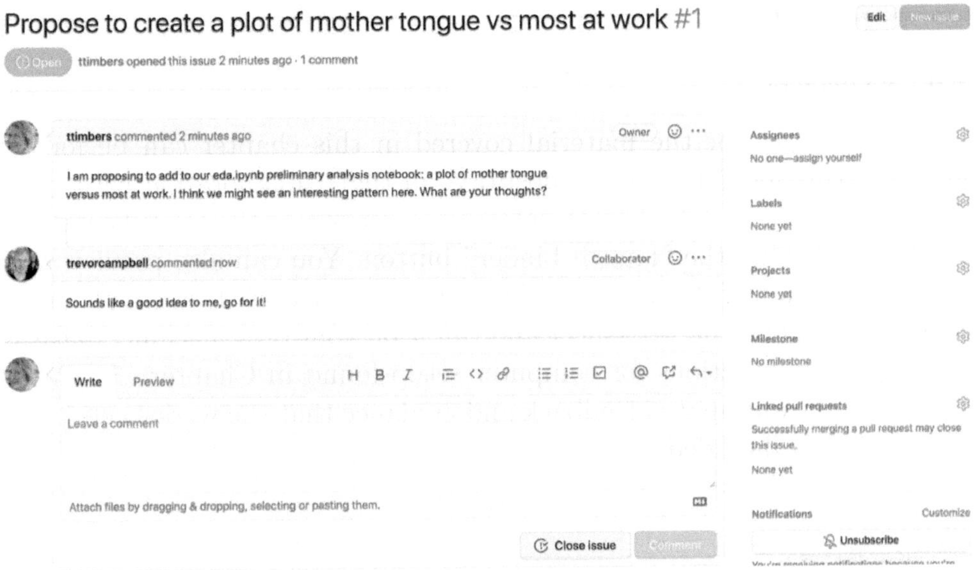

FIGURE 12.53: Dialog box for replying to GitHub issues.

When a conversation is resolved, you can click "Close issue". The closed issue can be later viewed by clicking the "Closed" header link in the "Issue" tab (Figure 12.54).

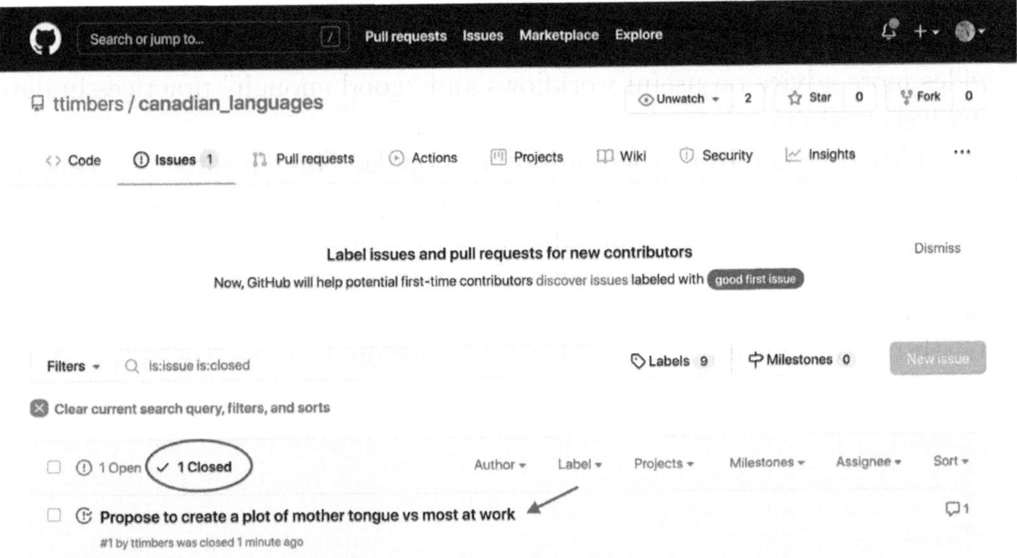

FIGURE 12.54: The "Closed" issues tab on the GitHub web interface.

12.9 Exercises

Practice exercises for the material covered in this chapter can be found in the accompanying worksheets repository[2] in the "Collaboration with version control" row. You can launch an interactive version of the worksheet in your browser by clicking the "launch binder" button. You can also preview a non-interactive version of the worksheet by clicking "view worksheet." If you instead decide to download the worksheet and run it on your own machine, make sure to follow the instructions for computer setup found in Chapter 13. This will ensure that the automated feedback and guidance that the worksheets provide will function as intended.

12.10 Additional resources

Now that you've picked up the basics of version control with Git and GitHub, you can expand your knowledge through the resources listed below:

- GitHub's guides website[3] and YouTube channel[4], and *Happy Git and GitHub for the useR*[5] are great resources to take the next steps in learning about Git and GitHub.
- Good enough practices in scientific computing[6] [Wilson et al., 2017] provides more advice on useful workflows and "good enough" practices in data analysis projects.
- In addition to GitHub[7], there are other popular Git repository hosting services such as GitLab[8] and BitBucket[9]. Comparing all of these options is beyond the scope of this book, and until you become a more advanced user, you are perfectly fine to just stick with GitHub. Just be aware that you have options!
- GitHub's documentation on creating a personal access token[10] and the

[2] https://github.com/UBC-DSCI/data-science-a-first-intro-worksheets#readme
[3] https://guides.github.com/
[4] https://www.youtube.com/githubguides
[5] https://happygitwithr.com/
[6] https://journals.plos.org/ploscompbiol/article?id=10.1371/journal.pcbi.1005510#sec014
[7] https://github.com
[8] https://gitlab.com
[9] https://bitbucket.org
[10] https://docs.github.com/en/authentication/keeping-your-account-and-data-secure/creating-a-personal-access-token

Happy Git and GitHub for the useR personal access tokens chapter[11] are both excellent additional resources to consult if you need additional help generating and using personal access tokens.

[11] `https://happygitwithr.com/https-pat.html`

13

Setting up your computer

13.1 Overview

In this chapter, you'll learn how to install all of the software needed to do the data science covered in this book on your own computer.

13.2 Chapter learning objectives

By the end of the chapter, readers will be able to do the following:

- Install the Git version control software.
- Install and launch a local instance of JupyterLab with the R kernel.
- Download the worksheets that accompany the chapters of this book from GitHub.

13.3 Installing software on your own computer

This section will provide instructions for installing the software required by this book on your own computer. Given that installation instructions can vary widely based on the computer setup, we have created instructions for multiple operating systems. In particular, the installation instructions below have been verified to work on a computer that:

- runs one of the following operating systems: Ubuntu 20.04, macOS Big Sur (version 11.4.x or 11.5.x), Windows 10 Professional, Enterprise or Education (version 2004, 20H2, or 21H1),
- has a connection to the internet,
- uses a 64-bit CPU,
- uses English as the default language.

13.3.1 Git

As shown in Chapter 12, Git is a very useful tool for version controlling your projects, as well as sharing your work with others. Here's how to install Git on the following operating systems:

Windows: To install Git on Windows, go to `https://git-scm.com/download/win` and download the Windows version of Git. Once the download has finished, run the installer and accept the default configuration for all pages.

MacOS: To install Git on Mac OS, open the terminal (how-to video[1]) and type the following command:

```
xcode-select --install
```

Ubuntu: To install Git on Ubuntu, open the terminal and type the following commands:

```
sudo apt update
sudo apt install git
```

13.3.2 Miniconda

To run Jupyter notebooks on your computer, you will need to install the web-based platform JupyterLab. But JupyterLab relies on Python, so we need to install Python first. We can install Python via the miniconda Python package distribution[2].

Windows: To install miniconda on Windows, download the latest Python 64-bit version from here[3]. Once the download has finished, run the installer and accept the default configuration for all pages. After installation, you can open the Anaconda Prompt by opening the Start Menu and searching for the program called "Anaconda Prompt (miniconda3)". When this opens, you will see a prompt similar to `(base) C:\Users\your_name`.

MacOS: To install miniconda on MacOS, you will need to use a different installation method depending on the type of processor chip your computer has.

If your Mac computer has an Intel x86 processor chip you can download the latest Python 64-bit version from here[4]. After the download has finished, run the installer and accept the default configuration for all pages.

---

[1] https://youtu.be/5AJbWEWwnbY

[2] https://docs.conda.io/en/latest/miniconda.html

[3] https://repo.anaconda.com/miniconda/Miniconda3-latest-Windows-x86_64.exe

[4] https://repo.anaconda.com/miniconda/Miniconda3-latest-MacOSX-x86_64.pkg

If your Mac computer has an Apple M1 processor chip you can download the latest Python 64-bit version from here[5]. After the download has finished, you need to run the downloaded script in the terminal using a command like:

```
bash path/to/Miniconda3-latest-MacOSX-arm64.sh
```

Make sure to replace `path/to/` with the path of the folder containing the downloaded script. Most computers will save downloaded files to the `Downloads` folder. If this is the case for your computer, you can run the script in the terminal by typing:

```
bash Downloads/Miniconda3-latest-MacOSX-arm64.sh
```

The instructions for the installation will then appear. Follow the prompts and agree to accepting the license, the default installation location, and to running `conda init`, which makes `conda` available from the terminal.

Ubuntu: To install miniconda on Ubuntu, first download the latest Python 64-bit version from here[6]. After the download has finished, open the terminal and execute the following command:

```
bash path/to/Miniconda3-latest-Linux-x86_64.sh
```

Make sure to replace `path/to/` with the path of the folder containing the downloaded script. Most often this file will be downloaded to the `Downloads` folder. If this is the case for your computer, you can run the script in the terminal by typing:

```
bash Downloads/Miniconda3-latest-Linux-x86_64.sh
```

The instructions for the installation will then appear. Follow the prompts and agree to accepting the license, the default installation location, and to running `conda init`, which makes `conda` available from the terminal.

13.3.3 JupyterLab

With miniconda set up, we can now install JupyterLab and the Jupyter Git extension. Type the following into the Anaconda Prompt (Windows) or the terminal (MacOS and Ubuntu) and press enter:

```
conda install -c conda-forge -y jupyterlab
conda install -y nodejs
pip install --upgrade jupyterlab-git
```

To test that your JupyterLab installation is functional, you can type `jupyter`

[5] https://repo.anaconda.com/miniconda/Miniconda3-latest-MacOSX-arm64.sh

[6] https://repo.anaconda.com/miniconda/Miniconda3-latest-Linux-x86_64.sh

`lab` into the Anaconda Prompt (Windows) or terminal (MacOS and Ubuntu) and press enter. This should open a new tab in your default browser with the JupyterLab interface. To exit out of JupyterLab you can click `File -> Shutdown`, or go to the terminal from which you launched JupyterLab, hold `Ctrl`, and press `c` twice.

To improve the experience of using R in JupyterLab, you should also add an extension that allows you to set up keyboard shortcuts for inserting text. By default, this extension creates shortcuts for inserting two of the most common R operators: `<-` and `|>`. Type the following in the Anaconda Prompt (Windows) or terminal (MacOS and Ubuntu) and press enter:

```
jupyter labextension install @techrah/text-shortcuts
```

13.3.4 R, R packages, and the IRkernel

To have the software used in this book available to you in JupyterLab, you will need to install the R programming language, several R packages, and the IRkernel. To install versions of these that are compatible with the accompanying worksheets, type the command shown below into the Anaconda Prompt (Windows) or terminal (MacOS and Ubuntu).

```
conda env update --file https://raw.githubusercontent.com/UBC-DSCI/data-science-a-first-intro-worksheets/main/environment.yml
```

This command installs the specific R and package versions specified in the `environment.yml` file found in the worksheets repository[7]. We will always keep the versions in the `environment.yml` file updated so that they are compatible with the exercise worksheets that accompany the book.

> You can also install the *latest* version of R and the R packages used in this book by typing the commands shown below in the Anaconda Prompt (Windows) or terminal (MacOS and Ubuntu) and pressing enter. **Be careful though:** this may install package versions that are incompatible with the worksheets that accompany the book; the automated exercise feedback might tell you your answers are not correct even though they are!

```
conda install -c conda-forge -y \
  r-base \
  r-cowplot \
  r-ggally \
  r-gridextra \
  r-irkernel \
```

[7] https://ubc-dsci.github.io/data-science-a-first-intro-worksheets

```
  r-kknn \
  r-rpostgres \
  r-rsqlite \
  r-scales \
  r-testthat \
  r-tidymodels \
  r-tidyverse \
  r-tinytex \
  unixodbc
```

13.3.5 LaTeX

To be able to render `.ipynb` files to `.pdf` you need to install a LaTeX distribution. These can be quite large, so we will opt to use `tinytex`, a light-weight cross-platform, portable, and easy-to-maintain LaTeX distribution based on TeX Live.

MacOS: To install `tinytex` we need to make sure that `/usr/local/bin` is writable. To do this, type the following in the terminal:

```
sudo chown -R $(whoami):admin /usr/local/bin
```

Note: You might be asked to enter your password during installation.

All operating systems: To install LaTeX, open JupyterLab by typing `jupyter lab` in the Anaconda Prompt (Windows) or terminal (MacOS and Ubuntu) and press Enter. Then from JupyterLab, open an R console, type the commands listed below, and press Shift + Enter to install `tinytex`:

```
tinytex::install_tinytex()
tinytex::tlmgr_install(c("eurosym",
                         "adjustbox",
                         "caption",
                         "collectbox",
                         "enumitem",
                         "environ",
                         "fp",
                         "jknapltx",
                         "ms",
                         "oberdiek",
```

```
                  "parskip",
                  "pgf",
                  "rsfs",
                  "tcolorbox",
                  "titling",
                  "trimspaces",
                  "ucs",
                  "ulem",
                  "upquote"))
```

Ubuntu: To append the TinyTex executables to our `PATH` we need to edit our `.bashrc file`. The TinyTex executables are usually installed in `~/bin`. Thus, add the lines below to the bottom of your `.bashrc` file (which you can open by `nano ~/.bashrc` and save the file:

```
# Append TinyTex executables to the path
export PATH="$PATH:~/bin"
```

Note: If you used `nano` to open your `.bashrc` file, follow the keyboard shortcuts at the bottom of the nano text editor to save and close the file.

13.4 Finishing up installation

It is good practice to restart all the programs you used when installing this software stack before you proceed to doing your data analysis. This includes restarting JupyterLab as well as the terminal (MacOS and Ubuntu) or the Anaconda Prompt (Windows). This will ensure all the software and settings you put in place are correctly sourced.

13.5 Downloading the worksheets for this book

The worksheets containing practice exercises for this book can be downloaded by visiting `https://github.com/UBC-DSCI/data-science-a-first-intro-worksheets`, clicking the green "Code" button, and then selecting "Download ZIP". The worksheets are contained within the compressed zip folder that will be downloaded. Once you unzip the downloaded file, you can open the folder and run each worksheet using Jupyter. See Chapter 11 for instructions on how to use Jupyter.

Bibliography

Jeffrey Arnold. *ggthemes*, 2019. URL `https://jrnold.github.io/ggthemes/`.

Evelyn Martin Lansdowne Beale, Maurice George Kendall, and David Mann. The discarding of variables in multivariate analysis. *Biometrika*, 54(3-4): 357–366, 1967.

Thomas Cover and Peter Hart. Nearest neighbor pattern classification. *IEEE Transactions on Information Theory*, 13(1):21–27, 1967.

Murray Cox. Inside Airbnb, n.d. URL `http://insideairbnb.com/`.

Sameer Deeb. The molecular basis of variation in human color vision. *Clinical Genetics*, 67:369–377, 2005.

David Diez, Mine Çetinkaya Rundel, and Christopher Barr. *OpenIntro Statistics*. OpenIntro, Inc., 2019. URL `https://openintro.org/book/os/`.

Norman Draper and Harry Smith. *Applied Regression Analysis*. Wiley, 1966.

M. Eforymson. Stepwise regression—a backward and forward look. In *Eastern Regional Meetings of the Institute of Mathematical Statistics*, 1966.

Evelyn Fix and Joseph Hodges. Discriminatory analysis. nonparametric discrimination: consistency properties. Technical report, USAF School of Aviation Medicine, Randolph Field, Texas, 1951.

Kristen Gorman, Tony Williams, and William Fraser. Ecological sexual dimorphism and environmental variability within a community of Antarctic penguins (genus *Pygoscelis*). *PLoS ONE*, 9(3), 2014.

Garrett Grolemund and Hadley Wickham. Dates and times made easy with lubridate. *Journal of Statistical Software*, 40(3):1–25, 2011.

Wolfgang Hardle. *Smoothing Techniques with Implementation in S*. Springer, New York, 1991.

Lionel Henry and Hadley Wickham. *tidyselect R package*, 2021. URL `https://tidyselect.r-lib.org/`.

Ronald Hocking and R. N. Leslie. Selection of the best subset in regression analysis. *Technometrics*, 9(4):531–540, 1967.

Allison Horst, Alison Hill, and Kristen Gorman. *palmerpenguins: Palmer Archipelago penguin data*, 2020. URL https://allisonhorst.github.io/palmerpenguins/. R package version 0.1.0.

Chester Ismay and Albert Kim. *Statistical Inference via Data Science: A ModernDive into R and the Tidyverse.* Chapman and Hall/CRC Press, 2020. URL https://moderndive.com/.

Gareth James, Daniela Witten, Trevor Hastie, and Robert Tibshirani. *An Introduction to Statistical Learning.* Springer, 1st edition, 2013. URL https://www.statlearning.com/.

Michael Kearney. *rtweet R package*, 2019. URL https://github.com/ropensci/rtweet.

Max Kuhn and David Vaughan. *parsnip R package*, 2021. URL https://parsnip.tidymodels.org/.

Max Kuhn and Hadley Wickham. *recipes R package*, 2021. URL https://recipes.tidymodels.org/.

Jeffrey Leek and Roger Peng. What is the question? *Science*, 347(6228): 1314–1315, 2015.

Thomas Leeper. *rio R package*, 2021. URL https://cloud.r-project.org/web/packages/rio/index.html.

Stuart Lloyd. Least square quantization in PCM. *IEEE Transactions on Information Theory*, 28(2):129–137, 1982. Originally released as a Bell Telephone Laboratories Paper in 1957.

Donald R. McNeil. *Interactive Data Analysis: A Practical Primer.* Wiley, 1977.

Albert Michelson. Experimental determination of the velocity of light made at the United States Naval Academy, Annapolis. *Astronomic Papers*, 1:135–8, 1882.

Kirill Müller. *here R package*, 2020. URL https://here.r-lib.org/.

Erich Neuwirth. *RColorBrewer: ColorBrewer Palettes*, 2014. URL https://cran.r-project.org/web/packages/RColorBrewer/index.html.

Roger D Peng and Elizabeth Matsui. *The Art of Data Science: A Guide for Anyone Who Works with Data.* Skybrude Consulting, LLC, 2015. URL https://bookdown.org/rdpeng/artofdatascience/.

Dmytro Perepolkin. *polite R package*, 2021. URL https://dmi3kno.github.io/polite/.

R Core Team. *R: A Language and Environment for Statistical Computing*. R Foundation for Statistical Computing, 2021. URL https://www.R-project.org/.

Real Time Statistics Project. Internet live stats: Google search statistics, 2021. URL https://www.internetlivestats.com/google-search-statistics/.

Vitalie Spinu, Garrett Grolemund, and Hadley Wickham. *lubridate R package*, 2021. URL https://lubridate.tidyverse.org/.

Stanford Health Care. What is cancer?, 2021. URL https://stanfordhealthcare.org/medical-conditions/cancer/cancer.html.

Statistics Canada. Population census, 2016a. URL https://www12.statcan.gc.ca/census-recensement/2016/dp-pd/index-eng.cfm.

Statistics Canada. The Aboriginal languages of First Nations people, Métis and Inuit, 2016b. URL https://www12.statcan.gc.ca/census-recensement/2016/as-sa/98-200-x/2016022/98-200-x2016022-eng.cfm.

Statistics Canada. The evolution of language populations in Canada, by mother tongue, from 1901 to 2016, 2018. URL https://www150.statcan.gc.ca/n1/pub/11-630-x/11-630-x2018001-eng.htm.

William Nick Street, William Wolberg, and Olvi Mangasarian. Nuclear feature extraction for breast tumor diagnosis. In *International Symposium on Electronic Imaging: Science and Technology*, 1993.

Pieter Tans and Ralph Keeling. Trends in atmospheric carbon dioxide, 2020. URL https://gml.noaa.gov/ccgg/trends/data.html.

Tiffany Timbers. *canlang: Canadian Census language data*, 2020. URL https://ttimbers.github.io/canlang/. R package version 0.0.9.

Truth and Reconciliation Commission of Canada. *They Came for the Children: Canada, Aboriginal Peoples, and the Residential Schools*. Public Works & Government Services Canada, 2012.

Truth and Reconciliation Commission of Canada. *Calls to Action*. 2015. URL https://www2.gov.bc.ca/assets/gov/british-columbians-our-governments/indigenous-people/aboriginal-peoples-documents/calls_to_action_english2.pdf.

Nick Walker. Mapping indigenous languages in Canada. *Canadian Geographic*, 2017. URL https://www.canadiangeographic.ca/article/mapping-indigenous-languages-canada.

Hadley Wickham. Tidy data. *Journal of Statistical Software*, 59(10):1–23, 2014.

Hadley Wickham. *Advanced R.* CRC Press, 2019. URL https://adv-r.hadley.nz/.

Hadley Wickham. *The Tidyverse Style Guide.* 2020. URL https://style.tidyverse.org/.

Hadley Wickham. *rvest R package*, 2021a. URL https://rvest.tidyverse.org/.

Hadley Wickham. *tidyverse R package*, 2021b. URL https://tidyverse.tidyverse.org/.

Hadley Wickham and Garrett Grolemund. *R for Data Science: Import, Tidy, Transform, Visualize, and Model Data.* O'Reilly, 2016. URL https://r4ds.had.co.nz/.

Hadley Wickham, Mara Averick, Jennifer Bryan, Winston Chang, Lucy D'Agostino McGowan, Romain François, Garrett Grolemund, Alex Hayes, Lionel Henry, Jim Hester, Max Kuhn, Thomas Lin Pedersen, Evan Miller, Stephan Milton Bache, Kirill Müller, Jeroen Ooms, David Robinson, Dana Paige Seidel, Vitalie Spinu, Kohske Takahashi, Davis Vaughan, Claus Wilke, Kara Woo, and Hiroaki Yutani. Welcome to the tidyverse. *Journal of Open Source Software*, 4(43):1686, 2019.

Hadley Wickham, Winston Chang, Lionel Henry, Thomas Lin Pederson, Kohske Takahashi, Claus Wilke, Kara Woo, Hiroaki Yutani, and Dewey Dunnington. *ggplot2 R package*, 2021a. URL https://ggplot2.tidyverse.org/.

Hadley Wickham, Romain François, Lionel Henry, and Kirill Müller. *dplyr R package*, 2021b. URL https://dplyr.tidyverse.org/.

Claus Wilke. *Fundamentals of Data Visualization.* O'Reilly Media, 2019. URL https://clauswilke.com/dataviz/.

Greg Wilson, Jennifer Bryan, Karen Cranston, Justin Kitzes, Lex Nederbragt, and Tracy Teal. Good enough practices in scientific computing. *PLoS Computational Biology*, 13(6), 2017.

Kory Wilson. *Pulling Together: Foundations Guide.* BCcampus, 2018. URL https://opentextbc.ca/indigenizationfoundations/.

Index